THE WILEY GUIDE TO PROJECT ORGANIZATION & PROJECT MANAGEMENT COMPETENCIES

T0344841

THE WILEY GUIDES TO THE MANAGEMENT OF PROJECTS

Edited by

Peter W. G. Morris and Jeffrey K. Pinto

The Wiley Guide to Project, Program & Portfolio Management
978-0-470-22685-8

The Wiley Guide to Project Control
978-0-470-22684-1

The Wiley Guide to Project Organization & Project
Management Competencies
978-0-470-22683-4

The Wiley Guide to Project Technology, Supply Chain &
Procurement Management
978-0-470-22682-7

THE WILEY GUIDE TO PROJECT ORGANIZATION & PROJECT MANAGEMENT COMPETENCIES

Edited by

Peter W. G. Morris and Jeffrey K. Pinto

JOHN WILEY & SONS, INC.

Library of Congress Cataloging-in-Publication Data:

ISBN: 978-0-470-22683-4

Printed in the United States of America

V10006418_112718

CONTENTS

THE WILEY GUIDE TO PROJECT ORGANIZATION & PROJECT MANAGEMENT COMPETENCIES: PREFACE AND INTRODUCTION

Peter W. G. Morris and Jeffrey Pinto

In 1983, Dave Cleland and William King produced for Van Nostrand Reinhold (now John Wiley & Sons) the *Project Management Handbook,* a book that rapidly became a classic. Now over twenty years later, Wiley is bringing this landmark publication up to date with a new series *The Wiley Guides to the Management of Projects,* comprising four separate, but linked, books.

Why the new title—indeed, why the need to update the original work?

That is a big question, one that goes to the heart of much of the debate in project management today and which is central to the architecture and content of these books. First, why "the management of projects" instead of "project management"?

Project management has moved a long way since 1983. If we mark the founding of project management to be somewhere between about 1955 (when the first uses of modern project management terms and techniques began being applied in the management of the U.S. missile programs) and 1969 / 70 (when project management professional associations were established in the United States and Europe) (Morris, 1997), then Cleland and King's book reflected the thinking that had been developed in the field for about the first twenty years of this young discipline's life. Well, over another twenty years has since elapsed. During this time there has been an explosive growth in project management. The professional project management associations around the world now have thousands of members—the Project Management Institute (PMI) itself having well over 200,000—and membership continues to grow! Every year there are dozens of conferences; books, journals, and electronic publications abound; companies continue to recognize project management as a core business discipline and work to improve company performance through it; and, increasingly, there is more formal educational work carried out in university teaching and research programs, both at the undergraduate and, particularly, graduate, levels.

Yet, in many ways, all this activity has led to some confusion over concepts and applications. For example, the basic American, European, and Japanese professional models of project management are different. The most influential, PMI, not least due to its size, is the most limiting, reflecting an essentially execution, or delivery, orientation, evident both in its *Guide to the Project Management Body of Knowledge, PMBOK Guide, 3rd Edition* (PMI, 2004) and its *Organizational Project Management Maturity Model, OPM3* (PMI, 2003). This approach tends to under-emphasize the front-end, definitional stages of the project, the stages that are so crucial to successful accomplishment (the European and Japanese models, as we shall see, give much greater prominence to these stages). An execution emphasis is obviously essential, but managing the definition of the project, in a way that best fits with the business, technical, and other organizational needs of the sponsors, is critical in determining how well the project will deliver business benefits and in establishing the overall strategy for the project.

It was this insight, developed through research conducted independently by the current authors shortly after the publication of the Cleland and King *Handbook* (Morris and Hough, 1987; Pinto and Slevin, 1988), that led to Morris coining the term "the management of projects" in 1994 to reflect the need to focus on managing the definition and delivery of *the project itself* to deliver a successful outcome.

These at any rate are the themes that we shall be exploring in this book (and to which we shall revert in a moment). Our aim, frankly, is to better center the discipline by defining more clearly what is involved in managing projects successfully and, in doing so, to expand the discipline's focus.

So, why is this endeavor so big that it takes four books? Well, first, it was both the publisher's desire and our own to produce something substantial—something that could be used by both practitioners and scholars, hopefully for the next 10 to 20 years, like the Cleland and King book—as a reference for the best-thinking in the discipline. But why are there so many chapters that it needs four books? Quite simply, the size reflects the growth of knowledge within the field. The "management of projects" philosophy forces us (i.e., members of the discipline) to expand our frame of reference regarding what projects truly *are* beyond of the traditional *PMBOK/OPM3* model.

These, then, are not a set of short "how to" management books, but very intentionally, resource books. We see our readership not as casual business readers, but as people who are genuinely interested in the discipline, and who is seek further insight and information— the thinking managers of projects. Specifically, the books are intended for both the general practitioner and the student (typically working at the graduate level). For both, we seek to show where and how practice and innovative thinking is shaping the discipline. We are deliberately pushing the envelope, giving practical examples, and providing references to others' work. The books should, in short, be a real resource, allowing the reader to understand how the key "management of projects" practices are being applied in different contexts and pointing to where further information can be obtained.

To achieve this aim, we have assembled and worked, at times intensively, with a group of authors who collectively provide truly outstanding experience and insight. Some are, by any standard, among the leading researchers, writers, and speakers in the field, whether as academics or consultants. Others write directly from senior positions in industry, offering their practical experience. In every case, each has worked hard with us to furnish the relevance, the references, and the examples that the books, as a whole, aim to provide.

What one undoubtedly gets as a result is a range that is far greater than any individual alone can bring (one simply cannot be working in all these different areas so deeply as all these authors, combined, are). What one does not always get, though, are all the angles that any one mind might think is important. This is inevitable, if a little regrettable. But to a larger extent, we feel, it is beneficial for two reasons. One, this is not a discipline that is now done and finished—far from it. There are many examples where there is need and opportunity for further research and for alternative ways of looking at things. Rodney Turner and Anne Keegan, for example, in their chapter on managing innovation (*The Wiley Guide to Project Technology, Supply Chain & Procurement Management,* Chapter 8) ended up positioning the discussion very much in terms of learning and maturity. If we had gone to Harvard, to Wheelwright and Clark (1992) or Christensen (1999) for example, we would almost certainly have received something that focused more on the structural processes linking technology, innovation, and strategy. This divergence is healthy for the discipline, and is, in fact, inevitable in a subject that is so context-dependent as management. Second, it is also beneficial, because seeing a topic from a different viewpoint can be stimulating and lead the reader to fresh insights. Hence we have Steve Simister giving an outstandingly lucid and comprehensive treatment in *The Wiley Guide to Project Control,* Chapter 5 on risk management; but later we have Stephen Ward and Chris Chapman coming at the same subject (*The Wiley Guide to Project Control,* Chapter 6) from a different perspective and offering a penetrating treatment of it. There are many similar instances, particularly where the topic is complicated, or may vary in application, as in strategy, program management, finance, procurement, knowledge management, performance management, scheduling, competence, quality, and maturity.

In short, the breadth and diversity of this collection of work (and authors) is, we believe, one of the books' most fertile qualities. Together, they represent a set of approximately sixty authors from different discipline perspectives (e.g., construction, new product development, information technology, defense / aerospace) whose common bond is their commitment to improving the management of projects, and who provide a range of insights from around the globe. Thus, the North American reader can gain insight into processes that, while common in Europe, have yet to make significant inroads in other locations, and vice versa. IT project managers can likewise gather information from the wealth of knowledge built up through decades of practice in the construction industry, and vice versa. The settings may change; the key principals are remarkably resilient.

But these are big topics, and it is perhaps time to return to the question of what we mean by project management and the management of projects, and to the structure of the book.

Project Management

There are several levels at which the subject of project management can be approached. We have already indicated one of them in reference to the PMI model. As we and several other of the *Guides*' authors indicate later, this is a wholly valid, but essentially delivery, or execution-oriented perspective of the discipline: what the project manager needs to do in order to deliver the project "on time, in budget, to scope." If project management professionals cannot do this effectively, they are failing at the first fence. Mastering these skills is the *sine qua non*—the 'without which nothing'—of the discipline. Volume 1 addresses this basic view of the discipline—though by no means exhaustively (there are dozens of other books on the market that do this excellently—including some outstanding textbooks: Meredith and Mantel, 2003; Gray and Larson, 2003; Pinto, 2004).

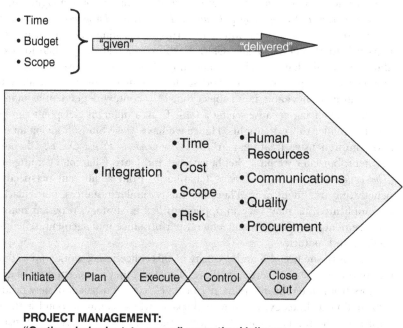

PROJECT MANAGEMENT:
"On time, in budget, to scope" execution/delivery

The overriding paradigm of project management at this level is a control one (in the cybernetic sense of control involving planning, measuring, comparing, and then adjusting performance to meet planned objectives, or adjusting the plans). Interestingly, even this model—for us, the foundation stone of the discipline—is often more than many in other disciplines think of as project management: many, for example, see it as predominantly oriented around scheduling (or even as a subset, in some management textbooks, of operations management). In fact, even in some sectors of industry, this has only recently begun to change, as can be seen towards the end of the book in the chapter on project management in the pharmaceutical industry. It is more than just scheduling of course: there is a whole range of cost, scope, quality and other control activities. But there are other important topics too.

Managing project risks, for example, is an absolutely fundamental skill even at this basic level of project management. Projects, by definition, are unique: doing the work necessary to initiate, plan, execute, control, and close-out the project will inevitably entail risks. These need to be managed.

Both these areas are mainstream and generally pretty well understood within the traditional project management community (as represented by the PMI *PMBOK*® *'Guide'* (PMI, 2004) for example). What is less well covered, perhaps, is the people-side of managing projects. Clearly people are absolutely central to effective project management; without people projects simply could not be managed. There is a huge amount of work that has been done on how organizations and people behave and perform, and much that has been written on this within a project management context (that so little of this finds its way into *PMBOK* is almost certainly due to its concentration on material that is said in *PMBOK* to be "unique" to project management). A lot of this information we have positioned in Volume 3, which deals more with the area of competencies, but some we have kept in the other volumes, deliberately to make the point that people issues are essential in project delivery.

It is thus important to provide the necessary balance to our building blocks of the discipline. For example, among the key contextual elements that set the stage for future activity is the organization's structure—so pivotal in influencing how effectively projects may be run. But organizational structure has to fit within the larger social context of the organization—its culture, values, and operating philosophy; stakeholder expectations, socio-economic, and business context; behavioural norms, power, and informal influence processes, and so on. This takes us to our larger theme: looking at the project in its environment and managing its definition and delivery for stakeholder success: "the management of projects."

The Management of Projects

The thrust of the books is, as we have said, to expand the field of project management. This is quite deliberate. For as Morris and Hough showed in *The Anatomy of Major Projects* (1987), in a survey of the then-existing data on project overruns (drawing on over 3,600 projects as well as eight specially prepared case studies), neither poor scheduling nor even lack of teamwork figured crucially among the factors leading to the large number of unsuccessful projects in this data set. What instead were typically important were items such as client changes, poor technology management, and poor change control; changing social, economic, and environmental factors; labor issues, poor contract management, etc. Basically, the message was that while traditional project management skills are important, they are often not *sufficient* to ensure project success: what is needed is to broaden the focus to cover the management of external and front-end issues, not least technology. Similarly, at about the same time, and subsequently, Pinto and his coauthors, in their studies on project success (Pinto and Slevin, 1988; Kharbanda and Pinto, 1997), showed the importance of client issues and technology, as well as the more traditional areas of project control and people.

The result of both works has been to change the way we look at the discipline. No longer is the focus so much just on the processes and practices needed to deliver projects "to scope, in budget, on schedule," but rather on how we set up and define the project to deliver stakeholder success—on how to manage projects. In one sense, this almost makes the subject impossibly large, for now the only thing differentiating this form of management from other sorts is "the project." We need, therefore, to understand the characteristics of the project development life cycle, but also the nature of projects in organizations. This becomes the kernel of the new discipline, and there is much in this book on this.

Morris articulated this idea in *The Management of Projects* (1994, 97), and it significantly influenced the development of the Association for Project Management's Body of Knowledge as well as the International Project Management Association's Competence Baseline (Morris, 2001; Morris, Jamieson, and Shepherd, 2006; Morris, Crawford, Hodgson, Shepherd, and Thomas, 2006). As a generic term, we feel "the management of projects" still works, but it is interesting to note how the rising interest in program management and portfolio management fits comfortably into this schema. Program management is now strongly seen as the management of multiple projects connected to a shared business objective—see, for example, the chapter by Michel Thiry (*The Wiley Guide to Project, Program & Portfolio Management*, Chapter 6.) The emphasis on managing for business benefit, and on managing projects, is exactly the same as in "the management of projects." Similarly, the recently launched *Japanese Body of Knowledge, P2M (Program and Project Management)*, discussed *inter alia* in Lynn Crawford's chapter on project management standards (*The Wiley Guide to Project Organization & Project Management Competencies*, Chapter 10), is explicitly oriented around managing programs and projects to create, and optimize, business value. Systems management, strategy, value management, finance, and relations management for example are all major elements in *P2M:* few, if any, appear in *PMBOK.*

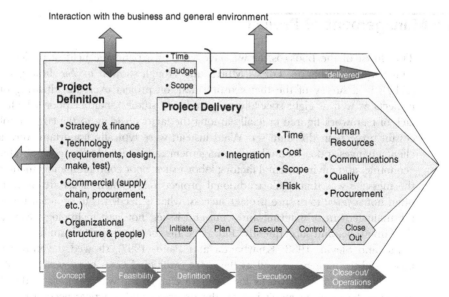

THE MANAGEMENT OF PROJECTS involves managing the definition and delivery of the project for stakeholder success. The focus is on the project in its context. Project and program management – and portfolio management, though this is less managerial – sit within this framework.

("The management of projects" model is also more relevant to the single project situation than *PMBOK* incidentally, not just because of the emphasis on value, but via the inclusion of design, technology, and definition. There are many single project management situations, such as Design & Build contracts for example, where the project management team has responsibility for elements of the project design and definition).

Structure of *The Wiley Guide to Project Organization & Project Management Competencies*

The Wiley Guides to the Management of Projects series is consists of four distinct, but interrelated, books:

- *The Wiley Guide to Project, Program & Portfolio Management*
- *The Wiley Guide to Project Control*
- *The Wiley Guide to Project Organization & Project Management Competencies*
- *The Wiley Guide to Project Technology, Supply Chain & Procurement Management*

This book, *The Wiley Guide to Project Organization & Project Management Competencies*, looks both at some familiar basics of project organization as well as new thinking, for example on organizational networks, knowledge management and project-based learning, and the whole area of competencies.

At bottom, project management represents a "people" challenge; that is, the ability to successfully manage a project from inception through delivery is predicated on our ability to appreciate and effectively employ the competencies of all those who are associated with the project development and delivery process. This volume covers a range of organizational and people-based topics that are occupying the project management world today. Foremost among these are issues to do with knowledge, learning, and maturity. Before addressing these topics, however, we look at structure, teams, leadership, power and negotiation, and competencies.

1. But first, in Chapter 1, Dennis Slevin and Jeff Pinto provide a broad overview of some key behavioral factors impacting successful project management. Drawing on original research on practicing project managers specially carried out for this book, they summarize these into twelve critical issues that impact on the performance of the project manager, ranging from the personal ("micro") to the organizational ("macro").

2. In order to effectively mobilize the resources needed to manage the project, a great deal needs to be understood about the organizational structures and systems, and roles and responsibilities that must be harnessed to undertake the project. Erik Larson, in Chapter 2, provides a solid overview of the principal forms of organization structure found in project management. Erik concentrates particularly on the matrix form, but shows the types of factors that will affect the choice of organization structure. Each of these choices,

deliberate or otherwise, will have a tremendous impact on the resulting likelihood for successful project management.

3. Connie DeLisle, in Chapter 3, brings us up to date with her chapter on "Contemporary Views on Shaping, Developing and Managing Teams." Connie begins in a character-istically arresting manner: "We have knowledge and wisdom to change, but why do we not do so, or even act in ways contradictory to successful team building and manage-ment?" She looks at team working in three sections: first, the forces shaping teams (crisis management, globalization, donated time, organizational anorexia, senior exec priori-ties); then team development (team characteristics, key responsibilities, knowledge com-petencies, personalities, resource negotiation, development itself, nature / nuture, and virtual working); and, finally, team management, where her focus is particularly on addressing team pathologies.

4. Peg Thoms and John Kerwin look in Chapter 4 at leadership in projects. They begin by reviewing some of the key concepts, particularly around the so-called charismatic and transformational schools of leadership theory. Then they focus on the importance of vision-creation as a key activity in leadership—not just having one but being able to communicate it (imaging). Visioning reflects personality, but also in part reflects one's future-looking orientation. Techniques are presented for helping create a vision and time orientation—past, present, and future—which are discussed in terms of project leadership. Finally, these ideas are illustrated by two cases, one a highway project, one a film production. Peg and John end by summarizing: "We can learn a great deal from leadership theories, but even more by observing effective leaders. Paying attention to how problems are solved, how innovative strategies are developed, and how great project leaders communicate and motivate are the best ways to improve our leadership ability."

5. Few project managers have much formal power. Much is informal and has to be ne-gotiated. Influencing skills are very important. In Chapter 5, Jeff Pinto and John Ma-genau review the different sources of power and forms of influence. Then they look at developing influencing and negotiating skills and how to prepare for, and conduct, ne-gotiations.

6. Martina Huemann, Rodney Turner, and Anne Keegan broaden the discussion in Chap-ter 6 of human resource management challenges in project-based organizations. Begin-ning with a review of the specific challenges posed by projects—essentially the lack of certainty over future work and career development—they move on to a central discus-sion of the importance of competences in project management, something that will now occupy much of the rest of this volume.

7. In Chapter 7, Andrew Gale looks in depth at organizational and personal competencies. He goes over our definitions of competencies, noting the basic normative idea that they are what are required to fulfill a specific role and concluding that "competence is con-cerned with the capacity to undertake specific types of action and can be considered as an holistic concept involving the integration of attitudes, skills, knowledge, performance and quality of application." Becoming, and even remaining, competent clearly involves learning and learning thus is important in any discussion of competency (hence Chapters 8 and 9). Andrew then looks at the idea of project management (professional) compe-tency frameworks which he sees as predominantly normative, reductionist, deterministic,

and restricting. He looks at alternative models (for example, actor network theory) and at studies on organizational competencies. He concludes by looking at experience in trying to measure competency: no one has yet found it possible to generate hard measures.

8. Knowledge management and organizational learning became subjects of considerable interest from the late 1990s onwards. In Chapter 8, Christophe Bredillet offers a stimulating *tour d'horizon* of the field applied to the management of projects. All the key ideas in both fields, KM and OL, are presented before looking at projects, which, in his view, are specially good places for "learning at the edge of organization."

9. In Chapter 9, Peter Morris poses the question, how do we know what is what in project management—how valid are the rules, practices, insights, and other knowledge that we may wish to pass on to, or even impose on, others? First, we have to define our frame of reference—this tells us at least what the ballpark looks like. Undoubtedly at some level there are insights that can be generalized. The question is, at what point do they become inapplicable or not useful? Much of the most valuable project management knowledge is process-based rather than substantive in a context-specific way. Risk management is a good example: the process rules are relatively straightforward, the substantive judgments often quite difficult. Best practice rules on project-based learning suggest that while it ought to be possible to support real role-specific learning (at a price), general competency standards and accreditation is more questionable.

10. Lynn Crawford, in Chapter 10's comprehensive review of global project management knowledge and standards, probably agrees. Lynn thoroughly surveys all the international project management professional standards as well as related national and international standards. She addresses first what she calls project-based standards—essentially the four major Bodies of Knowledge (PMI, APM, IPMA, ENAA); then people-based standards (the Australian, UK, and South African competency standards); and finally organizational standards—basically the maturity models and methodologies (OPM3, PRINCE2, etc.). Lynn then goes on to compare the contents of the different BOKs before revisiting the issue of competency assessment, noting, quite rightly, the lack of evidence on the attitudes and behaviors that several pm competency assessing bodies quite understandably rate as so important, as well as at pm qualifications. Touching briefly on international initiatives to form a global perspective on a PM BOK and standards, Lynn concludes on a cautionary note, pointing to the difficulties many feel (a) in translating 'hard' engineering-base project management to 'soft' organizational projects, and (b) the differences between program and project management. Also, "the process for standards development, which is largely a process of making explicit and codifying through consensus the tacit knowledge of experienced practitioners, ensures that standards will remain conservative and will lag behind the cutting edge of both research and practice."

11. In Chapter 11, David Frame brings us back to the world of real projects with his discussion of project evaluation (pre-, mid-, and post-project) and, in doing so, touches on many of the issues addressed above. David covers the basic principles of evaluation and what is needed to convert learning into action. He deals in particular with structured walkthroughs, EISA, and Customer Acceptance Tests. He concludes by tying all this back to the learning organization and HR management.

12. Bill Ibbs, Justin Reginato, and Young Hoon Kwak in Chapter 12 extend the discussion of how the organization can extract value from project management by discussing the program of research they have been doing for several years on the Return on Investment of project management. Their method is heavily based around the concept of organizational maturity: data on project management practices is benchmarked against other organizations', and the different levels of maturity are assessed against the organizations' abilities to achieve projects "on time and to cost."

13. Organizations are indeed now looking closely at how they can leverage value from project management, and to many the maturity concept is an appealing way to tackle this challenge. Terry Cooke-Davies in Chapter 13 carefully explains the difficulties with the concept, however. He traces the origins of maturity work from the Quality movement through software (SEI's CMM) into project management. He reviews the OPM3 and PMMM models before moving on to "untangle the vocabulary and distinguish the relevant concepts." As in several of the previous chapters, he also notes the importance of what is being measured, and the challenge of distinguishing between the maturity of processes and practices (in various different types of projects the organization might undertake) and its overall ability to perform. Project management maturity models make a lot of assumptions here. Like Lynn Crawford, Andy Gale, and Peter Morris, Terry is skeptical that uniform approaches of assessment will work for all project situations. In short, the approach needs applying with care, tailored to the organization's specific circumstances.

14. Finally, in Chapter 14, Lynn Crawford briefly reviews the rise and role of project management associations around the world since the mid 1960s. She looks at the various international initiatives that are now working in parallel with these and concludes by looking at the future. She pulls no punches: there is still much debate as to whether or not it is a practice or a discipline, and how it fits with other subjects and disciplines. Nevertheless, many now recognize—indeed, demand—that standards exist for the discipline. Defining these at national and international levels is an important responsibility for us all.

About the Authors

Dennis P. Slevin

Dennis P. Slevin is Professor of Business Administration at the Katz Graduate School of Business, University of Pittsburgh. He received his education in a variety of university settings, starting with a B.A. in Mathematics at St. Vincent College and continuing with a B.S. in Physics at Massachusetts Institute of Technology, an M.S. in Industrial Administration at Carnegie Mellon University, and a Ph.D. in Business Administration at Stanford University in 1969. Dr. Slevin's research interests focus on entrepreneurship, project management, and corporate governance. He has co-authored the *Total Competitiveness Audit* and the *Project Implementation Profile;* each instrument proposes a conceptual model and a diagnostic tool. He has published widely in a variety of professional journals including *Administrative*

Science Quarterly, Academy of Management Journal, Management Science, Sloan Management Review, Project Management Journal, and numerous other journals and proceedings. His book, *The Whole Manager: How to Increase Your Professional and Personal Effectiveness,* New York, AMACOM, 1989, (paperback, 1991) provides concrete tools for use by practicing managers. He was co-chair of PMI Research Conference 2000, Paris, France. This conference gathered project management researchers from around the world and resulted in the book, *The Frontiers of Project Management Research,* PMI 2002, of which he is co-editor. He was co-chair of PMI Research Conference 2002, Seattle, July, 2002. Since 1972, he has also been president of Innodyne, Inc., a management consulting firm specializing in the design and implementation of specially targeted management development programs. He has worked with numerous companies and organizations such as PPG Industries, General Electric, Alcoa, Westinghouse, GKN plc, IBM, and many other large and small firms.

Jeff Pinto

Dr. Jeffrey K. Pinto is the Samuel A. and Elizabeth B. Breene Professor of Management in the Sam and Irene Black School of Business at Penn State Erie. His major research focus has been in the areas of Project Management, Implementation of New Technologies, and the Diffusion of Innovations in Organizations. Professor Pinto is the author or editor of seventeen books and over one hundred and twenty scientific papers that have appeared in a variety of academic and practitioner journals, books, conference proceedings, and technical reports. Dr. Pinto's work has been translated into French, Dutch, German, Finnish, Russian, and Spanish, among other languages. He is also a frequent presenter at national and international conferences and has served as keynote speaker and as a member of organizing committees for a number of international conferences. Dr. Pinto served as Editor of the *Project Management Journal* from 1990 to 1996 and is a two-time recipient of the Project Management Institute's Distinguished Contribution Award. He has consulted widely with a number of firms, both domestic and international, on a variety of topics, including project management, new product development, information system implementation, organization development, leadership, and conflict resolution. A recent book, *Building Customer-Based Project Organizations,* was published in 2001 by Wiley. He is also the co-developer of SimProject™, a project management simulation for classroom instruction.

Erik Larson

Erik Larson is professor and chair of the management, marketing, and international business department at the College of Business, Oregon State University. He teaches executive, graduate, and undergraduate courses on project management and leadership. His research and consulting activities focus on project management. He has published numerous articles on matrix management, product development, and project partnering. He is co-author of a popular textbook, *Project Management: The Managerial Process, 2nd Ed,* as well as a professional book, *Project Management: The Complete Guide for Every Manager.* He has been a member of the Portland, Oregon chapter of the Project Management Institute since 1984. In 1995, he worked as a Fulbright scholar with faculty at the Krakow Academy of Economics on modernizing Polish business education. He received a B.A. in psychology from Claremont Mc-Kenna College and a Ph.D. in management from State University of New York at Buffalo.

Connie Delisle

Connie Delisle completed a Ph.D. in the Department of Civil Engineering in the Project Management Specialization, a M.Sc. in resource management at the University of Calgary, and two bachelors' degrees from the University of Victoria. The study of the psychology of human behavior and teamwork has been an integral part of her educational studies for the past fifteen years. Her recent practical work experience in the area of e-learning includes course design and delivery of online learning to executive MBAs working in virtual teams. Earlier roles include working as an environmental advisor and project team leader for local government in BC, and senior oil and gas companies in Alberta. Experience as a Canadian national team rower and triathlete provide Connie with a solid understanding of the principles and practical aspects of creating winning team experiences. She has published over thirty articles in the area of success, teams and communication, as well as a recent joint publication with Dr. Janice Thomas and Dr. Kam Jugdev that represents a three-year international research effort in the area of selling project management to senior executives. Connie recently moved to Ottawa, Ontario Canada and is currently employed as a Senior Consultant with Consulting and Audit Canada, a cost recovery based branch of Public Works and Government Services Canada.

Peg Thoms

Peg Thoms earned her Ph.D. in Organizational Behavior from the Fisher School of Business at The Ohio State University. She is currently an Associate Professor of Management and the Director of the MBA Program at Penn State Erie, The Behrend College. She also recently served as a visiting professor of project management at Umeå University in Umeå, Sweden. In addition, Dr. Thoms has sixteen years of business management experience. She conducts research and has published in the areas of leadership vision, time orientation, leadership development, and self-managed work teams. She has published articles in a number of journals including *Human Resource Development Quarterly*, *Journal of Organizational Behavior*, and *Journal of Management Inquiry*, as well as a chapter on the motivation of project teams in the *Project Management Handbook* (ed. J. Pinto). She co-authored a book entitled *Project Leadership from Theory to Practice*. Her new book, *Driven by Time: A Guide to Time Orientation and Leadership*, is published by Greenwood/Praeger (2003). She has consulted with various organizations in the manufacturing, health care, insurance, and banking industries. She has also worked with many not-for-profit groups. Dr. Thoms has won numerous teaching and research awards including The Walter F. Ulmer, Jr. Applied Research Award from the Center for Creative Leadership for her work on leadership vision and *Project Management Journal's* Paper of the Year for her article, "Project Leadership: A Question of Timing." She teaches management, human resources, and leadership.

John Kerwin

John Kerwin received his master's degree from the Edward R. Morrow School of Communications at Washington State University. Previously, he was producer, director, and writer at his production company in Los Angeles, California, Kerwin Communications, Inc. In that capacity, his company produced programs and commercials for NBC, CBS, ABC, and ESPN; he had his own show on ESPN for four years. Kerwin started in the production business with NBC and worked on Super Bowls, World Series, Wimbledon Tennis, the Nightly News, political campaign coverage, elections, and prime time news specials, one of which was honored with an Emmy. His company also worked on movies and theatrical productions. Kerwin's area of research is the application of visual production technology to creative and practical concepts in communications. He is currently assistant professor of communications at Penn State University and has produced and published several works for the marketing and public awareness of services and resources in the educational and corporate sectors. His latest work will be "The First Year Experience," a thirty-minute production which will be distributed internationally on the subject of the freshman experience in higher education.

John Magenau

John Magenau holds a Ph.D. from the State University of New York-Buffalo (1981). His research and teaching interests are in the areas of labor-management relations and negotiation. He is director of the Sam and Irene Black School of Business and serves as a trustee on the board of Lake Erie College of Osteopathic Medicine and as president of the board of the Enterprise Development Center of Erie County and as secretary of board of directors for The Center for eBusiness and Advanced Information Technology.

Martina Huemann

Dr. Martina Huemann holds a doctorate in project management of the Vienna University of Economics and Business Administration. She also studied business administration and economics at the University Lund, Sweden, and the Economic University Prague, Tech Republic. Currently, she is assistant professor in the Project Management Group of the Vienna University of Economics and Business Administration. There she teaches project management to graduate and postgraduate students. In research, she focuses on individual and organizational competences in project-oriented organizations and project-oriented societies. She is visiting fellow of The University of Technology Sydney. Martina has project management experience in organizational development, research, and marketing projects. She is certified Project Manager according to the IPMA—International Project Management Association—certification. Martina organizes the annual PM days research conference and the annual PM days student paper award to promote project management research. She contributed to the development of the pm baseline—the Austrian project management body of knowledge and is board member of Project Management Austria—the Austrian project management association. She is assessor of the IPMA Award and trainer of the IPMA advanced courses. Martina is trainer and consultant of Roland Gareis Consulting. She has experience with project-oriented organizations of different industries and the public sector. Martina is specialised in management audits and reviews of projects and programs, and human resource management issues like project management assessment centers for project and program managers.

Rodney Turner

Rodney Turner is Professor of Project Management at Erasmus University Rotterdam, in the Faculty of Economics. He is also an Adjunct Professor at the University of Technology Sydney, and Visiting Professor at Henley Management College, where he was previously Professor of Project Management, and Director of the Masters program in Project Management. He studied engineering at Auckland University and did his doctorate at Oxford University, where he was also for two years a post-doctoral research fellow. He worked for six years for ICI as a mechanical engineer and project manager, on the design, construction, and maintenance of heavy process plant, and for three years with Coopers and Lybrand as a management consultant. He joined Henley in 1989 and Erasmus in 1997. Rodney Turner is the author or editor of seven books, including *The Handbook of Project-based Management*, the best-selling book published by McGraw-Hill, and the *Gower Handbook of Project Management*. He is editor of *The International Journal of Project Management*, and has written articles for journals, conferences and magazines. He lectures on and teaches project management world wide. From 1999 and 2000, he was President of the International Project Management Association, and Chairman for 2001–2002. He has also helped to establish the Benelux Region of the European Construction Institute as foundation Operations Director. He is also a Fellow of the Institution of Mechanical Engineers and the Association for Project Management.

Anne Keegan

Anne Keegan is a University Lecturer in the Department of Marketing and Organisation, Rotterdam School of Economics, Erasmus University Rotterdam. She delivers courses in Human Resource Management, Organisation Theory, and Behavioural Science in undergraduate, postgraduate and executive-level courses. She has been a member of ERIM (Erasmus Research Institute for Management) since 2002. In addition, she undertakes research into the Project Based Organisation and is a partner in a European Wide Study into the Versatile Project Based Organisation. Her other research interests include HRM in Knowledge Intensive Firms, New Forms of Organising and Critical Management Theory. Dr. Keegan has published in *Long Range Planning* and *Management Learning* and is a reviewer for journals including the *Journal of Management Studies* and the International Journal of Project Management. She is a member of the American Academy of Management, the European Group for Organisation Studies (EGOS) and the Dutch HRM Network. Dr. Keegan studied management and business at the Department of Business Studies, Trinity College Dublin, and did her doctorate there on the topic of *Management Practices in Knowledge Intensive Firms*. Following three years post-doctoral research, she now works as a university lecturer and researcher. Dr. Keegan has also worked as a consultant in the areas of Human Resource Management and Organizational Change to firms in the computer, food, export and voluntary sectors in Ireland and the Netherlands.

Andrew W. Gale

Andrew Gale is Senior Lecturer in Project Management and Program Director for the MSc Project Management Professional Development Program in the Manchester Centre for Civil & Construction Engineering UMIST. He is a Chartered Civil Engineer and Chartered Builder specializing in construction project management. He teaches project management with emphasis on people and culture, equality and diversity and group and team process. He is actively involved in joint collaborative teaching of art and civil engineering students. He is leading the introduction of project management curriculum in the new British University in Dubai. He has managed many research and consultancy grants since 1990 and published over ninety papers and articles. He has extensive experience in working with Russian construction firms and academic institutions and led development training programs, distance learning and curriculum development in Russia, funded by UK and EU government agencies. He has over eighteen years experience in research on construction organization and project culture, with specific interests in diversity, equality and inclusion. Currently, he is developing a collaborative research program (with the Art and Design Faculty at The Manchester Metropolitan University) investigating how civil engineers and artists can learn from each other in the context of project management. He is an active member of the European sub-group of the ICE International Policy Committee with special responsibility for Russia.

Christophe Bredillet

Professor Christophe N. Bredillet, PgD in Project Management, MBA, Dr.Sc., M.Sc. Eng EC Lille, Certificated Program Director IPMA Level A, CMP, CCE, has eighteen years of experience in project and program management with several industries (banking, sporting goods, and IT). For the past ten years, he has been the program director of the MS, MBA, and Doctorate in Project & Program Management at ISGI-Groupe ESC Lille. He was appointed Professor of Project Management at UTS in 2001. The field of his research in PM is the design of learning and knowledge management systems, the development of standards, and the epistemology and evolution of the field. He is strongly involved in development of PM Professional Associations worldwide. He is PMI Global Accreditation Committee Member and has been for the past three years Research MAG, and in the co-lead team member of two standards development projects for PMI (Project Managers Competences Framework and Organizational Project Management Maturity Model). He is a steering committee member of the Global Project Management Forum, and of the international non-aligned "think tank" group named OLCI. He is Vice-President International

Affairs, member of the Board of Directors and Regional Manager, North of France, for AFITEP (the French professional project management Association). He is founder and President of the PMI Chapter "Hauts-de-France."

Peter W. G. Morris

Peter Morris is Professor of Construction and Project Management at University College London, Visiting Professor of Engineering Project Management at UMIST, and Director of the UCL/UMIST based Centre for Research in the Management of Projects. He is also Executive Director of INDECO Ltd, an international projects-oriented management consultancy. He is a past Chairman and Vice President of the UK Association for Project Management and Deputy Chairman of the International Project Management Association. His research has focused significantly around knowledge management and organizational learning in projects, and in design management. Dr. Morris consults with many major companies on developing enterprise-wide project management competency. Prior to joining INDECO, he was a Main Board Director of Bovis Limited, the holding company of the Bovis Construction Group. Between 1984 and 1989, he was a Research Fellow at the University of Oxford and Executive Director of the Major Projects Association. Prior to his work at Oxford, he was with Arthur D. Little in Cambridge, Massachusetts and previously with Booz, Allen & Hamilton in New York, and Sir Robert McAlpine in London. He has written approximately one hundred papers on project management as well as the books: *The Anatomy of Major Projects* (Wiley, 1988) and *The Management of Projects* (Thomas Telford, 1997). He is a Fellow of the Association for Project Management, Institution of Civil Engineers, and Chartered Institute of Building and has a Ph.D., M.Sc., and B.Sc., all from UMIST.

Lynn Crawford

With a diverse background as architect, project manager, regional planner, and policy adviser, and with qualifications in human resource management and business administration, Lynn Crawford has experience both as a project manager and an adviser to project-based organizations on human resources, strategic and business planning, and development issues. Ongoing research and practice include working with leading organizations that are developing their organizational project management competence by sharing and developing knowledge and best practices as members of a global system of project management knowledge networks. A particular area of research and expertise is the assessment and development of individual and corporate project and program management capability. Lynn was until recently Director of the postgraduate Project Management Programme at the University of Technology Sydney. She has been Project Director for several major research projects funded by the Australian Research Council in areas of project management competence and the management of multiple, interdependent and soft projects and continues to conduct research in these areas, working with industry partners. Lynn was a member of the Steering Committee for the development of Australian National Competency Standards in Project Management and is currently leading initiatives aimed at development of global standards for project management. These involve all major project management professional associations, recognized leaders in project management and representatives of global corporations.

J. Davidson Frame

Dr. Frame is Dean at the University of Management and Technology. Prior to joining UMT in 1998, he was Professor of Management Science at the George Washington University, where he served on the faculty from 1979 until 1998. At GWU, he was chairman of the Management Science Department, Director of the Program on Science, Technology, and Innovation, and founder of the project management master's degree program. Between 1973–1979, Dr. Frame was vice president at Computer Horizons, Inc. While at CHI, he headed the Washington office and directed some thirty software development and research projects.

David Frame has written eight books, including the business best-seller, *Managing Projects in Organizations* (3rd ed., 2003). His most recent book is *Managing Risk in Organizations,* published in July 2003. He has also written some forty scholarly articles in the area of technology management, the management of intellectual property, and project management. Dr. Frame is active in the project management professional arena. He served as Director of Certification at the Project Management Institute in 1990–1996. He then served as PMI's Director of Educational Services in 1997–98. He was elected to PMI's international Board of Directors and served in that capacity from January 2000 until December 2002. He was awarded PMI's Distinguished Contribution Award in 1993 and its Person of the Year Award in 1995. He is on the editorial board of *The International Journal of Project Management.*

William Ibbs

William Ibbs is the founding principal of The Ibbs Consulting Group. Dr. Ibbs is also Professor of Project Management at the University of California at Berkeley. He has authored more than one-hundred and fifty scholarly papers and received various awards including PMI's Presidential Citation and the National Science Foundation's Presidential Investigator Award. Active in the Project Management Institute, he has served in a number of positions including Research Director. PMI sponsored his research, which led to the Berkeley Project Management Process Model and new ways to measure Project Management's Return on Investment (the PM/ROI^{SM} concept). Two of his books resulted from that work and were published by PMI. Current research is investigating role of project management for managing innovation. This includes distinguishing characteristics and management styles of different project categories: derivative, platform, and breakthrough. Dr Ibbs holds BS and MS degrees from Carnegie Mellon University and a Ph.D. from the University of California at Berkeley, all in civil engineering.

Justin Reginato

Justin Reginato is currently a candidate for a Ph.D. degree in Engineering and Project Management, emphasizing in the Management of Technology, at the University of California, Berkeley. The focus of his research is determining project management's role in the development of innovative and research-intense projects and its impact on the market value of corporations. Additionally, Mr. Reginato is an adjunct professor of Engineering at Santa Clara University where he teaches courses on project management. Justin has over five years of project management experience with URS Corporation, as well as consulting experience with several high technology companies. He is co-author, with C. William Ibbs, of *Quantifying the Value of Project Management*, published in 2002 by PMI.

Young Hoon Kwak

Dr. Young Hoon Kwak is a faculty member of the project management program at the management science department at the George Washington University (GWU), Washington, D.C. He received his B.S. (1991) in civil engineering from Yonsei University in Seoul, Korea, and M.S. (1992) and Ph.D. (1997) in engineering and project management from the University of California, Berkeley. Before joining GWU, he taught at the Florida International University in Miami and was a post doctoral scholar at the Massachusetts Institute of Technology. Dr. Kwak published and presented numerous papers in engineering and project management/construction management/technology management related areas at peer reviewed journals and conferences. He is serving as the member of the Editorial Review Board for *Project Management Journal*. Dr. Kwak has also consulted for U.S. Naval Facilities Engineering Command, DAEWOO Engineering and Construction, Construction, and Economy Research Institute of Korea (CERIK) and other Fortune 500 companies. He was the co-principal investigator of Project Management Institute's nationwide research "Benefits of Project Management: Financial and Organizational Rewards to Corporations." Dr. Kwak's major interests include project management and control, project risk management, construction management, technology management and international project management.

Terry Cooke-Davies

Terry Cooke-Davies has been a practitioner of both general and project management continuously since the end of the 1960s. He is the Managing Director of Human Systems Limited, which he founded in 1985 to provide services to organizations in support of their innovation projects and ventures. Through the family of project management knowledge networks created and supported by Human Systems, he is in close touch with the best project management practices of more than seventy leading organizations globally. The methods developed in support of the networks are soundly based in theory, as well as having practical application to members, and this was recognized by the award of a Ph.D. to Terry by Leeds Metropolitan University in 2000 for a thesis entitled, "Towards Improved Project Management Practice: Uncovering the evidence for effective practices through empirical research." He is now an Adjunct Professor of Project Management at the University of Technology, Sydney and an Honorary Research Fellow at University College, London. Terry is a regular speaker at international project management conferences in Europe, North America, Australia, and Asia and has published more than thirty book chapters, journal and magazine articles, and research papers. He has a bachelor's degree in Theology, and qualificiations in electrical engineering, management accounting and counselling in addition to his doctors degree in Project Management.

The Wiley Guides to the Management of Projects series offers an opportunity to take a step back and evaluate the status of the field, particularly in terms of scholarship and intellectual contributions, some twenty-four years after Cleland and King's seminal *Handbook*. Much has changed in the interim. The discipline has broadened considerably—where once projects were the primary focus of a few industries, today they are literally the dominant way of organizing business in sectors as diverse as insurance and manufacturing, software engineering and utilities. But as projects have been recognized as primary, critical organizational forms, so has recognition that the range of practices, processes, and issues needed to manage them is substantially broader than was typically seen nearly a quarter of a century ago. The old project management "initiate, plan, execute, control, and close" model once considered the basis for the discipline is now increasingly recognized as insufficient and inadequate, as the many chapters of this book surely demonstrate.

The shift from "project management" to "the management of projects" is no mere linguistic sleight-of-hand: it represents a profound change in the manner in which we approach projects, organize, perform, and evaluate them.

On a personal note, we, the editors, have been both gratified and humbled by the willingness of the authors (very busy people all) to commit their time and labor to this project (and our thanks too to Gill Hypher for all her administrative assistance). Asking an internationally recognized set of experts to provide leading edge work in their respective fields, while ensuring that it is equally useful for scholars and practitioners alike, is a formidable challenge. The contributors rose to meet this challenge wonderfully, as we are sure you, our readers, will agree. In many ways, the *Wiley Guides* represent not only the current state of the art in the discipline; it also showcases the talents and insights of the field's top scholars, thinkers, practitioners, and consultants.

Cleland and King's original *Project Management Handbook* spawned many imitators; we hope with this book that it has acquired a worthy successor.

References

Christensen, C. M. 1999. *Innovation and the General Manager*. Boston: Irwin McGraw-Hill.

Cleland, D. I., and King, W. R. 1983. *Project Management Handbook*. New York: Van Nostrand Reinhold.

Cleland, D. I. 1990. *Project Management: Strategic Design and Implementation*. Blue Ridge Summit, PA: TAB Books.

Gray, C. F., and E. W. Larson. 2003. *Project Management*. Burr Ridge, IL: McGraw-Hill.

Griseri, P. 2002. *Management Knowledge: a critical view*. London: Palgrave.

Kharbanda, O. P., and J. K. Pinto. 1997. *What Made Gertie Gallop?* New York: Van Nostrand Reinhold.

Meredith, J. R., and S. J. Mantel. *Project Management: A Managerial Approach*. 5th Edition. New York: Wiley.

Morris, P. W. G., and G. H. Hough. 1987. *The Anatomy of Major Projects.* Chichester: John Wiley & Sons Ltd.

Morris, P. W. G. 1994. *The Management of Projects.* London: Thomas Telford; distributed in the USA by The American Society of Civil Engineers; paperback edition 1997.

Morris, P. W. G. 2001. "Updating The Project Management Bodies Of Knowledge" *Project Management Journal* 32(3):21–30.

Morris, P. W. G., H. A. J. Jamieson, and M. M. Shepherd. 2006. "Research updating the APM Body of Knowledge 4th edition" *International Journal of Project Management* 24:461–473.

Morris, P. W. G., L. Crawford, D. Hodgson, M. M. Shepherd, and J. Thomas. 2006. "Exploring the Role of Formal Bodies of Knowledge in Defining a Profession—the case of Project Management" *International Journal of Project Management* 24:710–721.

Pinto, J. K. and D. P. Slevin. 1988. "Project success: definitions and measurement techniques," *Project Management Journal* 19(1):67–72.

Pinto, J. K. 2004. *Project Management.* Upper Saddle River, NJ: Prentice-Hall.

Project Management Institute. 2004. *Guide to the Project Management Body of Knowledge.* Newtown Square, PA: PMI.

Project Management Institute. 2003. *Organizational Project Management Maturity Model.* Newtown Square, PA: PMI.

Wheelwright, S. C. & Clark, K. B. 1992. *Revolutionizing New Product Development.* New York: The Free Press.

THE WILEY GUIDE TO PROJECT ORGANIZATION & PROJECT MANAGEMENT COMPETENCIES

CHAPTER ONE

AN OVERVIEW OF BEHAVIORAL ISSUES IN PROJECT MANAGEMENT

Dennis P. Slevin, Jeffrey K. Pinto

G laciers move. If the geological conditions are right, they move inexorably toward the sea. Anyone who has visited and stood on a glacier in areas such as Alaska's Inland Passage is impressed by the dynamism and fluidity of the process. Glacial moraines are amazing flowing rivers of rock. Analogously, project management has been engaged in an inexorable movement toward the human side of the enterprise. The field of project management is one that has always been characterized by its joint emphasis on a blend of technical elements (e.g., PERT charts, beta distributions, earned value analysis, resource leveling) coupled with its vital connection to behavioral and management concepts. While numerous tools, techniques, and quantitative aids were developed in the 1960s and 1970s, people and teamwork have become crucial issues at the turn of the millennium.

Our interest in this topic has been an enduring one, starting almost two decades ago (Slevin and Pinto, 1986, 1988; Slevin and Pinto, 1992). We have long noted that projects are not successful because of the use of the latest project management techniques; they occur as the result of understanding the role that people play in fostering an environment for success. Many of these issues are addressed in a recent collection of new research studies published by the Project Management Institute (Slevin, Cleland, and Pinto, 2002). Research continues to bear out this position as recent studies clearly show. Interest in the behavioral side of successful project management continues to be keen and will continue to grow in this decade (Kloppenborg and Opfer, 2002). The purpose of this chapter is to provide a broad overview of some key behavioral factors impacting successful project management. We sought clarity and guidance in this task by interviewing a number of practicing project managers, many of whom wished to remain anonymous. Their insights and observations were invaluable in helping shape our thinking on these issues and providing the framework for this chapter.

A 12-Factor Model

As one examines the current literature, a number of key behavioral factors emerge as central to successful project management. We have identified 12 factors that we believe are crucial in impacting behavioral issues of project management. We developed this list by doing a quick scan of a variety of recent texts in project management, selected journal articles and papers presented at 2000 and 2002 PMI frontiers of Project Management Research Conferences. Also, a review was conducted of our own recent project management course syllabi. We then sorted the list of factors on the micro–macro continuum in an attempt to provide a useful structure for comments provided to us by practitioners. While we do not argue that this is an exhaustive list, we do feel that it takes a big bite out of the universe of key behavioral issues impacting the project manager. Our analysis has been primarily from the perspective of the individual project manager (the person on the firing line), as opposed to broader perspectives, such as a project management office or general organizational structure. These factors are listed in Figure 1.1.

As one goes down the list, the focus transitions from more micro (individual) issues to more macro (organization wide) issues. Changes in the environment over the past decade have generated increasing challenges in each of these 12 areas for the modern project manager. It is likely, in fact, that the changes we are observing at this point are simply milestones in the overall movement of the field, much as the glacier's movements may be easy to track from point to point, though the eventual destination of the glacier will always remain in question.

Key Comments from Practitioners

In an attempt to make this chapter as pragmatic as possible, we have solicited comments from practitioners concerning these key issues. Via e-mail and telephone interview techniques, a number of practicing project managers were asked to share their opinions con-

FIGURE 1.1. TWELVE KEY BEHAVIORAL FACTORS FOR SUCCESSFUL PROJECTS.

M   1. Personal Characteristics of the Project Manager
I   2. Motivation of the Project Manager
C   3. Leadership and the Project Manager
R   4. Communications and the Project Manager
O   5. Staffing and the Project Manager
   6. Cross-Functional Cooperation and the Project Manager
M   7. Project Teams and the Project Manager
A   8. Virtual Teams and the Project Manager
C   9. Human Resource Policies and the Project Manager
R   10. Conflict and Negotiations and the Project Manager
O   11. Power and Politics and the Project Manager
   12. Project Organization and the Project Manager

cerning trends in project management over the past decade, and where they felt the field was going in the future. Among others, this panel represents an opportunistic selection of individuals who attended PMI Research Conference 2002 (July 14 to 17, 2002, Seattle, Washington; Conference Co-Chairs: Dennis P. Slevin, Jeffrey K. Pinto, and David I. Cleland).

We believe that selected practitioner comments provide some interesting insights into current trends in the field. They also serve as an important perspective to the academic view of these constructs. In conducting these interviews, we deliberately sought a mix of project managers from a variety of organizational or business settings, including traditional production organizations, service industries, and governmental agencies. The range of responses provided additional evidence that more and more organizations are becoming "totally projectized" as they attempt to cope with rapidly changing technology and turbulent business environments (Lundin and Hartman, 2000).

1. Personal Characteristics of the Project Manager

It has been suggested for some time that project management skills are related closely and directly to key general management skills (*PMBOK Guide*, 2000 Edition). While it is clear that general management is a challenging profession, it is also obvious that the project manager often faces special challenges (Meredith and Mantel, 2003). While the general manager often has formal authority and considerable power, the project manager often faces the challenge of working from a low-power, informal position. The unique setting of project management has given rise to a literature that attempts to identify the characteristics most conducive to running successful projects (Posner, 1987; Einsiedel, 1987; Petterson, 1991). Based on these studies, a number of perspectives, traits, and features of project leadership have begun to emerge. Though there is by no means a general consensus of the specific traits of successful project managers, our evaluation of the relevant characteristics from the universe of 32 behavioral dimensions would include the following list (Byham, 1981; Slevin, 1989):

Dimensions

- *Planning and organizing.* Establishing a course of action for self and/or others to accomplish a specific goal; planning proper assignments of personnel and appropriate allocation of resources.
- *Control.* Establishing procedures to monitor one's own job activities and responsibilities or to regulate the tasks and the activities of subordinates. Taking action to monitor the results of delegated assignments or projects.
- *Technical/professional knowledge.* Level of understanding of relevant technical/professional information.
- *Oral communication.* Effective expression in individual or group situations (includes organization, gestures, and nonverbal communication).
- *Listening.* Use of information extracted from oral communication. The ability to pick out the essence of what is being said.
- *Written communication.* Clear expression of ideas in writing in good grammatical form; includes the plan or format of the communication.

- *Sensitivity.* Actions that indicate a consideration for the feelings and needs of others. Awareness of the impact of one's own behavior on others.
- *Group leadership.* Utilization of appropriate interpersonal styles and methods in guiding a group with a common task or goal toward task accomplishment, maintenance of group cohesiveness, and cooperation. Facilitation of group process.
- *Job motivation.* The extent to which activities and responsibilities available in the job overlap with activities and responsibilities that result in personal satisfaction; the degree to which the work itself is personally satisfying.
- *Analysis.* Identifying issues and problems, securing relevant information, relating and comparing data from different sources, and identifying cause-and-effect relationships.
- *Judgment.* Developing alternative courses of action and making decisions that reflect factual information, are based on logical assumptions, and take organizational resources into consideration.
- *Initiative.* Originating action and maintaining active attempts to achieve goals; self-starting rather than passively accepting. Taking action to achieve goals beyond what is necessarily called for.

Two things come to mind when selecting personal characteristics of the project manager:

- *The list tends to be long and diverse.* Project management is an intellectually and physically challenging profession. It requires a wide range of capabilities.
- *Technical skills are important.* While a general manager might surround him- or herself with technical experts, the project manager must be intimately involved with the technology concerning his or her project. As technology advances, this technical proficiency challenge becomes even more significant.

It has been suggested that successful project managers are both born and made (Melymuka, 2000). Some have suggested that project managers have key management styles that account for success, focusing primarily on their ability to function well as facilitators and communicators (Montague, 2000). Others have suggested that full-time leadership skills are essential (Schulz, 2000). In fact, though many of these theories offer some face validity and surface appeal, they also reinforce the problem of trying to isolate the type of person who makes an effective project manager.

Key Comments from Practitioners

"For the most part, I find that today's project managers tend to be very achievement-oriented with strong characteristics towards working together with others in a cooperative manner, as opposed to the classical superior/subordinate relationship that was characterized ten years ago."

"Organizations that recognize this are beginning to put more investment in team building and teamwork training."

"In our organization, we don't look for the ideal project manager. Too much of this job is learned as you go. We find it better to select likely candidates and work with them, giving them small assignments and testing their abilities. Can they make decisions? Are they intelligent enough to ask the right questions? Do they know what they don't know?"

2. Motivation and the Project Manager

The project manager must be a motivational genius. The project manager must have a high level of self-motivation and also be quite skillful at motivating the project team, often under situations of insufficient resources, low team member commitment and morale, and little formal authority. The self-motivation of the project manager is often an intrinsic thing. NASA has been quoted as saying "we don't work very hard on motivating astronauts, but we certainly are extremely careful in selecting astronauts." In that sense, self-motivated project managers are born, not made. However, the organization can do a variety of things to make sure that the motivational structures for project managers are as well developed and carefully executed as those for functional managers (Dunn, 2001). Job satisfaction can obviously be enhanced through appropriate motivational techniques. Further, having clear career ladders for project managers is essential. In many organizations, project managers form a subclass of manager. Because they do not belong to any department, it is easy for their careers to be overlooked in favor of functional standouts. As one wag put it to us, "In our organization, there are two career ladders, but only one has rungs!" Project managers will be self-motivated to the degree they perceive that their performance is likely to earn them advancement or other positive reinforcers.

Concerning the motivation of the project team, often highly creative and unusual techniques must be exploited. Feedback to the team is often long in coming in terms of project success. A typical salesperson receives information concerning sales progress every month; however, in complex projects, there are often unclear measures of progress. While information on schedule, budget, earned value analysis, and other dimensions of project progress may be available, the typical global feedback concerning project success occurs after it is completed, transferred, and used. This presents a particular challenge to a project manager concerning motivation. Recently, one of the authors attended a Christmas party at which the president of an entrepreneurial IT consulting company presented awards to ten employees, many of whom manage projects off-site. Each person received a very nice bronze plaque indicating that he or she had been awarded the President's Excellence in Performance Award for the year. In addition, each individual was given a very stylish briefcase. The president then said, "Each of you should open your briefcase. Who knows? One of them might have $1,000 in it." To the astonishment of the recipients, each briefcase contained 1,000 one-dollar bills—not an inexpensive approach to motivation, but in a competitive IT world where turnover is often a problem, this had a major impact on these lucky employees.

Another area of motivational import for the project manager concerns the management of risk. In a rapidly changing technological world, risk management becomes increasingly important. The importance of risk has been identified by a number of researchers in the field (Turner, 1993; Chapman and Ward, 1997; Wideman, 1998). The *PMBOK Guide*, 2000 Edition contains a significantly revised chapter on project risk management. A new edition of a major textbook in the field contains a substantially enhanced treatment of risk (Meredith and Mantel, 2003). It is important that the organization develop an open and cooperative attitude toward risk, along with approaches that reduce the motivation for concealing risks (Schmidt and Dart, 1999).

Key Comments from Practitioners

"While organizations that recognize project management as a valid discipline and process have put into place motivational inducers such as specific job descriptions and career paths, this is not the norm in the corporate environment but rather the exception. Hence, I see more motivation coming from the individual project manager's need to achieve any personal satisfaction than from the classical motivational means. While this serves well in most cases, the lack of tangible motivational rewards in many organizations leads to disappointment and apathy when the position of project manager is not recognized and does not lead to a specific career objective."

"When we ask practitioners, 'what is your motivation to stay in project management?' the comments often include a passion, a challenge, opportunity to influence, growth, finding better ways, the variety, ability to impact, being a change agent, able to achieve results, et cetera. The people who stay in project management are often self-motivated, at least until a cumbersome management grinds them down."

3. Leadership and the Project Manager

Leadership is crucial for effective project management in two ways:

- Leadership determines the effectiveness of the project planning process.
- Leadership style has a crucial impact on the effectiveness of the project team.

Leadership is important at the onset of the project because it provides key inputs to the project planning process. For example, leadership is crucial in definition of the project scope and the development of the project plan (Globerson and Zwikael, 2002). The implications of this finding are key: The leadership of the project manager immediately sets the stage for not only project team development but the metamorphosis of the project. Hence, effective organizations attempt to develop a positive leadership environment to enhance project success (Jiang, Klein, and Chen, 2001).

As Peg Thoms and John Kerwin explore in their chapter, leadership style issues present a particular problem for the project manager (Slevin, 1989; Slevin and Pinto, 1991). One of the key challenges of the project manager is the need to use consensus leadership approaches in working with the project team. One approach to the clarification of consensus issues is the two-dimensional Bonoma-Slevin Leadership Model. The two dimensions are *information input* and *decision authority*. Information input is represented by the subordinate groups' degree of information inputted into the decision-making process. The decision authority dimension determines whether the leader makes the decision solely by him- or herself or shares the decision making with the group. The grid below helps to define four leadership decisions (see Figure 1.2).

The four extremes of leaders (depicted in the four corners of the grid) are the following:

- *Autocrat* (100, 0). Such managers solicit little or no input from their groups and make the managerial decisions solely by themselves.
- *Consultative autocrat* (100, 100). In this managerial style, intensive input is elicited from the members, but these formal leaders keep all substantive decision-making authority to themselves.

FIGURE 1.2. BONOMA-SLEVIN LEADERSHIP MODEL.

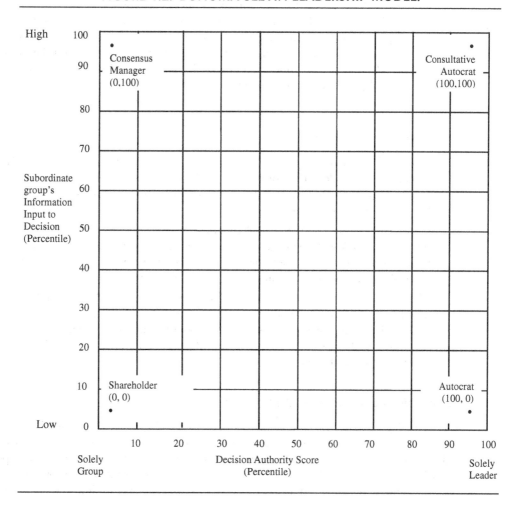

- *Consensus manager* (0, 100). Purely consensual managers throw open the problem to the group for discussion (input) and simultaneously allow or encourage the entire group to make the relevant decision.
- *Shareholder manager* (0, 0). This position is literally poor management. Little or no information input and exchange take place within the group, while the group itself has ultimate authority for the final decision.

The leadership style challenge for the project manager concerns heavy pressures, driving the decision making in the consensus area (northwest corner) of the grid, while often time pressures induce one to behave in an autocratic fashion. The good news from this leadership model is that leadership behaviors are flexible. It would be a mistake to assume that once identified as possessing a certain style, it is impossible to alter that style for different circumstances or situations. In fact, successful project managers have been shown to employ a great deal of flexibility in their use of leadership approaches. They may employ more authoritative practices when dealing with troublesome team members, consensus styles when working with technical people to collectively solve a problem, or a consultative autocrat style when developing project plans, duration estimates, or PERT (program evaluation and review technique) schedules. The typical realities of the project environment suggest that maximum returns derive from a flexible, thoughtful approach to project leadership styles. The reactive, one-dimensional project manager will find his or her leadership style may work well under some situations but is totally unsuited for others (Kangus and Lee-Kelly, 2000).

Key Comments from Practitioners

"The project manager is becoming more of a leader than the classical view of a manager. As a leader, the project manager is beginning to meld together the leadership and management responsibilities into a well-rounded capability to not only see the trees but also see the forest . . . and what's on the other side of the forest."

"Leadership in our company usually consists of being out in front, being able to talk the technical talk with the engineers, the financial talk with the accountant, and the management talk with the administrators. Our project leaders have to have the respect of the rest of the team, and it is never given easily—it has to be earned!"

"The best leader I ever saw seemed to instinctively understand how to relate to different groups and different situations. I've watched him go from a sweet person to an SOB and back to a sweet person again in about five minutes, depending upon the situation and the person he was dealing with. It wasn't an act—he just knew the tune he had to play with each person."

4. Communications and the Project Manager

As technology advances, communications challenges will increase. Each time information is exchanged, time is expended and project resources are consumed (Back, 2001). Communication of the project vision to all affected stakeholders can be a tremendously important step in the process (Reed, 2002). The Internet can provide a major tool for project management communications but also can be a significant consumer of time and energy (Giffen,

2002). As a result, we are faced with a conundrum: Research demonstrates the importance of maintaining clear lines of communication with all project stakeholders to improve the chances of success, and yet, because of the manner in which many communication mediums operate, it is not always clear how to generate the most effective messages and communicate them for maximum benefit.

A number of factors have emerged that make communications a greater challenge every year for the project manager, including the following:

- Increased project complexity
- Globalization of projects
- The Internet and all of its ramifications
- E-mail
- Virtual teams

Example: A pharmaceutical company assembles a virtual project team of 30 scientists and managers in six different countries. They interact using the most advanced teleconferencing technologies. The e-mail load is enormous. Even though the technology is marvelous, scheduling and executing a meeting is a challenge because of global time zone differences. As a result, the communications network and initially approved modes for communication do not perform nearly as well as hoped, causing project delays and numerous face-to-face meetings to clarify differences: something the project manager had hoped to avoid by investing in state-of-the-art networking!

The *PMBOK Guide,* 2000 Edition suggests four major communication processes:

- Communications planning
- Information distribution
- Performance reporting
- Administrative closure

The manner in which each of these communications steps is approached can have a huge impact on the viability of project team and stakeholder communications. Each of these steps must be carefully considered, its strengths and weaknesses assessed, and fallback positions identified. Done well, project communications processes are a hugely important factor in project success. Done poorly, they may result in conflicting messages, priorities, isolated pockets within the team, and an information vacuum.

Key Comments from Practitioners

"Mulenburg's 4th Law: All people problems are problems of communication. And all (at least most) project leader problems are people problems—the team, the line manager, upper management, the customer, suppliers, et cetera—or involve people that have to be convinced, persuaded, stroked, or put on the right path."

"The project manager has become more of a communicator and facilitator than experienced years ago. With the increase in technological changes, the higher emphasis on 'better, quicker, cheaper,' and more complex projects, this has occurred more out of necessity than from design. However, project managers who are good communicators are becoming more and more difficult to find."

"There are two keys to communication: speaking well and writing well. I see some project managers that simply cannot write coherently. In fact, their writing is embarrassing. On the other hand, some project managers lack the ability to put two sentences together once they are up in front of people. We don't have the luxury of hiding project managers who can't communicate."

"The key to our business is keeping the customer happy. The person who is responsible for that is the project manager. Communication skills are essential!"

5. Staffing and the Project Manager

Careful staffing of organizations has long been known as a secret to success. There are two common problems found in staffing project teams: (1) taking the first available resource regardless of their level of motivation, skill, or background in the project being considered or (2) having functional managers use project teams as a dumping ground for their poorer performers. Unfortunately, while successful project teams should be staffed by the best and brightest available, often the reverse is true. It is not difficult to see the end result from creating a team made up of personnel with low motivation or the suddenly discovered news that their boss considers them expendable. Alternatively, research suggests that when care is taken to staff project teams from available talent pools, the end result is much more promising for creating an environment for success. It has been suggested that the interview process can be made more effective by following these ten steps:

1. Write the job description.
2. Conduct the job analysis.
3. Select the behavioral dimensions.
4. Construct an interview form.
5. Recruit qualified candidates.
6. Study the résumés or applications.
7. Interview the applicants, record the data, defer judgment.
8. Score the interviews.
9. Use multiple interview consensus.
10. Make the hiring decision (Slevin, 1989, p. 287).

The implications are clear: Project teams that are staffed carefully, based on the hunt for the best talent available, are more likely to perform well. At a time when turnover in many critical knowledge-based industries is high (Abdel-Hamid, 1992), it would be a mistake to approach project team staffing in a way that will turn off potential valuable contributors to the project. Research suggests that individuals should be selected not only for their skills but also for the interpersonal capabilities and diversity that they bring to the team (McDowell, 2001; Melymuka, 2002).

Key Comments from Practitioners

> "I have not seen many changes. Most project managers within organizations indicate that they have little, if any, impact on staffing of the project. In most cases, PMs inherit their staff and simply have to live with who they have. While this works in some cases, in some cases it does not work. High success has been shown in organizations who involve the project manager in determining, negotiating, and in some cases actually hiring staff for the project."

> "As mentioned, project management is still pretty much an accidental profession in [our organization], which appears consistent with much of industry from the literature."

> "Project managers (and team members) are selected based on technical expertise, not on managerial skills, especially not on communication skills, unless they have already shown capability in project management."

6. Cross-Functional Cooperation and the Project Manager

Most projects have long required a team that includes members of different functional groups or members with diverse backgrounds. The cultures of their departments and differentiated manner in viewing the world often combine to make it extremely difficult to achieve cross-functional cooperation. Because cross-functional teams can greatly facilitate the successful implementation of projects, it is critical to better understand the mechanisms and motivations by which members of different functional groups are willing to collaborate on projects. Research suggests that four antecedent constructs can be important in accomplishing cross-functional team effectiveness (Pinto, Pinto, and Prescott, 1993):

- *Superordinate goals.* The need to create goals that are urgent and compelling, but whose accomplishment requires joint commitment and cannot be done by any individual department.
- *Accessibility.* Project team members from different functional departments cooperate when they perceive that other team members are accessible, either in person or over the telephone or e-mail system.
- *Physical proximity.* Project team members are more likely to cooperate when they are placed within physically proximate locations. For example, creating a project office or "war room" can enhance their willingness to cooperate.
- *Formal rules and procedures.* Project team members receive formal mandates or notification that their cooperation is required.

Cross-functional/multifunctional members of the project team can present a challenge for harmonious and enthusiastic teamwork, but able leadership can overcome the challenge (Rao, 2001). Cross-functional teams have been found particularly useful the greater the novelty or technical complexity of the project (Tidd and Bodley, 2002).

Key Comments from Practitioners

"Project managers tend to be very cooperative with cross-functional relationships. I do not believe this is a problem. The problem I have seen, and continue to see, is just the reverse. The cross-functional individuals—especially functional managers and personnel—tend not to cooperate with the project managers. This creates serious problems, since cooperation is a two-way street."

"In our organization, marketing and engineering do not get along. We have developed this mentality where these two groups actively work to discredit each other. Of course, no one stops to think about who the real loser in this situation is!"

7. Project Teams and the Project Manager

Organizations of the future are relying more and more on project teams for success. This movement implies that the team-building processes themselves may be a subobjective of the project (Bubshait and Farooq, 1999). One important discovery in team research in recent years has been the work of Gersick (1988; 1989), who investigated the manner in which groups evolve and adapt to each other and to the problem for which they were formed. Her research suggests that the old heuristic of "forming, storming, norming, performing, and adjouning" (Tuchman, 1965) that has been used to guide group formation and development for decades does not stand close scrutiny when examined in natural settings. Rather, coining a term from the field of biology, "punctuated equilibrium," she found that groups tend to derive their operating norms very quickly, working at a moderate pace until approximately the midpoint of the project, at which time a sense of urgency, pent-up frustrations, and a desire to re-address unacceptable group norms lead to an internal upheaval. The result is to create a better-performing project team. Gersick's work has been important for helping project managers understand how to better and more proactively manage the process by which their teams develop.

The chapter by Larson refers to the strengths and weaknesses of organizing projects as dedicated teams, an important issue.

Another crucial element to team success is knowledge management (Drew, 2003). Assembling knowledge management teams and distributing knowledge across the players can be extraordinarily important. The use of project management offices (PMOs) has been a useful tool in maintaining this center for knowledge management. Though the transition to PMOs is not always a smooth one, the ability to apply a centralized base of project knowledge to ongoing problems makes this process a useful one for promoting project success and helping develop the expertise of project team members.

Key Comments from Practitioners

"A proliferation of books and articles address the issue of teams without much emphasis on the project manager's role for how to become the leader of these teams.

Everyone is supposed to 'get along' somehow. The only well-oiled teams I have seen are that way because of a project leader with the skills to make it work."

8. Virtual Teams and the Project Manager

The world is going virtual. Two major universities recently received a multimillion-dollar grant to perfect the use of supercomputers in an application that enables people several thousand miles apart to work jointly together as if they were in the same room. With the increasing globalization of project management, teams comprising individuals who may never directly interact with each other are becoming commonplace. Their primary means of communication is through Internet, e-mail, and virtual meetings. This increase in virtual teamwork creates an entirely new level of complexity to the challenge of team building for project success (Adams and Adams, 1997; Townsend, DeMarie, and Henrickson, 1998).

Issues of cost, transportation, globalization, skill distribution, and a variety of other pressures have hastened the movement toward virtual team use in recent years (Elkins, 2000). Likewise, more and more global companies are experimenting with the process of partnering, which implies additional pressures on the virtual team (Bresnen and Marshall, 2002). However, because of a variety of factors beyond the control of project managers, such as organization structure or corporate culture, the degree to which project teams quickly acclimate to the virtual environment is quite variable. Some organizations have been able to develop and employ effective virtual teams, including adopting quite effective virtual team-building processes, while others have continued to find the technology difficult to master (Delisle, 2002). (See the chapter by Delisle.)

Key Comments from Practitioners

"With improved conferencing technology, virtual teams have become a reality (no pun intended). Distance and cost combine to make virtual teams a necessity in many instances. A major change that seems obvious is that to work, they need face time together initially, and periodically throughout the project."

"The big challenge we face is trying to make virtual teams act and work just like real teams. When you lose the sense of proximity to others working on the project, there is a feeling of disconnect. We require our virtual teams to make up in frequency of communications what they lack in proximity of communications."

9. Human Resource Policies and the Project Manager

For decades, human resource policies have been designed primarily to fulfill the needs of line management activities. Recent experiences have shown that the human resource function can become a full business partner with a project management process without losing integrity to line managers (Clark, 1999). In other words, the HR function is being designed more carefully to expedite project team development and staffing. The *PMBOK Guide,* 2000 Edition suggests the following major processes concerning project human resource management.

- Organizational planning
- Staff acquisition
- Team development

Another way in which HR is becoming more attuned to project management needs is through legal issues, compliance, and safety and health in the workplace. As projects are occasionally created in less than optimum work conditions, such as harsh environmental conditions or to work on projects with health or safety risks, human resource expertise has been tapped by project managers so they have a clearer understanding of issues of corporate liability and due diligence regarding safety and hiring practices.

Key Comments from Practitioners

> "In general, organizations have not responded to the needs of project management with respect to human resource policies. Most HR policies address the organizational needs from an ongoing, functional aspect with few addressing the particular aspects of project management. Key 'holes' exist in addressing the temporary nature of projects, matrix management, and the classical problem in accounting of accruals (organization) versus committed (project) costs."

10. Conflict and Negotiations and the Project Manager

The project manager is in a constant environment of conflict and negotiations (Kellogg, Orlikowski, and Yates, 2002). As Jeff Pinto and John Magenau explain in their chapter, the need to exercise the influential side of project management occurs for a number of reasons. For example, many organizations run projects within structures where departmental heads retain all control over project resources, requiring project managers to negotiate for their team resources. Other reasons for conflict and negotiation occur within projects where it is vital that the project manager and key team members understand important terms and conditions of contracts. The result is a circumstance in which project managers routinely exercise influence, deal with conflict, and negotiate with parties both inside and outside their own organization. Consequently, negotiation skills are considered to be an important part of the project manager's tool kit (Pravda and Garai, 1995). Project managers face a constant dilemma of determining how they are to acquire the authority to overrule resource and line managers in order to accomplish project objectives (Pinto et al., 1998; Vandersluis, 2001).

Key Comments from Practitioners

> "Project managers continue to be very involved with conflict management and negotiations. However, very little formal or informal training and development in these areas is prevalent. There still tends to be the concept of the 'accidental project manager,' resulting in throwing individuals into situations for which they are really not prepared."

"As with most elements of project management, organizations need to have specific and structured training and development programs that address project management in general and the specific areas of conflict management and negotiations."

"Our most productive project managers are the ones who instinctively understand that their job does not start and stop with the scheduling and administrative duties. They have to handle the hard duties, like negotiation and conflict, every day."

"Our brand-new project managers (the people who have never run projects before) are always shocked at how little power they have in this company. If they want to succeed, they learn that they better sharpen up their negotiation skills real fast!"

11. Power and Politics and the Project Manager

One of the least-talked-about aspects of project management duties involves the necessity of mastering the art of influence and political behavior. Attitudes regarding the use of politics, in this sense, point to an interesting dichotomy among managers in organizations. On the one hand, by a margin of almost 4 to 1, successful mid-level managers acknowledge that politics and influence are vital to performing their jobs effectively. On the other hand, by the same margins, these managers routinely affirm that the use of politics wastes company resources, is unpleasant to engage in, and is personally repugnant to them. The implications are interesting: On the one hand, managers do not like to use politics in their jobs, and yet on the other hand, they recognize that in order to successfully manage their projects, it is a vital skill to master.

Often negotiating the political terrain can be a greater challenge than the technical details of the project itself. All projects have numerous stakeholders. The political processes that characterize interactions between project managers and top managers are becoming evermore important in the success of new forms of organizations such as the project management office (Vandersluis, 2002). One solution to enhancing the project management process from the power and politics perspective is the institutionalization of an executive champion (Wreden, 2002). Champions can often serve to alleviate some of the political headaches that project managers accrue by serving as the point man for the project with key stakeholder groups. Champions exert their own kind of influence on behalf of the project. The difference is that because of the authority of status of champions, they are in a better position to help the project along.

Key Comments from Practitioners

"Many organizations are beginning to focus on the results of the two types of power: organizational power resulting in either compromise or compliance, and portable power resulting in commitment or loyalty. When faced with the question "would you rather have a project team that is committed and loyal or one that compromises and complies?" most recognize that the former creates a stronger project team and leads to a higher success rate. By focusing on this analysis, power and authority concerns tend to become resolved."

"I get a real kick out of the reaction of people who join our organization right out of college and are confronted with their first real taste of company politics. They can argue until they are blue that their opinion should win out because their way is 'the best,' but until they learn how to get things accomplished around here through the back door, they will never really be successful. All our successful project managers are successful politicians."

12. Project Organization and the Project Manager

The *PMBOK Guide,* 2000 Edition suggests two extremes of organizational form to a project:

- *The functional organization.* People and positions are grouped together according to the work they perform.
- *The projectized organization.* People are grouped together by project commitments, regardless of the functional background or expertise they possess.

As one moves toward the projectized organization, *PMBOK Guide,* 2000 Edition suggests four levels of matrix:

- *Weak matrix organization.* Limited project manager authority; 0 to 25 percent of personnel time dedicated to project management work; part-time project management administrative staff.
- *Balanced matrix organization.* Low to moderate project manager authority; 15 to 60 percent of personnel time dedicated to project management work; part-time project management administrative staff.
- *Strong matrix organization.* Moderate to high project manager authority; 50 to 95 percent of personnel time dedicated to project management work; full-time project management administrative staff.
- *Projectized organization.* High to almost total project manager authority; 85 to 100 percent of personnel time dedicated to project management work; full-time project management administrative staff.

The argument regarding the optimal type of organization suggests that there is a transition period in which organizations move from less-than-optimal structures, such as the functional structure, to those that are better able to support and sustain a project focus. Wheelwright and Clark (1992) refer to this movement as the drive toward "heavyweight" project organizations, in which power and decision-making authority are no longer shared between project and function, but rest solely in the hands of project managers. Research has clearly demonstrated the benefits to project success from crafting an organization form that supports these activities (Gobeli and Larson, 1987). Experience in interacting with senior managers indicates that more and more organizations are moving in the "projectized" direction (Lundin and Soderholm, 1995; Lundin and Midler, 1998).

Key Comments from Practitioners

"More and more I see project organizations aligned towards managing multiple projects as opposed to the single, large stand-alone project. This is especially true in internal corporate IT organizations, product development, and internal organizational support functions. It is also more prevalent in organizations that do projects for profit for external customers. Unfortunately, the concept of the project or program office in support of this is just now starting to catch on but suffers from a lack of documented experience, research, literature, and project management software tools that focus on multiple project management."

Conclusions

We have spent the better part of the past two decades researching, teaching, and consulting in project management and project organizations. Over that time, we have had the opportunity to witness the advent of a number of important innovations in the project management field in a variety of areas: scheduling, project monitoring and control, structural changes, and so forth. While all of these ideas have doubtlessly had a positive affect on the way projects are being run today, we find ourselves, in some sense, coming full circle as we note that the "true" determinants of successful project management are in many ways as clear today as they were two decades ago. Successful projects are those in which the "people side" has been well managed. All the technology in the world cannot overcome poor leadership, motivation, communications skills, team building, and so forth. On the other hand, project managers who take the time to perfect their skills in these critical areas continue to demonstrate that successful project management depends first and foremost on our ability to effectively manage the human resources for which we have been made responsible.

This chapter has offered a brief overview of some of the important themes in managing projects and the behavioral challenges that this process involves. As the chapter makes clear, the challenges are diverse; they broadly cover the gamut of individual and interpersonal relationships all the way to larger, organization theory issues of organization structure and cultural processes. As a result, it should be apparent that the types of skills needed to master the discipline of project management, whether from a practitioner or academic research perspective, requires both a depth of understanding and a breadth of knowledge that makes project management a truly unique undertaking. Successful project managers must learn first an appreciation of the myriad behavioral challenges they are going to face, as well as develop a commitment to pursuing knowledge in these diverse areas.

Acknowledgment

The authors are indebted to Walter Bowman, Gerald Mulenburg, and a number of other anonymous, practicing project managers for substantial insights and helpful comments in the preparation of this chapter.

References

Abdel-Hamid, T. K. 1992. Investigating the impacts of managerial turnover/succession on software project performance. *Journal of Management Information Systems* 9:127–145.

Adams, J. R., and L. L. Adams. 1997. The virtual projects: managing tomorrow's team today. *PM Network,* 11(1):37–41.

Back, W. E. 2001. Information management strategies for project managers. *Project Management Journal* 32:10–20.

Bresnen, M., and N. Marshall. 2002. The engineering or evolution of co-operation? A tale of two partnering projects. *International Journal of Project Management* 20:497–505.

Bubshait, A. A., and G. Farooq. 1999. Team building and project success. *Cost Engineering* 41:37–42.

Byham, W. C. 1981. *Targeted selection: A behavioral approach to improved hiring decisions.* Pittsburgh: Development Dimensions International.

Chapman, C. B., and S. Ward. 1997. *Project risk management: Processes, techniques and insights.* Chichester, UK: Wiley.

Clark, I. 1999. Corporate human resources and "bottom line" financial performance. *Personnel Review.* 28:290–307.

Delisle, C. 2002. Success and communication in virtual project teams. PhD diss., Department of Civil Engineering, University of Calgary. Calgary, Alberta.

Drew, R. 2003. Assembling knowledge management teams. *Information Strategy: The Executive's Journal* 19:37–42.

Dunn, S. C. 2001. Motivation by project and functional managers in matrix organizations. *Engineering Management Journal* 13:3–10.

Einsiedel, A. A. 1987. Profile of effective project managers. *Project Management Journal* 18(5):51–56.

Elkins, T. 2000. Virtual teams. *IIE Solutions.* 32:26–32.

Gersick, C. 1988. Time and transition in work teams: Towards a new model of group development. *Academy of Management Journal.* 31:9–41.

———. 1989. Making time predictable transitions in task groups. *Academy of Management Journal.* 32: 274–309.

Giffin, S. D. 2002. A taxonomy of internet applications for project management communication. *Project Management Journal.* 33:32–47.

Globerson, S. and O. Zwikael. 2002. The Impact of the *Project* Manager on *Project Management* Planning Processes. *Project Management Journal.* 33:58–65.

Gobeli, D. H., and E. W. Larson. 1987. Relative effectiveness of different project management structures. *Project Management Journal.* 18(2):81–85.

Jiang, J. J., G. Klein, and H. Chen. 2001. The relative influence of IS project implementation. *Project Management Journal.* 32(3):49–55.

Kangis, P., and L. Lee-Kelley. 2000. Project leadership in clinical research organizations. *International Journal of Project Management.* 18:393–342.

Kellogg, K., W. Orlikowski, and J. Yates. 2002. Enacting new ways of organizing: Exploring the activities and consequences of post-industrial work. *Academy of Management Proceedings.*

Kloppenborg, T. J. and W. A. Opfer. 2002. Forty years of project management research: Trends, interpretations, and predictions. In *The frontiers of project management research,* ed. D. P. Slevin, D. I. Cleland, and J. K. Pinto. Newtown Square, PA: Project Management Institute.

Loo, R. 2002. Journaling: A learning tool for project management training and team-building. *Project Management Journal.* 33:61–68.

Lundin, R. A., and F. Hartman. 2000. Pervasiveness of projects in business. In *Projects as business constituents and guiding motives,* ed. R. A. Lundin and F. Hartman. Dordrecht, Germany: Kluwer Academic Publishers.

Lundin, R. A., and C. Midler. 1998. *Projects as arenas for renewal and learning processes*. Norwell, MA: Kluwer Academic Publishers.

Lundin, R. A., and A. Soderholm. 1995. A theory of the temporary organization. *Scandinavian Journal of Management*. 11(4):437–455.

McDowell, S. W. 2001. Just-in-time project management. *IIE Solutions*. 33:30–34.

Melymuka, K. 2000. Born to lead projects. *Computerworld*. 34:62–64.

———. 2002. Who's in the house? *Computerworld*. 36.

Meredith, J. R. and S. J. Mantel. 2003. *Project Management: A Managerial Approach*. New York: Wiley.

Montague, J. 2000. Frequent, face-to-face conversation key to proactive project management. *Control Engineering*, Vol. 47:16–17.

Petterson, N. 1991. What do we know about the effective project manager? *International Journal of Project Management*. 9:99–104.

Pinto, J. K. and D. P. Slevin. 1988. Critical success factors across the project life cycle. *Project Management Journal* 67–75.

———. 1992. *Project implementation profile (PIP)*, Tuxedo, NY: XICOM INC.

Pinto, J. K., P. Thoms, P., J. Trailer, T. Palmer, and M. Govekar. 1998. *Project leadership from theory to practice*. Newtown Square, PA: Project Management Institute.

Pinto, M. B., J. K. Pinto, and J. E. Prescott. 1993. Antecedents and consequences of project team cross-functional cooperation. *Management Science*. 39:1281–1298.

PMBOK Guide 2000 *A guide to the project management body of knowledge*. Newtown Square, PA: Project Management Institute.

Posner, B. Z. 1987. What it takes to be a good project manager. *Project Management Journal*. 18(1):51–54.

Pravda, S. and G. Garai. 1995. Using skills to create harmony in the cross-functional team. *Electronic Business Buyer*. 21:17–18.

Rao, U. B., 2001. Managing cross-functional teams for project success. *Chemical Business*. 5:8–10.

Reed, B. 2002. Actually making things happen. *Information Executive*. 6:10–12.

Schmidt, C., and P. Dart. 1999. Disincentives for communicating risk: A risk paradox. *Information and Software Technology*. 41:403–412.

Schulz, Y. 2000. Project teams need a qualified full-time leader to succeed. *Computing Canada*. 26:11.

Slevin, Dennis P. 1989. *The whole manager*. Innodyne, Inc., Pittsburgh, PA.

Slevin, D. P., D. I. Cleland, and J. K. Pinto, eds. 2002. *The frontiers of project management research*. Newtown Square, PA: Project Management Institute.

Slevin, D. P. and J. K. Pinto. 1987. Balancing strategy and tactics project implementation. *Sloan Management Review*. 29(1):33–41.

———. 1991. Project leadership: understanding and consciously choosing your style. *Project Management Journal*. 22(1):39–47.

Tidd, J. and J. Bodley. 2002. The influence of project novelty on the new product development process. *R&D Management*. 32:127–139.

Townsend, A. M., S. DeMarie, and A. R. Hendrickson. 1998. Virtual teams: technology and the workplace of the future. *Academy of Management Executive*. 12(3):17–29.

Tuchman, B. W. 1965. Developmental sequence of small groups. *Psychological Bulletin*. 63:384–399.

Turner, J. R. 1993. *The handbook of project-based management*. New York: McGraw-Hill.

Vandersluis, C. 2001. Projecting your success. *Computing Canada*. 27:14–16.

Wheelwright, S. C., and K. Clarke. 1992. Creating project plans to focus product development. *Harvard Business Review*. 70(2):70–82.

Wideman, R. M. 1998. Project risk management. In *The Project Management Institute project management handbook*, ed. J. K. Pinto. Jossey-Bass Publishers and Project Management Institute.

Wreden, N. 2002. Executive champions: Vital links between strategy and implementation, *Harvard Management Update*. 7:3–6.

CHAPTER TWO

PROJECT MANAGEMENT STRUCTURES

Erik Larson

A project management structure provides a framework for launching and implementing project activities within a parent organization. A good structure appropriately balances the needs of both the organization and the project by defining the interface between the project and parent organization in terms of authority, allocation of resources, and eventual integration of project outcomes into mainstream operations (Gray and Larson, 2003).

In the past, many business organizations have struggled to create an effective system for implementing projects. One of the major reasons for this struggle is that projects contradict fundamental design principles embedded in traditional organizations. Projects are unique, one-time efforts with a distinct beginning and end. Most organizations are designed to efficiently manage ongoing activities. Efficiency is achieved primarily by breaking down complex tasks into simplified, repetitive activities, as characterized by assembly line production methods. Projects by their very nature are not routine and are therefore an anomaly in these work environments.

A second reason businesses find it difficult to effectively organize projects is that most projects are multidisciplinary in nature. For example, a new-product development project will likely involve the combined efforts of people from design, marketing, manufacturing, and finance. However, most organizations have been departmentalized according to functional expertise, with specialists from design, marketing, manufacturing, and finance residing in different units. Many researchers have noted that these groupings naturally develop unique customs, norms, values, and working styles that inhibit integration across functional boundaries (Lawrence and Lorsch, 1969; Harrison and Beyer, 1993; Majchrzak and Wang, 1996). Not only are there departmental silos, but managing projects poses the additional dilemma of who is in charge of the project. In most organizations, authority is distributed hierarchically across functional lines. Because projects span functional areas, identifying and legitimizing project management authority is often problematic.

In recent years there has been a dramatic shift in the management of projects. Global competition and technological advances have led to the compression of the product life cycle, the emergence of speed as a competitive advantage, and corporate downsizing that has placed a premium on implementing projects. Projects are no longer the exception but are central to the success of firms. As a result, organizations are being reengineered to support project implementation. Businesses are taking advantage of the latest information technology to coordinate and track the efforts of professionals both within and across organizations.

Four different approaches to project management organization are discussed here: functional organization, dedicated project teams, matrix structure, and network organization. Although not exhaustive, these structures and their variant forms represent the major approaches for organizing projects. The advantages and disadvantages of each of these structures are discussed, as well as some of the critical factors that might lead a firm to choose one form over others.

Organizing Projects within the Functional Organization

One approach to organizing projects is to simply manage them within the existing functional hierarchy of the organization. Once management decides to implement a project, the different segments of the project are delegated to the respective functional units, with each unit responsible for completing its segment of the project (see Figure 2.1). Coordination is maintained through normal management channels. For example, an optical scope manufacturing firm decides to differentiate its product line by offering high-end binoculars designed for avid bird watchers. Senior management authorizes the project, and different segments of the project are distributed to appropriate areas. The industrial design department is responsible for designing the binoculars. The production unit is responsible for devising the means for producing binoculars according to these new design specifications. The marketing department is responsible for gauging demand and price, as well as identifying distribution outlets. The overall project will be managed within the normal hierarchy, with the project being part of the working agenda of top management.

The functional organization is also commonly used when, given the nature of the project, one functional area plays a dominant role in completing the project or has a dominant interest in the success of the project. Under these circumstances, a high-ranking manager in that area is given the responsibility for coordinating the project. For example, the transfer of equipment and personnel to a new office would be managed by a top-ranking manager in the firm's facilities department. Likewise, a project involving the upgrading of the management information system would be managed by the information systems department. In both cases, most of the project work would be done within the specified department and coordination with other departments would occur through normal channels.

There are advantages and disadvantages for using the existing functional organization to administer and complete projects (Stuckenbruck, 1981; Youker 1977; Verma, 1995). The major advantages are the following:

FIGURE 2.1. FUNCTIONAL ORGANIZATION.

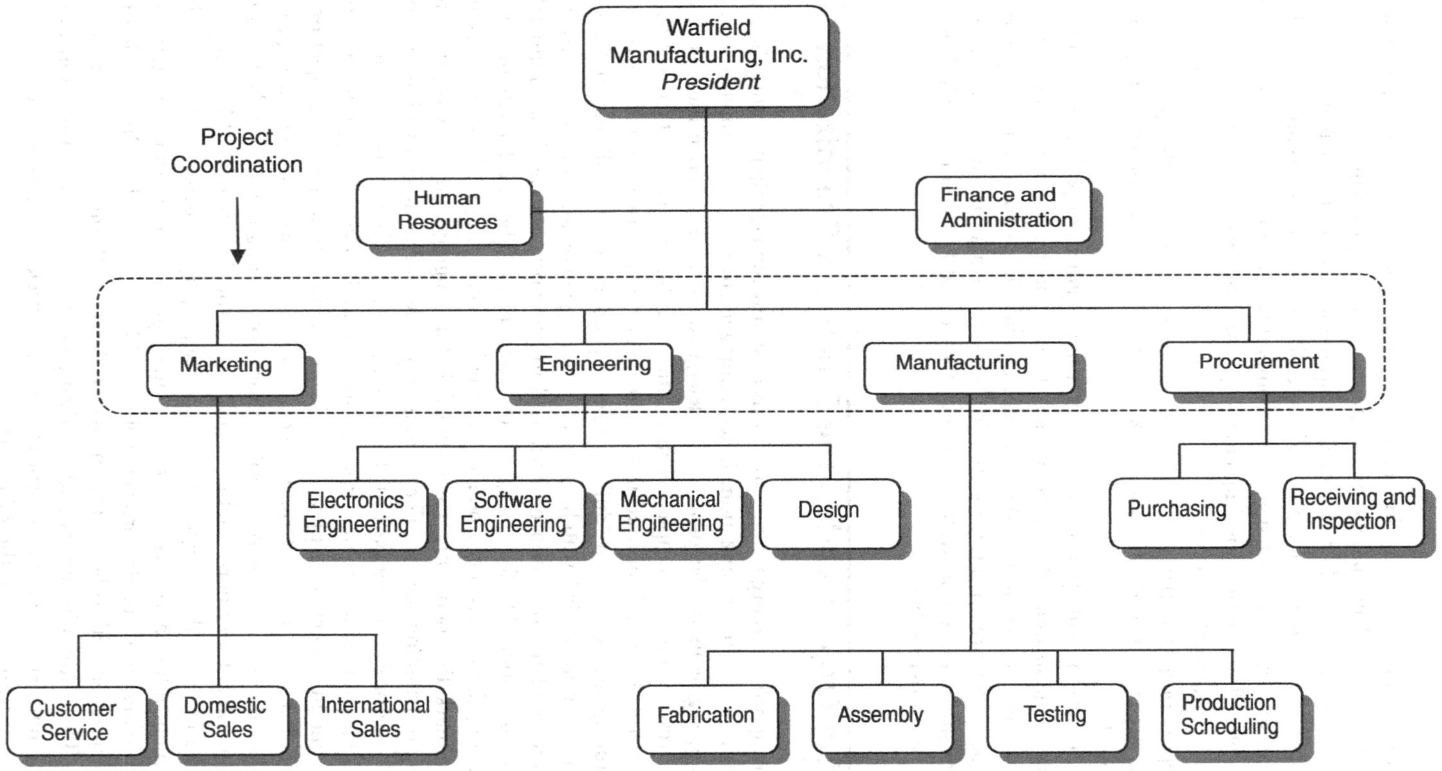

- *No change.* Projects are completed within the basic functional structure of the parent organization. There is no radical alteration in the design and operation of the parent organization.
- *Flexibility.* There is maximum flexibility in the use of staff. Appropriate specialists in different functional units can temporarily be assigned to work on the project and then return to their normal work. With a broad base of technical personnel available within each functional department, people can be switched among different projects with relative ease.
- *In-depth expertise.* If the scope of the project is narrow and the proper functional unit is assigned primary responsibility, in-depth expertise can be brought to bear on the most crucial aspects of the project.
- *Easy post-project transition.* Normal career paths within a functional division are maintained. While specialists can make significant contributions to projects, their functional field is their professional home and the focus of their professional growth and advancement.

Just as there are advantages for organizing projects within the existing functional organization, there are also disadvantages. These disadvantages are particularly pronounced when the scope of the project is broad and one functional department does not take the dominant technological and managerial lead on the project:

- *Low commitment.* Each functional unit has its own core routine work to do; sometimes project responsibilities get pushed aside to meet primary obligations. This difficulty is compounded when the project has different priorities for different units. For example, the marketing department may consider the project urgent, while the operations people consider it only of secondary importance.
- *Poor integration.* There may be poor integration across functional units. Cross-functional communication and coordination are slow, and limited at best, in most hierarchical organizations. Furthermore, there is a tendency to suboptimize the project, with respective functional specialists being concerned only with their segment of the project and not the total project.
- *Slow.* It generally takes longer to complete projects through this functional arrangement. This is in part attributable to slow response time—project information and decisions have to be circulated through normal management channels. Furthermore, the lack of horizontal, direct communication among functional groups contributes to rework as specialists realize the implications of others' actions after the fact.
- *Lack of ownership.* The motivation of people assigned to the project can be weak. The project may be seen as an additional burden that is not directly linked to their professional development or advancement. Furthermore, because they are working on only a segment of the project, professionals do not identify with the project. Lack of ownership discourages strong commitment to project-related activities.

Organizing Projects as Dedicated Teams

At the other end of the spectrum is the creation of dedicated project teams. These teams operate as independent units from the rest of the parent organization. Usually a full-time project manager is designated to recruit a core group of specialists who work full-time on the project. The project manager recruits necessary personnel from both within and outside the parent company. The subsequent team is physically separated from the parent organization and given marching orders to complete the project (see Figure 2.2).

The interface between the parent organization and the project teams will vary. In some cases, the parent organization prescribes administrative and financial control procedures over the project. In other cases, firms allow the project manager maximum freedom to get the project done. Such was the case of the original *Skunk Works* established by Kelly Johnson at Lockheed Martin. Kelly and a small, isolated band of Lockheed mavericks developed the revolutionary U-2 spy plane during the early 1950s. Lockheed went on to use this approach to develop a series of high-speed planes, including the F-117 Stealth Fighter (Johnson, Smith and Geary, 1990; Miller, 1996).

In the case of firms where projects are the dominant form of business, such as a construction firm or a consulting firm, the entire organization is designed to support project teams. Instead of one or two special projects, the organization consists of sets of quasi-independent teams working on specific projects. The main responsibility of traditional functional departments is to assist and support these project teams. For example, the marketing department is directed at generating new business that will lead to more projects, while the human resources department is responsible for managing a variety of personnel issues, as well as recruiting and training new employees. This type of organization is referred to in the literature as a *project* organization and is graphically portrayed in Figure 2.3.

FIGURE 2.2. DEDICATED PROJECT TEAM.

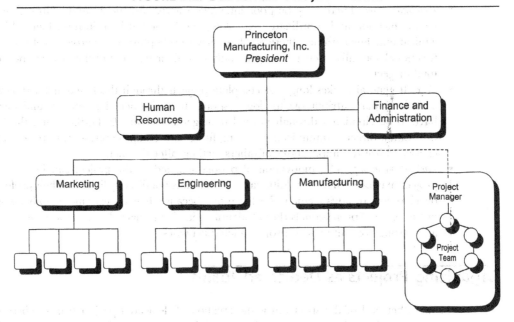

FIGURE 2.3. PROJECT ORGANIZATION STRUCTURE.

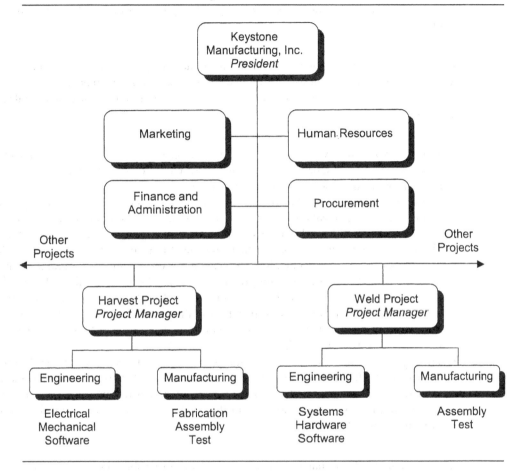

As in the case of functional organization, the dedicated project team approach has strengths and weaknesses (Stuckenbruck, 1981; Youker, 1979, Verma, 1995). The following are recognized as strengths:

- *Simple.* Does not directly disrupt ongoing operations. Other than taking away resources in the form of specialists assigned to the project, the functional organization remains intact with the project team operating independently.
- *Speed.* Perhaps the main reason for this is that participants devote their full attention to the project and are not distracted by other obligations and duties. Furthermore, response time tends to be quicker under this arrangement because most decisions are made within the team and are not deferred up the hierarchy.
- *Cohesion.* Participants share a common goal and personal responsibility toward the project and the team.

- *Cross-functional integration.* Specialists from different areas work closely together and, with proper guidance, become committed to optimizing the project, not their respective areas of expertise.

In many cases, the project team approach is the optimum approach for completing a project, when you view it solely from the standpoint of what is best for completing the project. Its weaknesses become more evident when the needs of the parent organization are taken into account:

- *Expensive.* Not only have you created a new management position (project manager), but resources are also assigned on a full-time basis. This can result in duplication of efforts across projects and a loss of economies of scale.
- *Internal strife.* A strong we/they divisiveness sometimes emerges between the project team and the parent organization. This divisiveness can undermine not only the integration of the eventual outcomes of the project into mainstream operations but also the assimilation of project team members back into their functional units once the project is completed.
- *Limited technological expertise.* Creating self-contained teams inhibits maximum technological expertise being brought to bear on problems. Technical expertise is limited somewhat to the talents and experience of the specialists assigned to the project. While nothing prevents specialists from consulting with others in the functional division, the we/they syndrome and the fact that such help is not formally sanctioned by the organization discourage this from happening.
- *Post-project assimilation.* Assigning full-time personnel to a project creates the dilemma of what to do with personnel after the project is completed. If other project work is not available, the transition back to their original functional departments may be difficult because of their prolonged absence and the need to catch up with recent developments in their functional areas.

A good example of internal strife is the saga of the successful Macintosh project team at Apple Computer. Steve Jobs, who at the time was both the chairman of Apple and the project manager for the Mac team, pampered his team with perks including at-the-desk massages, coolers stocked with freshly squeezed orange juice, a Bosendorfer grand piano, and first-class plane tickets. No other employees at Apple got to travel first-class. Jobs considered his team to be the elite of Apple and had a tendency to refer to everyone else as "bozos" who "didn't get it." Engineers from the Apple II division, which was the bread and butter of Apple's sales, became incensed with the special treatment their colleagues were getting.

One evening at Ely McFly's, a local watering hole, the tensions between Apple II engineers seated at one table and those of a Mac team at another boiled over. Aaron Goldberg, a long-time industry consultant, watched from his barstool as the squabbling escalated. "The Mac guys were screaming, 'We're the future!' The Apple II guys were screaming, 'We're the money!' Then there was a geek brawl. Pocket protectors and pens were flying. I was waiting for a notebook to drop, so they would stop and pick up the papers" (Carlton, 1997).

Although comical from a distance, the discord between the Apple II and Mac groups severely hampered Apple's performance during the 1980s. John Sculley, who replaced Steve Jobs as chairman of Apple, observed that Apple had evolved into two "warring companies" and referred to the street between the Apple II and Macintosh buildings as "the DMZ" (demilitarized zone) (Sculley, 1987).

Organizing Projects within a Matrix Arrangement

Matrix management is a hybrid organizational form in which a horizontal project management structure is overlaid on the normal functional hierarchy. In a matrix system, there are usually two chains of command, one along functional lines and the other along project lines. Instead of delegating segments of a project to different units or creating an autonomous team, project participants report simultaneously to both functional and project managers.

Companies apply this matrix arrangement in a variety of different ways. Some organizations set up temporary matrix systems to deal with specific projects, while matrix may be a permanent fixture in other organizations. Let us first look at its general application and then proceed to a more detailed discussion of finer points. Consider Figure 2.4. There are three projects currently under way: Silver, Gold, and Rust. All three project managers report to a director of project management, who supervises all projects. Each project has an administrative assistant, although the one for the Rust project is only part-time.

The Silver project involves the design and expansion of an existing production line to accommodate new metal alloys. To accomplish this objective, the project has assigned to it 5.5 people from manufacturing and 6 people from engineering. These individuals are assigned to the project on a part-time or full-time basis, depending on the project's needs during various phases of the project. The Gold project involves the development of a new product that requires the heavy representation of engineering, manufacturing, and marketing. The Rust project involves forecasting changing needs of an existing customer base. While these three projects, as well as others, are being completed, the functional divisions continue performing their basic, core activities.

The matrix structure is designed to optimally utilize resources by having individuals work on multiple projects as well as being capable of performing normal functional duties. At the same time, the matrix approach attempts to achieve greater integration by creating and legitimizing the authority of a project manager. In theory, the matrix approach provides a dual focus between the functional/technical expertise and project requirements that is missing in either the project team or functional approach to project management. The project manager is responsible for integrating functional input and overseeing the completion of the project. Functional managers usually "own" the resources in their area and are responsible for overseeing the functional contribution to the project. See Table 2.1 for a further delineation of the two roles.

Different Matrix Forms

In practice there are really different kinds of matrix systems, depending on the relative authority of the project and functional managers (Larson and Gobeli, 1985; Smith and

FIGURE 2.4. MATRIX ORGANIZATION STRUCTURE.

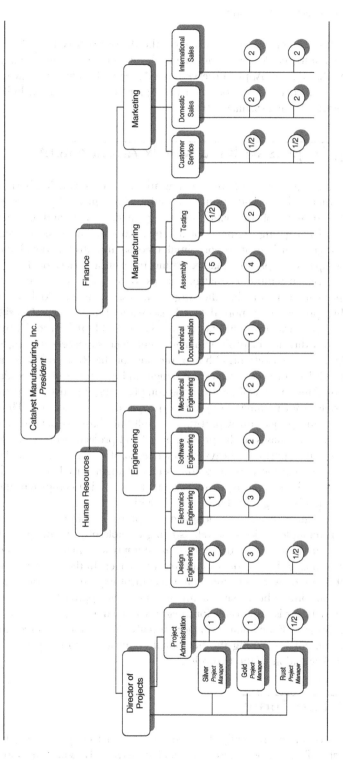

TABLE 2.1. DIVISION OF PROJECT MANAGER AND FUNCTIONAL MANAGER RESPONSIBILITIES IN A MATRIX STRUCTURE.

Project Manager	Negotiated Issues	Functional Manager
What has to be done?	Who will do the task?	How will it be done?
When should the task be done?	Where will the task be done?	How will the project involvement impact normal functional activities?
How much money is available to do the task?	Why will the task be done?	How well has the functional input been integrated?
How well has the total project been done?	Is the task satisfactorily completed.	

Reinertsen, 1995; Bowen, Clark, Holloway, and Wheelwright, 1994). *A weak matrix*, also called a lightweight or functional matrix, is one in which the balance of authority strongly favors the functional managers. A *balanced matrix*, also called a middleweight matrix, is used to describe the traditional matrix arrangement. *A strong matrix*, also called heavyweight or project, is one in which the balance of authority is strongly on the side of the project manager.

The relative difference in power between functional managers and project managers is reflected along a number of related dimensions. One such dimension is level of reporting relationship. A project manager who reports directly to the vice president of product development has more clout than a marketing manager who reports to a regional sales manager. Location of project activities is another subtle but important factor. A project manager wields considerably more influence over project participants if they work in his office than if they perform their project-related activities in their functional offices. Likewise, the percentage of full-time staff assigned to the project contributes to relative influence. Full-time status implies transfer of obligations from functional activities to the project.

One other significant factor is who is responsible for conducting performance appraisals and compensation decisions. In a weak matrix, the project manager is not likely to have any direct input in the evaluation of participants who worked on the project. This would be the sole responsibility of the functional manager. Conversely, in a strong matrix, the project manager's evaluation would carry more weight than the functional manager's. In a balanced matrix, either input from both managers is sought, or the project manager makes recommendations to the functional manager, who is responsible for the formal evaluation of individual employees. Often companies will brag that they use a strong, project-oriented matrix only to find upon closer examination that the project managers have very little say over the evaluation and compensation of personnel.

Ultimately, whether the matrix is weak or strong is determined by the extent to which the project manager has authority over participants. Authority may be determined informally by the persuasive powers of managers involved and the perceived importance of the project, or formally by the prescribed powers of the project manager. Here is a thumbnail sketch of the three kinds of matrices:

- *Weak matrix.* This form is very similar to a functional approach with the exception that there is a formally designated project manager responsible for coordinating project activities. Functional managers are responsible for managing their segment of the project. The project manager basically acts as a staff assistant who draws the schedules and checklists, collects information on status of work, and facilitates project completion. The project manager has indirect authority to expedite and monitor the project. Functional managers call most of the shots and decide who does what and when the work is completed.

- *Balanced matrix.* This is the classic matrix in which the project manager is responsible for defining what needs to be accomplished, while the functional managers are concerned with how it will be accomplished. More specifically, the project manager establishes the overall plan for completing the project, integrates the contribution of the different disciplines, sets schedules, and monitors progress. The functional managers are responsible for assigning personnel and executing their segment of the project according to the standards and schedules set by the project manager. The merger of "what and how" requires both parties to work closely together and jointly approve technical and operational decisions.

- *Strong matrix.* This form attempts to create the "feel" of a project team within a matrix environment. The project manager controls most aspects of the project, including scope trade-offs and assignment of functional personnel. The project manager controls when and what specialists do, and has final say on major project decisions. The functional manager has title over his/her people and is consulted on a need basis. In some situations, a functional manager's department may serve as a "contractor" for the project, in which case they have more control over specialized work. For example, the development of a new series of laptop computers may require a team of experts from different disciplines working on the basic design and performance requirements within a project matrix arrangement. Once the specifications have been determined, final design and production of certain components (i.e., power source) may be assigned to respective functional groups to complete.

Both matrix management in general and in its specific forms have unique strengths and weaknesses (Stuckenbruck, 1981; Youker, 1979; Larson and Gobeli, 1987; Verma, 1995). The advantages and disadvantages of matrix organizations in general are noted in the list that follows, which also briefly highlights specifics concerning different forms.

- *Efficient resource utilization.* Resources can be shared across multiple projects as well as within functional divisions. Individuals can divide their energy across multiple projects on an as-needed basis. This reduces duplication required in a pure project team structure.

- *Dual project/functional focus.* A strong project focus is provided by having a formally designated project manager who is responsible for coordinating and integrating contributions of different units. This helps sustain a holistic approach to problem solving that is often missing in the functional organization. At the same time, functional input reinforces rigor and high quality standards.

- *Post-project assimilation.* Because the project organization is overlaid on the functional divisions, the project has reasonable access to the entire reservoir of technology and expertise of functional divisions. Furthermore, unlike dedicated project teams, specialists maintain ties with their functional group, so they have a home port to return to once the project is completed.

- *Flexible.* Matrix arrangements provide for flexible utilization of resources and expertise within the firm. In some cases, functional units may provide individuals who are managed by the project manager. In other cases, the contributions are monitored by the functional manager.

The strengths of the matrix structure are considerable. Unfortunately, so are the potential weaknesses. In large part, this is because a matrix structure is more complicated and the creation of multiple bosses represents a radical departure from the traditional hierarchical authority system. Furthermore, one does not simply install a matrix structure over night. Experts argue that it takes three to five years for a matrix system to fully mature (Davies and Lawrence, 1977; Graham and Englund, 1997). So many of the weaknesses described in the following represent growing pains:

- *Dysfunctional conflict.* The matrix approach is predicated on creative tension between functional managers and project managers who bring critical expertise and perspectives to the project. Such tension is viewed as healthy and a necessary mechanism for achieving an appropriate balance between complex technical issues and unique project requirements. Unfortunately, sometimes legitimate conflict can spill over to a more personal level, resulting from conflicting agendas and accountabilities. Worthy discussions can degenerate into heated arguments that engender animosity among the managers involved.
- *Infighting.* Any situation in which equipment, resources, and people are being shared across projects and functional activities lends itself to conflict and competition for scarce resources. Infighting can occur among project managers, who are primarily interested in what is best for their project.
- *Stress.* Matrix management violates the management principle of unity of command. Project participants have at least two bosses—their functional head and one or more project managers. Working in a matrix environment can be stressful if you are being told to do three conflicting things by three different managers.
- *Slow.* In theory, the presence of a project manager to coordinate the project should accelerate the completion of the project. In practice, decision making can get bogged down, as agreements have to be forged across multiple functional groups. This may be especially true for the balanced matrix.

The advantages and disadvantages are not necessarily true for all three forms of matrix. The strong matrix is likely to enhance project integration, diminish internal power struggles, and ultimately improve control of project activities and costs. On the downside, technical quality may suffer because functional areas have less control over their contributions. Finally, internal strife may erupt as the members develop a strong team identity and relations outside the project become strained.

The weak matrix is likely to improve technical quality as well as provide a better system for managing conflict across projects because the functional manager assigns personnel to different projects. The problem is that functional control is often maintained at the expense of poor project integration. The balanced matrix can achieve better balance between technical and project requirements, but it is a very delicate system to create and manage and is more likely to succumb to many of the problems associated with the matrix approach (Larson and Gobeli, 1987).

Organizing Projects within Network Organizations

The turn of the century has seen a radical shift in the organizational architecture of business firms. Corporate downsizing and cost control have combined to produce what some have called "network organizations" (Miles and Snow, 1995; Miles, Snow, Mathews, Miles, and Coleman, 1997). In theory, a network organization is an alliance of several organizations for the purpose of creating products or services for customers. This collaborative structure typically consists of several satellite organizations beehived around a "hub" or "core firm." The core firm coordinates the network process and provides one or two core competences, such as marketing or product development. For example, Cisco Systems mainly designs new products and utilizes a constellation of suppliers, contract manufacturers, assemblers, and other partners to deliver products to their customers. Likewise, Nike, another prime example of this kind of organization, provides marketing expertise for its sports footwear and apparel. The key organizing principle is that, instead of doing everything in-house, a firm outsources key activities to other businesses with the requisite competencies.

The shift toward network organizations is easily apparent in the film industry. During the golden era of Hollywood, huge, vertically integrated corporations made movies. Studios such as MGM, Warner Brothers, and Twentieth Century Fox owned large movie lots and employed thousands of full-time specialists—set designers, camera people, film editors, directors, and even actors. Today, most movies are made by a collection of individuals and small companies who come together to make films project-by-project. This structure allows each project to be staffed with the talent most suited to its demands rather than choosing from only those people the studio employs. This approach also disperses financial risk across many organizations.

The network approach is now being applied to a wide range of projects. For example, see Figure 2.5.

Figure 2.5 depicts a situation in which a new reclining chair is being developed. The genesis for the chair comes from a mechanical engineer who suffers from a bad back. She developed the prototype in her garage. The inventor negotiates a contract with a furniture firm to develop and manufacture the recliner. The furniture company, in turn, creates a project team of manufacturers, suppliers, and marketing firms to create the new chair. Each participant adds requisite expertise to the project. The furniture firm provides its brand name and distribution network to the project. Tool-and-die firms provide customized parts, which are delivered to a manufacturing firm that will assemble the chairs. Marketing firms refine the design and test-market potential names and options. A project manager is assigned by the furniture firm to work with the inventor and the other parties to complete the project.

Collectively, the project is the summation of different structures. For example, the tool-and-die firm may assign a dedicated team to create the process for producing the custom

FIGURE 2.5. NETWORK PROJECT.

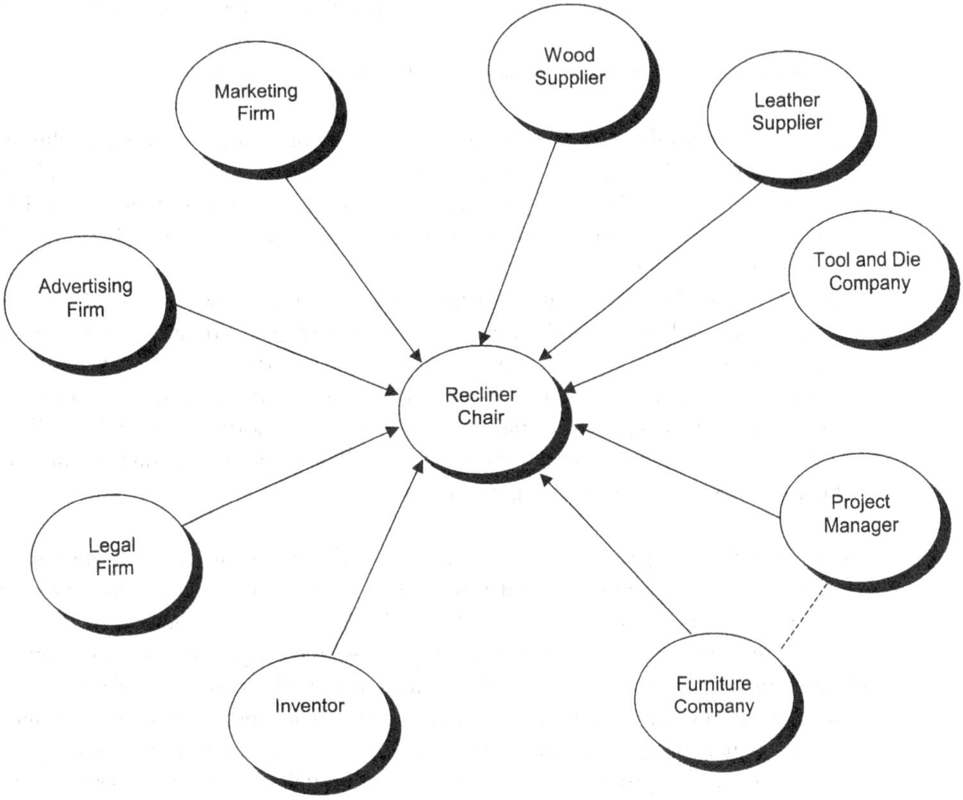

parts. The marketing work may be performed within functional departments, while the project manager works on this project part-time.

The advantages of network projects are many:

- *Cost reduction.* The most noteworthy is cost reduction. Companies can secure free-market prices for contracted services, especially if the work can be outsourced offshore. Furthermore, overhead costs are dramatically cut, since the company no longer has to maintain internally the contracted services.
- *High level of expertise.* A high level of expertise and technology can be brought to bear on the project. A company no longer has to keep up technological advances. Instead, it can focus on developing its core competencies and hire firms with the know-how to work on relevant segments of the project.

- *Increased flexibility.* Organizations are no longer constrained by their own resources but can pursue a wide range of projects by combining their resources with talents of other companies. Small companies can instantly go global by working with foreign partners.

The disadvantages of network projects are less well documented:

- *Breakdowns in coordination.* Synchronizing the work of professionals from different organizations can be challenging, especially if the project work requires close collaboration and mutual adjustment. This form of project management structure tends to work best when each party, as in the case of most construction projects, is responsible for a well-defined, independent deliverable.
- *Loss of control.* The core team depends on other organizations over which they do not have direct authority. While long-term survival of participating organizations depends upon performance, a project may falter when one partner fails to deliver.
- *Conflict.* Finally, networked projects are more prone to interpersonal conflict, since the different participants do not share the same values, priorities, and culture. Trust, which is essential to project success, can be difficult to forge when interactions are limited and people come from different organizations.

Most networked projects operate in a virtual environment in which people are linked by computers, faxes, computer-aided design systems, and video teleconferencing and sometimes rarely, if ever, see one another face-to-face. Many projects are being networked across time zones so that work never stops. For example, members from one organization work on a software project in New York and then pass their work at the end of day to another organization in Hawaii. The Hawaiian team passes their work to a team in India, which in turn passes their work to a Dutch firm. Although it is too early to say how applicable this 24-hour tag team approach to project management will be, it exemplifies the potential that exists given the information technology that is available today.

Within networked projects, people come and go as services are needed, much like a matrix structure. They are not formal members of one organization, just technical experts who form a temporary alliance with an organization, fulfill their contractual obligations, and then move on to the next project.

Choice of Project Management Structure

There is growing empirical evidence that project success is directly linked to the amount of autonomy and authority project managers have over their projects (Gray, Dworatschek, Gobeli, Knoepfel, and Larson, 1990; Gobeli and Larson 1987; Brown and Eisenhardt, 1995). For example, Larson and Gobeli (1989) studied 546 development projects and found that projects relying on functional organization or a functional matrix were less successful than those that used a balanced matrix, project matrix, or project team. Furthermore, the project matrix outperformed the balanced matrix in meeting schedules and outperformed the dedicated project team in controlling cost. However, this and other studies have focused only on what is best for managing specific projects. It is important to remember what was stated in the beginning of the chapter—that the best system balances the needs of the project with those of the parent organization. So what project structure should an organization use?

This is a complicated question with no precise answers. A number of issues need to be considered at both the organization and project level.

Organizational Considerations

At the organization level, the first question that needs to be asked is how important is project management to the success of the firm? What percentage of core work involves projects? If over 75 percent of work involves projects, an organization should consider a fully projectized organization. If an organization has both standard products and projects, a matrix arrangement would appear to be appropriate. If an organization has very few projects, a less formal arrangement is probably all that is required. Temporary task forces could be created on an as-needed basis and the organization could outsource project work.

A second key question is resource availability. Remember, the matrix evolved out of the necessity to share resources across multiple projects and functional domains, while at the same time creating legitimate project leadership. For organizations that cannot afford to tie up critical personnel on individual projects, a matrix or network system would appear to be appropriate. An alternative would be to create a dedicated team but outsource project work when resources are not available internally.

A third consideration is whether the organization has a firm grasp of its priorities and can effectively communicate the relative importance of different projects. This is particularly true for matrix structures in which resources are shared across projects and functions. An effective project priority system needs to be in place to guide resource assignments and avoid the infighting that can unravel a matrix into a chaos. If priorities are not established, dedicated or network project teams are advised to sidestep the resource contention issue. However, this can carry an expensive price tag for top management failing do to their job (Graham and Englund, 1997).

The final consideration is the culture of the firm. The metaphor that is used to describe the relationship between organizational culture and project management is that of a riverboat trip. Culture is the river and the project is the craft. Organizing and completing projects within an organization in which the culture is conducive to project management is like paddling downstream: Much less effort is required, and the natural force of the river generates progress toward the destination. In many cases, the current can be so strong that steering is all that is required. Such is the case for projects that operate in a project-friendly environment where teamwork and cross-functional cooperation are the norms, where there is a strong commitment to excellence, and where healthy conflict is voiced and dealt with quickly and effectively.

Conversely, trying to complete a project in an organization in which the dominant culture inhibits effective project management is like paddling upstream: Much more time, effort, and attention are needed to reach the destination. This would be the situation in cultures where cross-functional competition is high, where risks are to be avoided, and where getting ahead is based less on performance and more on cultivating favorable relationships with superiors. In such cases, the project manager not only has to overcome the natural obstacles of the project but also has to overcome the prevailing negative forces inherent in the culture of the organization.

The implications of this analogy are obvious but important. Greater project authority and resources are necessary to complete projects that encounter a strong, negative cultural current. Conversely, less formal authority and fewer dedicated resources are needed to complete projects in which the cultural currents generate the behavior and cooperation essential to project success. The key issue is the degree of interdependency between the parent organization and the project team and the corresponding need to create a unique project culture conducive to successful project completion (Kerzner, 1997; Brown, Grove, Kelly, and Rana, 1997; Jassawalla and Shashittalk, 2002).

Project Considerations

At the project level, the question is how much autonomy does the project need in order to be successfully completed. Hobbs and Menard (1993) identify seven factors that should influence the choice of project management structure:

- Size of project
- Strategic importance
- Novelty and need for innovation
- Need for integration (number of departments involved)
- Environmental complexity (number of external interfaces)
- Budget and time constraints
- Stability of resource requirements

Hobbs and Ménard advise that the higher the levels of these seven factors, the more autonomy and authority the project manager and project team needs to be successful. This translates into using either a dedicated project team or a project matrix structure. For example, these structures should be used for large projects that are strategically critical and are new to the company, thus requiring much innovation. These structures would also be appropriate for complex, multidisciplinary projects that require input from many departments, as well as for projects that require constant contact with customers to assess their expectations.

Shenhar and his colleagues have taken a slightly different approach through their efforts to develop a meaningful framework for classifying different kinds of projects (Shenhar, Dvir, Lechler, and Poli, 2002; Shenhar, 1998; Shenhar and Dvir, 1996—see also the chapter by Shenhar and Dvir toward the end of this book). They use three dimensions to distinguish different kinds of projects: *Un*certainty at the moment of project initiation, *C*omplexity as reflected in the number and variety of elements/disciplines required to complete the project, and *Pa*ce with regard to speed and criticality of time. Together, these dimensions form the *UCP model* for distinguishing different kinds of projects. Shenhar and his colleagues argue that, depending upon how a project is configured on these dimensions, different project management techniques and styles are required for success. For example, "critical/blitz projects," which are most urgent and essential to the success of the firm, require dedicated project teams and strong top management support. Conversely, projects that lack inherent time-to-market pressure tend to take longer than planned without tight management.

Evidence of this contingency approach can be found in firms that have created a flexible management system that organizes projects according to project requirements. For example, Chaparral Steel, a mini-mill that produces steel bars and beams from scrap metal, classifies projects into three categories: advanced development, platform, and incremental (Bowen, Clark, Holloway, and Wheelwright, 1994) Advanced-development projects are high-risk endeavors involving the creation of a breakthrough product or process. Platform projects are medium-risk projects involving system upgrades that yield new products and processes. Incremental projects are low-risk, short-term projects that involve minor adjustments in existing products and processes. At any point in time, Chaparral might have 40 to 50 projects under way, of which only one or two are advanced, three to five are platform projects, and the remainder are small, incremental projects. The incremental projects are almost all done within a weak matrix, with the project manager coordinating the work of functional subgroups. A strong matrix is used to complete the platform projects, while dedicated project teams are typically created to complete the advanced-development projects. It is anticipated that more and more companies will be using this mix-and-match approach to managing projects.

Summary

Four different kinds of project management structures have been described and their relative strengths and weaknesses discussed. No one structure is optimum. Choice depends upon the needs of the organization and the requirements of the project. The future should see more and more organizations adopt a flexible project management system in which different structures will be used for different projects. Sophisticated information systems will be used to provide real-time progress reports and optimally utilize resources across projects. Still, the importance of organizational culture cannot be underestimated. Matrix management can flourish in one organization and be a total disaster in another. The reason for this was not matrix per se, but differences in the culture of the two organizations.

References

Bowen, H. K., K. B. Clark, C. A. Hollaway, and S. C. Wheelwright, eds. 1994. *The perpetual enterprise machine*. New York: Oxford Press.

Brown, P., S. Grove, R. Kelly, and S. Rana. 1997. Is cultural change important in your project? *PM Network*. January: 48–51.

Brown, S., and K. M. Eisenhardt. 1995. Product development: Past research, present findings, and future directions. *Academy of Management Review* 20(2):343–378.

Carleton, J. 1997. *Apple: The inside story of intrigue, egomania, and business blunders.* New York: Random House.

Davies, S. M., and P. R. Lawrence. 1977. *Matrix.* Reading, MA: Addison-Wesley.

Gobeli, D. H., and E. Larson. 1987. The relative effectiveness of different project management structures. *Project Management Journal.* 18(2):81–85.

Graham, R. J., and R. L. Englund. 1997. *Creating an environment for successful projects: The quest to manage project management.* San Francisco: Jossey-Bass.

Gray C., and E. Larson. 2003. *Project management: A managerial approach.* New York: McGraw-Hill/Irwin.

Gray, C., S. Dworatschek, D. H. Gobeli, H. Knoepfel, and E. Larson. 1990. International comparison of project organization structures: Use and effectiveness. *International Journal of Project Management.* 8(1):26–32.

Harrison, M. T., and J. M. Beyer. 1993. *The culture of organizations.* Upper Saddle River, NJ: Prentice Hall.

Hobbs, B., and P. Ménard. 1993. Organizational choices for project management. In *The AMA handbook of project management,* ed. Paul Dinsmore. New York: AMACOM.

Jassawalla, A. R., and H. C. Sashittal. 2002. Cultures that support product-innovation processes. *Academy of Management Executive.* 15(3):42–54.

Johnson, C. L., M. Smith, and L. Geary. 1990. *Kelly: More than my share of it all,* Washington D.C.: Smithsonian Institute Publications.

Kerzner, H. 1997. *In search of excellence in project management.* New York: Van Nostrand Reinhold.

Larson, E., and D. Gobeli. 1985. Project management structures: Is there a common language? *Project Management Journal.* 16(2):40–44.

———. 1989. Significance of project management structure on development success. *IEEE Transactions in Engineering Management.* 36(2):119–125.

Lawrence, P. R., and J. W. Lorsch. 1969. *Organization and environment.* Homewood, IL: Irwin.

Majchrzak, A., and Q. Wang. 1996. Breaking the functional mind-set in process organizations. *Harvard Business Review.* (September–October): 93–99.

Miles, R. E., and C. C. Snow. 1995. The new network firm: A spherical structure built on a human investment philosophy. *Organizational Dynamics.* Spring: 5–18.

Miles, R. E., C. C. Snow, J. A. Mathews, G. Miles, and H. J. Coleman. 1997. Organizing in the knowledge age: Anticipating the cullular form. *Academy of Management Executive.* 11(4):7–24.

Miller, J. 1996. *Lockheed Martin's skunk works.* New York: Speciality Publications.

Sculley, J. 1987. *Odyssey: Pepsi to Apple.* New York: Harper & Row.

Shenhar A. J. 1998. From theory to practice: Toward a typology of project management styles. *IEEE Transactions in Engineering Management.* 41(1):33–48.

Shenhar, A. J., and D. Dvir. 1996. Toward a typological theory of project management. *Research Policy* 25:607–632.

Shenhar, A. J., D. Dvir, T. Lechler, and M. Poli. 2002. One size does not fit all: True for projects, true for frameworks. *Frontiers of Project Management Research and Application. Proceedings of PMI Research Conference,* 99–106. Seattle.

Stuckenbruck, L. C. 1981. *Implementation of project management.* Upper Darby, PA: Project Management Institute.

Smith, P. G., and D. G. Reinersten. 1995. *Developing products in half the time.* New York: Van Nostrand Reinhold.

Verma, V. K. 1995. *Organizing projects for success: The human aspects of project management.* Newtown Square, PA: Project Management Institute.

Youker, R. 1977. Organizational alternatives for project management. *Project Management Quarterly.* 8:24–33

CHAPTER THREE

CONTEMPORARY VIEWS ON SHAPING, DEVELOPING, AND MANAGING TEAMS

Connie L. Delisle

What defines a project team? What has really changed in nearly 20 years in team research and practice? How have Western scientific views influenced our understanding of team development and management? We have knowledge and wisdom to change, but why do we not do so, or even act in ways contradictory to successful team building and management? Why push technical solutions in a global business "game" where all the players more or less understand the rules? To help address these really quite deep questions, this chapter begins to challenge the very assumptions about behaviors and attitudes that dominate our thinking about teams.

The chapter employs a deductive approach in examining teams and team building in order to take the broader business context into consideration. Part I takes a high-level look at the business context that impacts teams in a variety of potential ways. Part II concentrates on the more micro-level aspects of team development to consider when growing teams in ways that can be creative as well as efficient. Specifically:

- It covers the identification of human resources and the understanding of competencies, and personalities, as primary considerations in selecting team members. Once a team is selected, it needs to "grow" or be developed;
- It provides a brief discussion of evolutionary models of team development, as well as current thinking about how teams perform over time. In doing so, it challenges some long-held beliefs about how team building interacts with core environmental (external) and behavioral factors that influence how teams are developed. Although many environmental factors exist, team composition, demographics, and how teams view their progress in time seem to have a significant influence on team development. From a behavioral

point of view, mental toughness, cohesion, and motivation appear as key forces that shape team development;

- It introduces virtual project teams as a context that reflects how team development has shifted since its inception, adding to our understanding of how teams develop.

Part III takes a brief look at team management, specifically at the most common team pathologies. Suggested strategies to mitigate or avoid pathologies are also provided for consideration.

Part I: Teams and Forces Shaping Teams

Team by Definition or Characteristics

The acronym TEAM means, to some people, "Together Everyone Achieves More," but what does "team" really mean? There seems to be an ever-increasing number of conflicting definitions where it once appeared that this simple term was commonly understood. Team members, according to the PMBOK Guide (PMI, 2000), are those who report to the project manager or leader; nothing is mentioned about who does the work. In contrast, the Association of Project Management (UK) (Dixon, 2000) presents a brief discussion on teams in Section 7, focusing on describing characteristics of effectiveness rather than being prescriptive on what a team does.

What about the term "project"? According to the PMBOK Guide, a *project* is a temporary endeavor whose purpose is in producing a unique product or services (PMI, 2000). Consider process teams (e.g., manufacturing), which are supposed to act in predictable, repetitive, standardized ways with project teams (e.g., on a software project), which will be working within a defined start and end period, where the work is unique or difficult to standardize and that may take a long time to conclude. Rather than focusing on the correctness of the definition, effort may be better spent in finding a set of core characteristics that reflect what is happening in the real world.

Trends Impacting Teams

The division of labor creates and is in turn impacted by trends concerning the selection, development, and management of project teams. Five of the most palpable trends are briefly introduced and are then followed by a few practical responses to consider. These trends do not fit into categories labeled "drivers and barriers" (Wilemon and Baker, 1988; Miller, 1988), because changes in constraints over time will determine in part whether a driver becomes a barrier or vice versa. For example, crisis management may be considered a barrier to organization learning or a driver because it may trigger a recognition concerning the need to change.

Crisis Management. Crisis management seems to be the norm rather than the exception. The self-perpetuating problem stems from organizations losing their ability to plan and prioritize because they practice crisis management to handle dramatically increased workloads, which in turn further increases workload (Duxbury and Higgens, 2001). Thomas, Delisle, and Jugdev (2002) identify that senior executives often react to the time pressure when in organizational crises by maintaining the status quo (deny the severity of the crisis) or purchase a solution from outside their organization. To maintain the status quo, they may appoint internal project members who do not have the qualifications or expertise to fix the problems, resulting in the creation of "accidental project managers" who often make the situation worse. When the crunch is on, people either "hide out" or engage in dysfunctional behaviors marked by withdrawing or banding together in nonproductive cliques or preparing themselves for exiting the organization (DeMarco and Lister, 1987). This limits communication and the flow of information even further.

The management level response is to do the following:

- Treat organizational crises as a symptom of a more serious and systemic internal problems.
- Avoid buying into fear campaigns and reflexively reacting to crises. Focus on creating adaptive triggers or situations where employees can provide input, ideas and ask deep questions about the organization's values and beliefs and the way it conducts business.
- Notify employees of changes early, be honest and communicate often (Blount and Janicik, 2001).

Globalization and the Highly Skilled Mobile Workforce. The blurring of lines between temporal, geographical, and organizational boundaries impacts teams in many organizations. This has not been well reflected in the literature on developing and managing teams over the past two decades. Different fads, tips, and strategies tend to recycle the same principles in more attractive packages. Doing the right things right, at the right time, continues to be the fundamental challenge regardless of the traditional or virtual nature of a project team. Like any resource, accelerated rates of change increase the complexity of addressing the challenge. Grazier (2002) identifies three major trends in the global workforce. First, high-intensity collaboration that cuts through functional power levels in an organization and its networks. Second, shifts in leadership competencies from technical to human skills (this might also be interpreted as requirements for technical *and* human skills). And third, enhanced/expanded role for frontline workers so that they operate more like independent project contractors within a team context moving from team to team on the basis of need rather than job security. (See also the chapter by Huemann, Turner, and Keegan.)

Workers have responded to these pressures by becoming multiskilled and mobile. Upgrading education entails time off without pay to engage in full-time, part-time, or evening study, or simultaneous work and study through (often virtual) post-secondary institutions. Workers focus on developing and maintaining current referral networks to allow them to quickly move on to other opportunities, refocusing and fully exploiting their skills, experience, and education at the first hint of crises related to merger or market downturns. The most pronounced change related to organizational expectations of teamwork relates to the need for a vigorous justification of the cost of training based on return on investment (ROI) and organizational impact (Brown, 2002).

The management level response here is to do the following:

- Recognize that individuals being trained often have more depth of knowledge and/or experience than the trainer or senior levels of management. Capitalize on their expertise (strategic minds) rather than seeing them as bodies charged with carrying out the work.
- Co-manage team members' careers by assisting then in acquiring the right skills and current education, and move them to positions that test their mettle.
- Fully account for the cost of training and return on investment to the organization for each team member. Take this into account in hiring and firing decisions.

Donated Time. The phenomenon of donating work time does not appear to be novel, but it is finally validated in research. Duxbury and Higgens (2001) find that over half of study respondents report working 3.5 to 5 unpaid overtime hours per month, equating to between 40 to 60 unpaid days of overtime per year. The bad news—as with Newton's second law of thermodynamics—is that for every action there is an equal and opposite reaction. The flip side of overtime is undertime. Undertime represents the compensatory days involved in workers catching up on their lives. DeMarco and Lister (1987) give a general rule of thumb that of a 1:1 ratio for over and undertime. Duxbury and Higgins (2001) conclude that the link between hours of work, role overload, work-life conflict, and health problems makes this practice nonsustainable over the long run (see Case Study 1).

Case Study 1

> The project team of experienced information technology (IT) professionals raced to meet the timeline for completion of a highly complex product in response to a major scope change nearly 3/4 of the way into the first test cycle. The team worked between 70 and 85 hours a week for the last 3 weeks of the project. They did not "count" the hours worked out of the office at home because being on "salary" implicitly meant "getting the job done" and being paid overtime was not a company policy. After the project was delivered, some team members took time off for various reasons including vacation and health. Other members sat in their offices and played "catch up," and others justified personal appointments as something that the company owed them.

The management level response here is to do the following:

- Investigate and make explicit the quantitative link between days off from work, health claims, and so on.
- Assess levels of job satisfaction, and open the door to discussion about unpaid overtime. Be honest about its impact on the organization and don't expect employees to suck it up!
- Aim for doable, not perfect, solutions by considering resource constraints and needs, not "wants" of clients.
- Triple time estimates for complex projects (rule of thumb), make sure the resources are in place, or be prepared to say "no" to unreasonable client demands late in the project life cycle.

Organizational Anorexia. Strassmann (1995) notes that outsourcing to cut labor costs shares attributes of the psychological disorder anorexia nervosa, where people refuse to eat to the point of starvation and have distorted self-images of being fat even when emaciated. Strassmann (1995) and later Duxbury and Higgins (2001) find that organizations with anorexia chose downsizing as a preferred method for restoring competitiveness even when they know they have too few resources and exponentially increasing workloads. The evidence is in a longer workweek, up from 37.5 hours to 50 hours in Canada, and higher job stress up from 13 percent to 35 percent from ten years ago (Duxbury and Higgins, 2001). A recent U.S. study of 750 employees and 250 employers shows that 44 percent of employees have more job stress than a year ago (Bureau of Labor Statistics, 1999–2000). Conditions appear similar across the European Union where nearly 30 percent of 21,500 workers interviewed cited stress as the second most serious problem (The European Foundation for the Improvement of Living and Working Conditions, 2000). Sixty percent of 19,000 Australian workers surveyed in a study by the Department of Industrial Relations stated having more stress than one year ago (Beder, 2001).

Why work so hard? In Canada, two incomes are necessary just to keep from losing ground and maintaining a family's standard of living (Statistics Canada, 2000). Employers blame global competition as the need to extend work hours to allow work across time zones, and compete and keep costs down by limiting the number of employees it deems feasible to hire (Duxbury and Higgens, 2001). The underlying fear of job layoff for refusal to work unpaid overtime continues to make branding an effective method. Johnson, Lero, and Rooney (2001) state that "job angst"—fear of losing one's job and being unable to find a comparable one—cropped up strongly in the 1990s, but it still seems to have plenty of momentum.

The management level response here is to do the following:

- Formally assess "perceived" stress levels related to work hours to achieve a baseline before any intervention: Talk to your employees and really listen to what they are saying.
- Establish a policy to have budgeted and actual working hours recorded. Examine the trends, use results to effect change, and reassess stress levels. An organization cannot afford not to know, and it cannot afford to penalize its employees for speaking the truth.

Senior Executive Priorities. CEO turnover is on the rise in all three major world markets. Drake, Beam, and Morin (2000) studied 476 public and private business organizations representing over 50 different industries in 25 countries to find that in just the past five years, nearly two-thirds of all major companies replaced their CEO. Almost half of CEO replacements occur because of mergers and acquisitions. Thomas, Delisle, and Jugdev (2002) also report that nearly 60 percent of international senior executives, project managers, and team member respondents were with their current employer for five years. In short, implementation of plans ends up with shorter time frames that focus on short-term business results (Drake, Beam, and Morin, 2000). Despite self-preservation priorities, corporate debt reduction, business ethics, and overreaching concerns for national security, attracting and retaining high-caliber employees is said to be second only to profitability as the most important priority (Meyer, 1988; Stephenson, 2000).

The management-level response here is to do the following:

- Educate top executives; they have to know the people behind the organization.
- Be involved in creating value-driven arguments for hiring that can be tied to quantitative outcomes.
- Reframe people information in executive terms; people are human resources that can deliver on value propositions related to saving the organization time, money, and so on.
- Do not leave the hiring to the human resources department needs; it is important to understand that the organization is hiring "minds" to make a strategic contribution, not just bodies to carry out the work.

Part II: Team Development

Understanding Teams

Not much has changed in the discussion about team selection and building from influential books such as the 1988 *Handbook on Project Management* (Clelend and King). What are some of the reasons why this should be so?

One reason may be that the concept of teams has shifted over time, but our understanding has not. Thus, what once looked liked a team may more aptly act as a group or committee. Table 3.1 presents a list of team-based literature. Rather than being exhaustive, the table serves as a starting point to help teams identify their dominant characteristics.

To add some structure, Miller (1988) suggests placing each characteristic in three larger categories to demonstrate their link to "task, result, or people factors" (p. 826). Keep in mind that results-oriented characteristics as described by Miller (1988) tend to look like success factors as described in project management literature (i.e., on-time performance). Decisions about what characteristics are most important in delivery of successful projects can assist a project manager in figuring out what kind of team can be built. However, this is certainly just a starting point! The tough work lies ahead in creating the conditions necessary for successful teamwork.

Another reason may be that the pervasive Western view has narrowed thinking about building a better team to finding ways to be more efficient and effective at tasks. Consider this line of wisdom in simple mechanical terms: To build a better or higher-performing machine, the wheels simply need to be tightened. In terms of people, simply make the same processes more efficient by tightening timelines can only improve efficiency a certain degree. Consider that over time, even a highly performing team may grind to a halt. Doing things differently means loosening the bolts on the "machine," letting the pieces fall, and reorganizing in a way that allows teams to do different things and to do things differently.

Once the team concept is more fully understood with respect to which characteristics are linked to successful teams, the business of identifying team members and the relationship between competencies and personalities may be more clearly understood.

Identify Key Resources

Miller (1988) comments that staffing the project is the "first milestone during the project formation phase" (p. 834). Project managers/leaders need to be involved in identifying and

TABLE 3.1. KEY CHARACTERISTICS of TEAMS.

Some Key Characteristics	Research Examples	Practice Examples
Celebrate, have fun, recognition of effort	Hoffman (2000); Delisle (2001)	Forsberg, Mooz, and Cotterman (2000; Hartman (2001); NASA (2002)
Common goals/cooperative team focus	Hoffman (2000); Chen and Barshes (2000); Delisle (2001)	Forsberg, Mooz, and Cotterman (2000); Kerzner (2001); McGannon (2001); NASA (2002)
Empowered	Hoffman (2000)	NASA 2002
Structured (establish clear boundaries around roles, conduct)	Hoffman (2000)	Forsberg, Mooz, and Cotterman (2000); Nasa (2002)
Cohesive (tribal culture)	Hoffman (2000); Delisle (2001)	Hartman (2000); Nasa (2002)
Interdependency	Hoffman (2000); Chen and Barshes (2000)	Forsberg, Mooz, and Cotterman (2000); Kerzner, (2001); NASA (2002)
Communicative (open communication and strong skills)	Chen and Barshes (2000); Hoffman (2000); Delisle (2001)	Miller (1988); Hartman (2000); Kerzner (2001); Nasa (2002)
Committed (to team and job)	Hoffman (2000); Delisle, (2001);	Nasa (2002); Kerzner (2001); Hartman (2000) Nasa (200)
Diversity	Hoffman (2000)	Miller (1988); McGannon (2001); Kerzner, (2001); Nasa (2002); Hartman (2000)
Results-oriented—Shared milestones/ performance criteria	Hoffman (2000)	Miller (1988); Hartman (2000); Kerzner (2001)
Trust	Crisp and Jarvanpaa (2000); Delisle (2001)	
Highly skilled/competent	Mills, Tyson, and Finn (2000); Dyrenfurth, (2000)	McGannon (2001); Kerzner (2001)
High spirit and *energy*	Next Step (2002); European Commission (2002)	Miller (1988); Forsberg, Mooz, and Cotterman (2000); Kerzner (2001)

negotiating for key resources to increase the probability of obtaining the necessary skilled resources to match the job requirements. Recruiting good people for projects is too important a job to be left to the human resources department (Meyer, 1998). In reality, the project manager/leader often has little influence on who will be a member of the project team. However, in an interview with Madigan (1998), Pinto suggests actively finding people to lessen the risk of having problem employees dumped on the project by department heads. Team leaders or self-directed teams need to communicate with decision makers about desired personal attributes, areas of competence (things a person will be required to do), and personality traits/behaviors that will likely result in a strong project team.

Knowing Competencies—Job-Related and Personal

Assessing skills and abilities and knowing personality traits and key behaviors of individuals increase the chances of choosing a team that has the potential to succeed (Sugarman, 1999). What are competencies and how do they fit in? Competencies differ from KSA (knowledge, skills, and aptitudes/abilities) in that competencies are based on the individual and their capability rather than on the job and its associated tasks (Kierstead, 1998). (See also the chapter by Gale.) Competencies generally fall into broad categories of intellectual, management, relationship, and personal (Kierstead, 1998). Key resources for project teams are those who have strong relationship competency (ability to build and maintain interpersonal relationships and communication ability) and personal competency (behavior flexibility or ability to be responsive to change, workload, transitions) (Kierstead, 1998).

Understanding competency as person-based rather than as job-based appears to be particularly important in a management-by-project environment, where jobs and associated tasks may not be narrowly defined. (Though Crawford, in her chapter, makes the case for professionally based competencies.) Decision makers can reduce the risk of making the wrong resource selection by identifying the areas of competence for the actual job or role descriptions. In an ideal situation, team member selection follows by assessing person-based competencies of each potential candidate and matching them against the job or role requirements (see Case Study 2).

Case Study 2

The project entailed developing a new book store as part of a national retail chain. The members of the team were brought together through human resources from different functional divisions within the same company. The project manager was inexperienced and did not have the influence or know-how to get involved in the team selection process.

Although the talent did exist within the organization, the human resources department did not clearly match the skill set of each person to the job requirements for the project. The mismatch was partly due to the project manager not communicating the requirements (roles and level of responsibility) needed for each position to HR.

The team therefore did not contain the competencies needed to get the project completed. In frustration, the project owner fired the project manager. The new project manager was more cognizant of the need to match skill level with the project requirements and thus was able to pull in a few replacements to strengthen the team and eventually complete the work.

Understanding Personalities

Personality tends to be enduring and, some argue, unchangeable. Thus, principles presented in Hill and Summers' (1988) chapter in the *Project Management Handbook* provide a solid grounding in personality and conflict dynamics, and these topics need not be redressed in detail. Social conflicts tend to be relational (between people), but triggers that ignite conflict reside in the individual (Hill and Summers, 1988). Thus, assessment of the way individuals gather and process information instruments such as the Myers-Briggs Type Indicator provide useful information about the strengths and weaknesses of individuals, as well as the potential triggers that might blindside the team. In addition, Hill and Summers (1988) make a strong case for assessing the emotional dynamics to gain insight into how individuals act in interpersonal relationships.

Frankel's (1999) work serves as a unique foundation from which to explore the role of emotions versus logical thinking on project teams. He asserts that an individual's cognition does play a role in elaborating and refining a repertoire of coping skills to help individuals adapt to their environment. However, emotions essentially govern how individuals continuously estimate how to avoid risks and exploit opportunities (Frankel, 1999). Although a simplified explanation, people tend to act as utility theorists by first seeking to reduce pain and then increase pleasure. Personality then is partly driven by the intensity of the positive or negative feelings as well as the valuation of the risk and opportunity, and not by pure rational thinking. Individuals practice what Frankel (1999) calls "emotional economics" to make sense of their world. If estimates are out of proportion with the situation (too low or high), a problem may erupt. When individuals estimate emotions accurately in a specific situation, responses are within the realm of their adaptive cognitive skill set. Thus, asking questions that test the adaptive coping skills of individuals on a cognitive as well as emotional level may help in the long run in avoiding intense conflicts.

Negotiate for Human Resources

Line managers or members of self-directed project teams need to know what problems senior executives face—what keeps them up at night. Knowing executive's top challenges makes it possible to negotiate for the resources that will provide the best fit, allowing the project team to deliver the level of value added needed to justify their costs. Step three and four of Belgard, Fisher, and Rayner's (2000) process provides the most opportunity for a project leader to influence in the negotiation process:

1. Size up the work to be done.
2. Establish selection criteria.
3. Find possible candidates.

4. Evaluate candidates and make selection.
5. Formulate the offer.
6. Orient the new member.

Once the offer is on the table, it is more difficult to do any negotiation. From the negotiator's perspective, backward planning might be a useful process. First, know what you need the outcome of the negotiation to be. Next, decide where you will draw the line—how far are you willing to go to get a resource? How willing is the candidate to join the team? (Miller, 1988) Finally, work out a "best" alterative to the negotiated agreement in the event the process goes further than you are willing to go.

Developing or Growing a Team

An obvious yet often overlooked step is to identify the purpose of the team. Traditionally, literature on project teams points to the attainment of a *common goal* as the hallmark of a high-performance team. However, consider that individuals make up the team and carry out the component tasks charged to the project team. Individuals attain goals that result in the team delivering a product or service. DeMarco and Lister's (1978) argument still makes sense: Align individual goals as a means of achieving a common goal. From a management point of view, it matters less that the team has the exact same concerns and more that the project team carefully focuses its energy on those concerns that matter to the project to make sure it will contribute to the overall business success.

Project managers and senior executives often make the naive assumption that teams accept the employer's goals as a condition of employment. Individuals tend to have many different reasons for involving themselves in the organizational or project goals, so it is naive to assume that simply refocusing the project team on an arbitrary company goal will bring success. DeMarco and Lister (1987) suggest that this may actually hinder the team's ability to gel. Project teams tend to gel much more readily around the challenge of achieving a goal that is related to the team itself. For example, Hartman (2001) describes a "tie tribe" (pp. 272 to 273). Deliverables are written on each tie stripe, starting with the last one at the top. As the team successfully achieves a deliverable, the tie is trimmed and the highest contributor wears the tie.

Evolutionary Model. How can we grow or develop a team once we know its purpose? Many practitioners and researchers still utilize a nearly 40-year-old model by Tuchman (1965). The model describes five sequential stages of team development (see Table 3.2, left). The traditional team-building model is grounded in the assumption that over time, a team moves toward better performance.

Tuchman's (1965) team building model assumes that teams develop sequentially. Once a stage has been mastered, a team moves to the next stage, ultimately leading to better performance. The leader gives direction, and the team carries out the work in a relatively predictable way. Social dynamics are dealt with in the early stages to minimize interference with the general task's focus. McGrath's (1990) model (see Table 3.2, right) is predicated on the same evolutionary principles, and even the labels for each stage appear similar.

TABLE 3.2. TEAM DEVELOPMENT MODELS.

Sequential Team Development Stages	Potential Activities by Combination
Model by Tuchman (1965)	Model by McGrath (1990)
Forming	**I. Inception**
Learn about members, learn about the environment, and size up members.	*Production focus:* Select goals, plan, and brainstorm.
	Social focus: Ensure inclusion, establish parameters for participation, and flesh out roles.
Norming	**II. Problem Solving**
Gain consensus about acceptable individual and group behaviors; clarify and align expectations; learn about roles, responsibilities; and set guidelines for operation.	*Production focus:* Identify problems (technical and task-related), provide alternatives, and solve problems.
	Social focus: Address power, status issues; clarify roles, expertise, and competencies.
Storming	**III. Conflict Resolution**
Express anxiety, defensiveness, engage in power struggles and interpersonal conflict, learn to deal with conflict and find resolution.	*Production focus:* Identify and resolve conflicts (actual or anticipated) related to the project methodology, processes, etc.
	Social focus: Address interpersonal relationships, styles of learning, cultural differences (organizational and ethnic).
Performing	**IV. Execution**
Achieve commitment and full attention of members, use of each other as resources, respect of boundaries; express cohesion (gel, bonding).	*Task focus:* Perform task and continually scan environment for barriers/opportunities.
	Social focus: Revisit participation, accountability, and communication competency.

49

McGrath's original model shows a team's activities as a set of three interlocking functions: production (task) function, member support functions, and group well-being function. Durarte and Tennant Synder (2001) combine the member support functions and group well-being functions and label it "social" focus which is reflected in figure 3.2 (right). They use the same labels for each activity and apply the model to virtual project teams, although McGrath (1990) does not make this distinction.

McGrath's (1990) model aptly reflects how teams grow and change, creating "patterns" that suit the needs of the situation (see Figure 3.1). As constraints change, the growth pattern of the team may change. For example, the introduction of new communication technology may shift the path and/or the balance of energy spent on social rather than task functions.

Pattern A indicates that teams move through stages or activities I to IV sequentially, most closely resembling Tuchman's model (teams may not spend the same amount of time in each phase). Pattern B (focus on activity I and IV) often occurs if teams are given a project to complete under extreme time pressure. They tend to skip social and production functions concerning problem solving and conflict activities, focusing on production functions in relation to the project's execution. This describes teams that end up stamping out crisis fires instead of making use of the midpoint change in direction to improve strategy (McGrath, 1990).

Pattern C (focus on activity I, II, IV) does not suggest that teams avoid conflict. However, teams tend to address problem solving at the task level that may not challenge earlier established strategy, thus avoiding personal types of conflict (McGrath, 1990). Pattern D (focus on activity I, III, IV) reflects project situations where teams receive direction rather than work together to solve problems (i.e., task force teams that are brought together to solve predetermined problems). Patterns B to D appear most susceptible to neglect of social functions, which increases the risk that collaboration will suffer over time and hamper the desire to work together again (McGrath, 1990).

Revolutionary Model. Evolutionary models, however, do not adequately explain why team performance varies over time in relationship to its evolution. A common observation is that teams, regardless of the stage or phase, tend to go through at least one major transformation about halfway through their growth cycle (Gersick and Hackman, 1990; McGrath, 1990; Duarte and Tennant Synder, 2001). Gersick and Hackman (1990) note it is during this

FIGURE 3.1. POTENTIAL TEAM GROWTH PATTERNS.

Activity	Pattern			
From table 39.2 (p. 993)	A	B	C	D
I				
II				
III				
IV				

process that teams "punctuate their equilibrium." That is, after midway into any length of a project, efficiency drops as teams enter a more revolutionary period of change. McGrath (1990) notes that this action does not support the kind of sequential linear patterning explained by Tuchman's development model. Instead, McGrath's model works more readily with a cyclical concept of time.

Reworking of the original plan, a sense of urgency to finish, and closer contact with sponsors and clients may trigger efforts to punctuate equilibrium, but the underlying cause-effect relationship is not known for certain (Duarte and Tennant Synder, 2001). Perhaps a more internal, fundamental process such as trust plays a role in the midpoint transition of a team's revolutionary development. Crisp and Jarvenpaa's (2000) study lends support to this idea, finding that the level of team trust drops sometime near the middle of a project. Few changes in trust levels occur until the end of the project. Their work shows that by the midpoint of team development, members have more substantive information about the project and thus, greater insight into the "real" personality and behavior of team members.

Team members tend to engage in relationship building early on in the formation to gain a sense of which person to trust. This will ultimately be measured by how much weight each appears to be pulling to get the project work completed. Project team members may be closest to their *true* nature when nearing or engaging in critical questioning about other's performance and contribution (see Case Study 3).

Case Study 3

> An external consultant was hired to create a post-project review process for a major oil and gas company. The project team was chosen by careful attention to personality traits, experience, and team work skills. Its members had agreed to work cooperatively together to produce a high-quality project. About a third of the way through the project when the first set of deliverables were due, internal grumbling began about having to pick up the slack because one member did not follow thorough on task or team processes as agreed. Not until midway through the project did two of the five members openly address this problem in a group discussion. The initial level of trust based on early agreements for performance had broken down at a critical point in the project, forcing the team to rework the plan, reorganize the workload, and shift roles and responsibilities to enable the successful completion of the project.

In consideration of time and its relationship to team development, no one pattern of growth is "right or wrong." Rather, team awareness of the features and benefits of each pattern may allow its members to bring production functions and social dynamics into sharper focus as the project moves through its life cycle. Overall, teams appear to mature by visiting and revisting the interlocking sets of tasks and social activities, with recognition that midpoint in any project may serve as an important platform to rethink current processes and practices. As such, resolutions may be partial, simultaneous, or even parallel in some or all stages of team development.

Nature or Nurture: What Influences Team Dynamics?

How can you make a team gel so they work through development more easily? DeMarco and Lister (1987) astutely note that nothing can make a team gel, only improve the odds of growing a team that *may* gel. A grow rather than build model recognizes that the process is inherently nonlinear and prone to chaotic processes. Thus, the initial point of becoming a "team" cannot be known in enough detail to predict its evolution (Duarte and Tennant Snyder, 2001). Sommerville and Dalziel (1988) find evidence that teams also don't know how they will handle their end point and often engage in a period of "mourning" after performing. In reality, it may take teams until the end of the cycle to gel, or they may never gel. A technique called "inversion" developed by DeBono (1977) has been used with success to increase the odds that they will gel. Inversion simply refers to thinking of ways to achieve the opposite of a team—to make team formation impossible. DeMarco and Lister (1987) point to phony deadlines, defensive management, and fragmentation of people's time to be surefire ways of failing, or committing "teamicide."

Environmental Factors Affecting Team Dynamics. Many factors act to influence whether a project team gels or not. Team composition, demographics, and temporal influences appear to be important dynamics that influence team development and performance. Each dynamic is briefly introduced in the following subsections as a starting point for further investigation.

Team Composition. Decision makers need to consider the size and level of expertise on the project team. A team of approximately ten core or central members seems to work effectively in many project-based organizations (Delisle, 2001). Considering that projects utilize a blend of physical and virtual contributors, determining the actual size of the team may be extremely difficult. Deciding who belongs to the core, mid-layer, and outer stakeholder group may in itself be an important exercise in establishing desired role and authority relationships. More important, how many teams should any one individual be part of at one time?

Team Demographics. Not enough can be said here about the changes to the labor force and cultural composition to justify its importance to understanding the dynamics of a modern-day project team. More women than men are entering the labor force in both the United States and Canada. The Bureau of Labor Statistics (1999 to 2000) in the United States projects that the share of the workforce by each gender will be almost equally divided by the year 2008. Statistics Canada Labor Force Survey shows the same trends, with women accounting for 46 percent of the workforce in 2001 (Ministry of Industry, 2002). In terms of race, projected growth rates of over 3 percent are expected for the Asian and Hispanic labor force by 2010, resulting in the strongest presence of ethnic groups in the U.S. labor market (Fullerton and Toosi, 2001). In Canada, immigration is expected to account for almost all of the net growth in the Canadian labor force by the year 2011 (HRDC, 2002).

How are team dynamics impacted? Individuals are members of multiple cultures that include national/ethnic origin, profession, function (within organizational departments), and business or corporate cultures. Individuals are also expected to become part of or establish a team's culture when selected for a project. Project leaders and managers need to consider the range of diversity that enables the team to manage differences as the project progresses. Understanding the extent of each team member's cultural memberships offers a starting point or common ground from which to grow a strong yet diverse team.

Temporal Orientation. Time or "temporal" orientations of the organization, project leader, and team members may also influence a team's dynamics and play a significant role in project performance (Thoms, forthcoming; Thoms and Pinto, 1999, and Thoms and Greenberger, 1995). Project leaders and member's *thoughts* may be aligned mostly in the past, present, or future (Thoms, forthcoming; Thoms and Greenburger, 1995). For example, past-oriented individuals may think that strategic planning about the future as a waste of time. Future-oriented individuals may have a difficult time in conducting project reviews because they don't see the value in going over what has already been accomplished. Present-oriented individuals may be so caught up in day-to-day operations that they cannot see the big picture and how past lessons can be applied to avoid future pitfalls.

As well, individuals' *behaviors* may be described in terms of the "temporal skills" they use to make sense of the past, present, and future orientation (Thoms and Pinto, 1999). Thoms and Greenburger (1995) propose that a new class of management skills based on temporal consideration needs to be taken into consideration, because overall adaptability is tied to the individual's time orientation (past, present, future). For example, a temporal skill such as "chunking" involves the creation of time units (i.e., workweek = 5 days) to allow for the team to attack the project in "doable" units rather than as one time frame (Thoms and Greenburger, 1995). Furthermore, "time warping" by constantly reaffirming that the team is moving toward the project milestones is a skill that helps team members stay motivated.

Thus, *attunement* or match between the temporal skill and thought pattern alignment in the past, present, or future orientation seems largely controlled by an individual's choice of and prowess in using a temporal skill (Thoms, 2003). Consider that as individuals become attuned, they may be drawn to members that have a similar resonance (i.e., they seem to "click") in relation to the flow of the work. McGrath (1990) finds evidence that by imposing an initial tight project deadline, teams switch to a high task focused work at a fast rate, and deliver poorer quality. This carries over to subsequent projects even when they do not have the same restricted time deadlines. Teams become "entrained" or work in synch much like the cardiac rhythm of two individuals synchronizing when their chest cavities touch, for example.

Behavioral Factors Affecting Team Dynamics. Although arguments can be made about many significant team behaviors, three internal factors identified by Sugerman (1999) make sense in terms of distinguishing high from poorly performing teams. A brief discussion about mental toughness, cohesion, and motivation follows.

Mental Toughness. Mental toughness refers to the ability to bounce back positively from an adverse situation or event (Sugerman, 1999). The majority of study and research has focused on sport and military teams. These lessons are sometimes taken and applied to the business world. Mentally tough team members, according to Brown (2002b) are those that

- seek feedback about performance, even when it is not guaranteed to be positive;
- choose tasks that are challenging but add value rather than taking the path of least resistance;
- trust gut feel and ability instead of procrastinating when they need to make critical decisions;
- face conflict by quickly taking action, engaging it as required or resolving it quickly;
- focus on solutions rather than reacting negatively and defensively to perceived threats; and
- contribute ideas that could make a solid impact on the company's bottom line even if fearful of failure or rejection.

From a personal trait point of view, a major barrier that individuals on a project team face relates to learning how to respond by technique and not emotion in the face of adversity. This does not mean becoming emotionally hardened or cold. Rather, detaching oneself from taking an issue personally helps to avoid destructive blame tendencies that erode a sense of shared loyalty and trust.

Cohesion. Cohesion refers to the closeness or "glue" that holds a team together. Popular literature is rife with examples of fire, police, and sport teams pulling together to beat the odds. Sugerman (1999) asserts that building cohesion is a matter of assessing and aligning expectations, establishing open communication, and engaging in direct assessments of team members. However, does cohesion need to occur for top performance? Does cohesion also have a flip side?

Cohesive teams may or may not visibly show signs of intense social commitment such as "high fives" often seen in professional hockey games. Professional sport teams spend a great deal of time together on and off the playing field, and socialization is part of the culture. In contrast, the Royal Canadian Mounted Police Emergency Response Team (ERT) does not spend a great deal of time together on or off duty. The team consists of a two highly trained and specialized tactical units. Daily activity is mostly individually planned, although tactical and technical training sessions are group-based. Training and conditioning is primarily done on an individual basis, unlike highly organized group training of elite teams in the armed forces. As well, individuals do not spend much "social" time together, such as eating meals. However, when called into action, they work seamlessly together on each project they undertake. How can both types of teams be highly cohesive?

At one end of the continuum, an underlying force driving cohesion of ERT stems from overcoming threats to physical safety by cooperatively uniting as a team. At the other end of the continuum, elite sport teams understand the feelings associated with overcoming the "odds" and reaching self-actualized goals through cooperative game play, creating strong cohesion. Consider the paradox of individual needs of ERT versus group needs of an elite sport team. The ERT appears driven to satisfy individual physiological *self-interests* related to safety, whereas the elite sport team is driven to satisfy higher-order or *socially related interests*. Thus, cohesion looks different to different teams.

Motivation. Sugerman (1999) characterizes motivation as the intensity of behavior (arousal) or psychological force that pushes a person to perform. Theories of motivation range widely, including instinct theory (we are genetically predetermined), drive theory (seek to reduce tension, gain balance); arousal theory (seek to calm down or psych up), and incentive theory (perform for external reward). Perhaps the whole question of whether motivation is extrinsic (learned in response to reward) or intrinsic (because you love something) needs to be reframed as subquestions that we can answer. An important subset of the motivation question relates to perceived lack of control. Psychology studies consistently find a relationship between lack of control over self, events, and situations; increased burden of responsibility; and stress associated with lack of motivation.

Teams members may experience intensified feelings of loss of control if they receive little or no acknowledgment of or responses to their contributions. This may even be more intense in a virtual environment because of the lack of physical cues. Eventual withdraw from participation in online collaboration may be taken as a lack of motivation when the problem stems from loss of awareness or incomplete awareness of other's activities in the project or inability to extract salient information from an overwhelming mass of information (Conklin et al., 1998). Whether virtual or not, the leader's task is to ensure team members control their effort levels and not hold them responsible and accountable for external events that they *cannot control.* Consider as well that, as utility theory, people are engaging in what Frankel (1999) calls "emotional economics." They calculate the "net interaction value" partly as a function of personal value less personal cost (Conklin et al., 1988). Cost and value differ by individual—depending on whether they consider and how they weight related emotional, intellectual, and social impacts.

Another important subset of the motivation question relates to an individual's perception of themselves within the team. Swann, Polzer, Seyle, and Ko's (2002) research challenges commonsense thinking about the way people's self-views influence their responses to motivating feedback. Their study finds that participants with positive self-views worked hardest if they believed they would receive positive feedback, and participants with negative self-views worked *least* when they received positive feedback. As predicted by self-verification theory, people with negative self-views withdrew effort when they thought that they would be receiving positive feedback because they felt undeserving (Swann et al., 2002). If the group verifies this view, the member feels detached and may quit. If the group does not verify this view, the member feels committed to the team, yet alienated. These findings point to the need for personal transformation where individuals on a team first get in touch with their own views to enable a clearer view of the group and the overall organization. Signs that point to negative self-views include perfectionist tendencies, taking on too much work, and taking blame for events beyond one's control.

Influence of Virtual Teams

The evolutionary and revolutionary models help us to understand the changes in team composition and dynamics over time. However, shifts in division of labor over time essentially helps to explain the bigger picture of how teams relate to their environment, especially as we continue to hear discussion about challenges of working in virtual teams—those teams who work primarily using telecommunications devices and who may seldom or never meet face to face. Figure 3.2 depicts the division of labor by organization type, showing a rough time frame for each division (Delisle, 2001).

A review of the organizational and management literature shows that organizations tend to move through cycles of more formal controls to periods of loose connections (Delisle, 2001). Lipnack and Stamps (1996) do not suggest that any one organizational structure disappears; rather, its dominance shifts over time. For example, pre-1900s was marked by the division of labor into small tribal groups that relied on nomadic wisdom or storytelling. Moving to the right in the figure, cottage industries roughly emerged in the 1940s, borne out of the shift to an agriculture focus.

This agriculture era transitioned into bureaucracies where gains in labor use came from the mass production of goods and services in the 1970s. In the 1980s packets or nodes of workers typically organized by groups or teams began to emerge in workplace. Nodes provided more autonomy, although control remained centralized and expectations were to follow organizational norms and rules (Delisle, 2001). The continuing surge of communication technology noticeably changed the division of labor in the 1990s, providing choices such as distributed work through telecommuting and hot desking (sharing desks between office workers).

Most recently, the literature notes the emergence of virtual teams or "tribes" composed of a more highly educated and mobile workforce driven by the need for organizations to increase profits/cut costs and fulfill individual needs of workplace autonomy. Virtual workers may act independently as knowledge contractors, as well as be part-time members of one or more traditional organizational teams. Early indications are that the workforce of the

FIGURE 3.2. DIVISION OF LABOR OVER TIME.

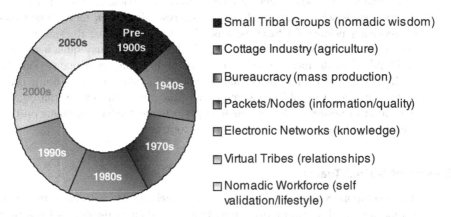

Adapted from Delisle, 2001.

future may become even more fragmented and driven by independent subject matter "nomads" who value their lifestyle sometimes ahead of work. Distinctions between the division of labor may be primarily around differentiating levels of technology sophistication, more fluid leadership sharing arrangements, and unique linkages. The linkages refer to interconnections between people, activities, and knowledge, rather than formal control (Lipnack and Stamps, 1996).

Challenges in Virtual Teams. What specific challenges do we face in building functional virtual teams? Relationships networks are becoming increasingly important to enabling virtual project teams. Thus, workers outside of large city centers may have difficulty gaining access to professional networks to allow them to select work. Therefore, organizations are beginning to focus more on building relationships with competitors and external suppliers that normally fall outside the commonly accepted "definition" of a project team (Delisle, 2001). On a micro level, virtual teams face specific challenges, of which two key ones are briefly touched on in this section. These challenges are in defining roles and responsibilities whose clarity is needed to assist in managing communications.

Given that virtual project teams emerge to solve a problem (task-driven), it might at first not seem necessary to define roles the way traditional teams do (Ott and Nastansky, 1997). To make matters more complex, traditional rules-based team design does not fit well with the reality of a virtual team. The most important issue is that responsibility tends to shift by task, *regardless of role definition*. Delisle, Thomas, Jugdev, and Buckle (2002) note that team members take on the tasks they believe they feel most skilled to do rather than the ones they feel obligated to do.

The Types of Work Index (TOW) by Margerison and McCann (1994) serves as a useful guide to helping virtual teams explore and understand their own internal capabilities (see Table 3.3). If the time is not afforded from this activity, the team can become frustrated (i.e., when timelines tighten), and they are not able to quickly identify who to draw from in solving problems related to the work and team function.

In terms of the TOW Index, a member may engage in one task (i.e., organizing) more often than another because he or she has had previous successes with the preferred behav-

TABLE 3.3. TYPES OF WORK INDEX.

Task	Description
Advising	Gathering and reporting information
Innovating	Creating and experimenting with ideas
Promoting	Exploring and presenting opportunities
Developing	Assessing and testing the applicability of new approaches
Organizing	Establishing and implementing ways of making things
Producing	Concluding and delivering outputs
Inspecting	Controlling and auditing the working of systems
Maintaining	Upholding and safeguarding standards and processes

iors. This, in turn, reinforces the likelihood of similar future behavior (Delisle, 2001). One area that appears to be consistently weak in virtual teams is the role of "maintaining," because unlike face-to-face teams, the administrative function is not tied to any specific role. Thus, meeting notes, coordinating work to be done, and following up on action items often falls off the team's radar screen unless someone volunteers. This essentially shifts a burden to one or two individuals, which can create resentment over time because of the added workload.

A closely related issue is of responsibility, because it tends to shift throughout a task depending on the work and workload demands. The use of a tool such as the RACI (Responsibility, Accountability, Consult and Inform) chart seems to elevate confusion specifically between who has a stake in contributing to the work (responsible) versus who has to deliver the work at the end of the day (accountable). As well, this tool when used consistently seems to resolve problems around what consult means (active involvement) versus inform (rubber-stamping near-final draft), which seem to be particularly problematic in virtual teams.

Part III: Team Management

Common Team Pathologies

Part III takes a practical look at ways to overcome common team pathologies. Research literature, popular books, magazines, and Web articles are filled with tips, tools, practices, and methods for managing teams. Rather than trying to cover off every tip or practice invented, this section focuses on a few key areas to focus efforts to do things differently and do different things in managing teams.

Arguably, teams go through some type of dysfunction or cycle of pathology at one point or another over the project life cycle. Pathologies are typically tied to dysfunctional behaviors. Lack of mental toughness, cohesion, and motivation as discussed appear to underlay many team dysfunctions. This section provides a brief introduction to common team pathologies in these three areas and suggests possible courses of action and tips to help overcome these difficulties.

Lack of Mental Toughness. Project manager/leaders often face the challenge of handling personalities that play a major role in shaping a team. Lack of mental toughness may be expressed in many forms, particularly tied to situations were people "give up" or are take the situation personally.

Two simple strategies can be used to increase metal toughness. First, all review and feedback from the project manager to team members should be couched in terms of "I" talk not "you" talk, especially concerning emotionally sensitive topics. For example, "I feel that your performance is not as strong as it was last month—I would like to spend some time hearing about what you think and finding ways that I can help." This is directly contrasting typical communication that begins with "you are not working to your potential; what are you going to do about it?"

Second, have the team agree on a code of conduct/guidelines that includes an agreement not to take things personally. This simple step will help the team go a long way in keeping blame from crippling the team's effectiveness.

From a behavioral point of view, Seligman's research in the mid-1960s on "learned helplessness" sheds light on mental toughness. For example, an individual fails at a task and believes he is at fault. Over time, the belief that he is incapable of doing anything in order to improve performance spills over to similar tasks. Perhaps the most damaging aspect is that this belief interferes with the ability to learn in a new situation where avoidance or alternate solutions are possible. For example, if conditions in the workplace change, the perceived lack of controllability holds employees in a motivational paralysis, and job burnout may occur over time (Potter, 1998). As a team, the first step is to become aware of learned helplessness and then work to create an environment to help members relearn adaptive behaviors.

Ineffective Cohesion. How does one go about holding together teams that work in the regular business world when cohesion is not based on fulfilling basic survival needs or achieving dizzying heights of self-actualization? Ironically, in the business context, there is greater incentive to *compete* than cooperate as a team. The practice of individual reward and recognition tends to reinforce individualistic tendencies.

One strategy may be in revisiting the message that is being sent: Does it reinforce team or individual behavior? Are the team member's choices driven by team member loyalty friendships rather than what is in the best interest of getting the work completed? An example from the popular television show *Survivor* where members try to outwit, outlast, and outplay their counterparts over the course of 39 days serves as a useful example. The team's survival (safety, food, shelter) is not ultimately threatened; thus, team member selection is mostly based on relationships formed among those "conforming" to the majority-driven social norms that emerge as the game evolves. Members who do not follow emergent social norms and form loyalty alliances are often quickly voted off the team. So are those who show innovation, strength, or high levels of motivation because they are seen as a "threat." In a corporate setting this behavior may be expressed as "groupthink," a condition where team members make decisions on the basis of social relationships or preserving self-esteem, not on business outcomes.

Strategies to reward and recognize team behavior are effective when tied to performance and are peer-driven (not manager-determined). Monetary rewards are not always the most effective; often a simple and informal recognition venue or events will suffice.

Lack of Motivation. The properties of the team itself in increasing or decreasing motivation deserve mention. Lack of motivation may be expressed symptomatically in many ways, including showing up late for work, producing poor quality, leaving work unfinished, complaining about other team members or policy, and so on.

However, root causes such as "social loafing," identified by Ringleham in 1913, are also critical to understand. Ringleham's psychological research identified that the greater the number of people working together on a task, the less effort each person decides to contribute (Ingham, Levinger, Graves, and Peckham, 1974). Studies of physical strength in rope-pulling contests reveal an individual's behavior is also affected by group dynamics. In group situations, the same individuals tended to pull less vigorously. Ultimately, the pressure to perform well is shared by the group, perhaps at an unconscious level. Awareness of social loafing is important in improving the quality of the team, because members are able to identify the signs and work on prevention rather than cure. Specifically, steps can be taken to identify individuals' inputs, evaluate inputs, involve all members in the task, and diffuse responsibility among members.

Another strategy is a modified "brainstorming" called the nominal group technique (Gersick and Hackman, 1990). Individuals first tackle the problem or question individually on paper. Contributions are collected and recorded on a flip chart like conventional brainstorming. Discussion continues until the alternatives are exhausted, and then individuals are asked to prioritize alternatives. This information is recorded by a facilitator and displayed to the group for further discussion. However, creation of a psychologically safe environment is as important as picking a method that will stimulate the team's thinking. Consider that if the body itself on a purely physiological level is far less responsive to any exercise routine beyond a six- to eight-week cycle, the mind is also likely to become bored by repeated use of any one method.

Misunderstanding of Success. Studies about success tend to talk about project- or business-level success without often considering the link between the two. Do businesses succeed when they make a profit? Do teams succeed when they deliver a successful product or service? Research shows that perceptions about success differ greatly, depending on who is asked and what context they are judging success. Success is a whole lot more complex than just meeting "priority triangle" criteria related to cost, time, and quality. Overall, project managers need to identify what role their team has in creating value.

First, the project teams need to establish what success means. Much confusion exists in practice and in the literature about what critical success factors (CFS) really mean. Initial research into success in the 1980s lumped success in one box and popularized CSF (Pinto and Slevin, 1988). To me, the CSF represents an overarching concept that includes both critical success indicators (CSIs) and critical success criteria (CSCs). Critical success indicators refers to the "softer" processes and markers organizations/teams agree to heed as a way to increase their chances of delivering successful projects (i.e., top management support). They are not often measured during or at the end of the project (Delisle, 2001). Critical success criteria (CSC) refers to "harder or more quantitative dimensions" that a project will be judged against as being successful (Delisle, 2001). Both CSIs and CSCs are often talked about in reference to the success of project management process. Outcome success then refers to the actual product/service itself that may be measured in relation to how well the project met the business goals/objectives (Delisle, 2001). (See the chapters by Brandon and by Cooke-Davies on this topic.)

Next, project teams need to determine what is most critical to their success. Part 1 of a study by Delisle (2001) asked participants what indicators of success they considered most critical. Respondents chose six CSIs from a list of 29 compiled from an extensive literature review (see Figure 3.3). As shown by the frequency of responses, participants chose open communication, communication skills, fun, trust, commitment, and culture. Crisp and Jarvenpaa's (2000) research on virtual project teams reveal a positive correlation between early team communication and increasing trust levels.

FIGURE 3.3. TEAM-LEVEL SUCCESS.

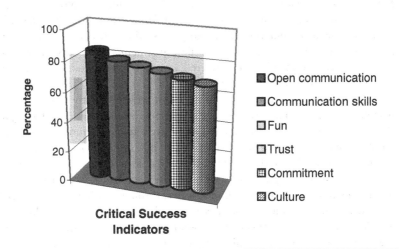

For example, if teammates communicated early on in the project, members of the team were perceived as more competent and worth trusting. As well, teams that communicated actively in the project tended to keep communicating more readily than those that did not as the project progressed (Crisp and Jarvenpaa, 2000).

These points underscore the importance of establishing predictable communication patterns early on in a project. Hartman (2000) found trust, fun, and open communication as three of seven elements for effective teams, *regardless* of whether they are virtual or not. Reaching team agreement about what could be done, and aligning this with what can be done in light of resource and project constraints, will help to identify CSIs that make sense for the project team rather than identifying one "right" list. Managing team success as an ongoing process will encourage the team to make adjustments to the CSIs and CSCs (add or drop) during the project life cycle as constraints change and the overall outcome success shifts.

Finally, project teams need to determine what they will pay attention to and what they will measure their success by during the project. Delisle's (2000) research identifies that only 45 percent of respondents report identifying CSIs or CSCs, and only 20 percent of these respondents report that their teams define and measure both. There is no magic formula that will ensure a successful team experience or guarantee a successful project. From a practical point of view, it may be more difficult for a team to agree whether they have successfully arrived at their destination if they have not taken the time to identify what signposts they expect to pass along the road to a successful project.

Misaligned Learning Styles. People differ in the way they take in and process information. For example, a team member may learn by listening (auditory), seeing a demonstration (visual), or touching or interacting with a model or real product (tactile). More likely, people learn by using a combination of learning styles. Many theories try to account for how people learn, although this has generated little agreement on the actual process. Learning style inventories simply identify the preferred way a person encodes information and changes it to make sense of a problem or situation.

Often, individuals are not even aware of which style they rely on and under pressure default to patterns that may not be as effective as possible to get the job done. Gardner (1993) advanced thinking about learning styles to accommodate multisensory abilities or "multiple intelligence" (MI). The term MI should not be confused with IQ. Individuals have varying levels of skill and can enhance their ability to learn by becoming more self-aware in terms of their strengths and weaknesses in relation to each of the MIs shown in Figure 3.4. For example, individuals who learn best through *words* use verbal/linguistic intelligence, those who learn best though *questioning* use logical/mathematical intelligence, and individuals who learn best through *pictures and images* rely on visual and spatial intelligence. Intuition as a way to learn seems to be guided by instinctive gut feel or "vibes."

Consider using the learning wheel to assess team member's MI at the start of the project. Sharing the results among the team may help avoid frustration and misunderstanding that often results from assuming people will readily adopt standard communication protocols (see Case Study 4).

FIGURE 3.4. MULTIPLE SENSORY LEARNING.

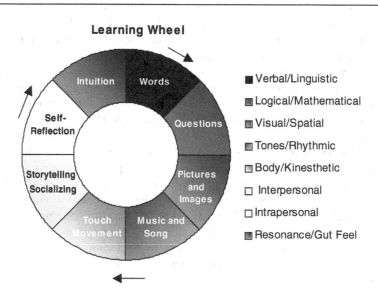

Case Study 4

An instructional design project team that produced interactive online learning tools met to brainstorm ideas to solve a technical problem. The team leader and the senior designer were engaged in a discussion that was going in circles. The leader kept saying "can you see what I am saying?" in response to the designer saying "I hear you, but I need you to answer these three questions." Another team member was fiddling with the actual tool, which created additional frustration because other members thought she was not paying attention. Had team members known what their natural preferences were for learning, their may have spent less time being right and moved on to solving the problem they could deal with.

Looking at One Side of the Equation. To breathe more life into project team that has lost its ingenuity, reintroduce small amounts of disorder. Why bother, who has time anyway? Research shows the importance of learning how to recognize the patterns of interaction that undermine learning because patterns of team interaction quickly become routinized (Conklin, 2001; Cha and Edmondson, 2003). Consider that what we may value as "good" also has an equal and "opposite" side, given the nature of the laws of thermodynamics. Cha and Edmondson (2003) identify individual vigilance, maximizing efficiency, and empowerment as three "good" patterns that can also *inhibit* learning.

First, *individual vigilance* encourages people to take personal responsibility to solve problems as they arise. Counterintuitively, this may create barriers to team success because it encourages independence. Next, business culture most often rewards and expects people to take *personal responsibility*. Consider the flip side. Cha and Edmondson (2003) report that 70 percent of nurses believe their managers expect them to work through the "daily disruptions" on their own. Speaking up about a problem or asking for help is perceived as incompetence. Finally, *empowerment* is often touted as a solution to productivity problems. Empowerment allows for the removal of mid-level managers or use of self-directed (leaderless) teams, leaving team members on their own to resolve problems that may stem from parts of the organization. Often, quick-fix solutions are applied, making a project look balanced. However, problems may resurface elsewhere, because the balance is artificial. The consequences then are even worse than before the problem was "solved."

Consider examining the lessons learned from a previous or similar project; look for patterns that do not work that once did make sense. Take steps to feed this information back into initiation phase of a new project. As well, a simple yet often overlooked step is to meet and ask for "permission" (at a social level, not in terms of authority) to change a process—don't assume that members all think the same as you do. Above all, keep an eye out for shifting constraints that may wreak havoc on things if monitoring and follow-up does not become part and parcel of actions that encourage learning.

Lack of a Common Vocabulary. Education and experience has predominantly prepared project teams to solve routine or "tame" problems. Teams set out to complete projects better and faster, yet projects failure rates remain high. Fragmentation of effort may be one cause linked to difficulties teams have in identifying and solving complex problems (Conklin, 2001). How do you know if fragmentation of effort is hurting your project? Blame is one visible symptom. Conklin (2001) characterizes blame as a kind of persistent white noise that becomes normal over time. High team personnel turnover is another symptom.

Developing a common vocabulary may be one antidote to reduce blame and turnover. On a micro level, developing a shared glossary of terms before the project starts often heads off problems as the project progresses, particularly if the project team training is multidisciplinary. Assuming that terminology in documents to assist in managing projects such the Project Management Institute's (PMI) *A Guide to the Project Management Body of Knowledge* is *generally accepted* and will be commonly understood appears to be a faulty assumption.

Herein lies the wicked problem: The term "generally accepted" according to PMI "means that knowledge and practices exists that apply to most projects most of the time, and that widespread consensus endures about their value and usefulness" (PMI, 2000, p. 3). Studies show that experts in any profession can only ever share parts of terminology and conceptual systems (Gaines and Shaw, 1989). Practical evidence is in the number of entries in a popular online project management glossary by Max Wideman soaring to over 5,400 terms/definitions from 100 sources, containing over 50,000 internal cross-referencing links (Delisle and Olson, 2004). The hundreds of project management glossaries filled with conflicting terms and definitions have paradoxically increased fragmentation instead of helping to create a shared social understanding. (See also the chapter by Crawford on standards in this regard.)

The first step lies in raising awareness about the extent of the problem. With this aim, a recent study asked 51 participants who on average had just over 15 years of project and business experience to define 20 "common" project management terms without using outside sources (Olson, 2001). The results are presented in Figure 3.5.

Only 37 percent consensus was achieved as an aggregate for the 20 common terms, disconfirming the belief that people-managing projects naturally have a high level of shared understanding. The problem is not easy to address. For example, 47 percent defined the term "competency" differently (high level of correspondence). Thirty-seven percent of respondents defined "project management maturity" totally differently. As well, they used different labels to describe it (i.e., age, experience), pointing to a high level of correspondence.

I have subsequently validated these findings by repeating the exact exercise using the same type of participants in the same conditions over the past two years. Consensus has never reached beyond 40 percent for all 20 terms. What level of shared understanding can we realistically achieve? Cha and Edmonson's (2003) study provides some insight into the why it is difficult to achieve a shared understanding (see Case Study 5).

Case Study 5

> The CEO of a young marketing company strongly espoused and clearly articulated his vision of the core company values (i.e., one was unpretentiousness). Employees initially embraced his *sent* or *stated* values, but over time as the company grew, employees made increasingly negative attributions about the CEO. The way that employees made sense of the CEO's stated values did not map exactly onto the

FIGURE 3.5. DIFFERENCE IN SHARED UNDERSTANDING OF 20 TERMS.

Definitions

	Same	**Different**
Terminology — Same	**37% Consensus** Use terminology and definitions in the same or very similar way.	**20% Correspondence** Use different definitions for the same terminology.
Terminology — Different	**28% Conflict** Use the same terminology but define concepts differently.	**13% Contrast** Totally differ in the use of terminology and definition.

actual words, nor did their meaning fit exactly. The employee's *elaborated* values were broader and more ideological than the CEO's sent values. For example, the CEO's sent value of unpretentiousness (meaning being informal, not egotistical) was elaborated on by employees to encompass "no hierarchy and elimination of rank" and labeled "equality." When they finally raised concerns about the decision to name shareholders within the company, it was limited to asking the CEO for change and not in seeking alternate explanations for the problem or findings solutions. In short, employees harshly judged the CEO as hypocritical because they believed the company's "shared" values were violated.

In the initial honeymoon or appraisal period, positive or negative attributions are made about a leader's actions (Cha and Edmonson, 2003). If subsequent actions are inconsistent with the espoused values, people engage in "sensemaking" where they critically analyze and judge the situation. On a biological level, physical stress increases and results in more simplified responses if actions are deemed to be inconsistent (Weick, 1995). On a cognitive level, people seek to blame someone that is highly visible for the problem, disregarding any role that they many had or have in order to preserve self-esteem (Cha and Edmonson, 2003).

What can a project team do about creating a shared understanding?

- Avoid assuming anything and seek clarification.
- Create a safe environment that invites questions and clarification without judgment.
- Raise awareness and acknowledge differences and similarities among member's training.

- Identify how large a gap exists in shared social understanding on the team up front in the project.
- Capture the output using a display system (mind map or equivalent) to make the issues and ideas "visible" and reusable.
- Engage the team in a process of articulating issues/ideas as a way of getting to a reasonable level of consensus on terms that everyone needs to understand. This is a process of making knowledge "explicit" and visible so that it can be reused by the team (including stakeholders).
- Aim to identify and agree on the key characteristics of complex concepts rather than trying to nail down one catchall definition.
- Understand that meaning is socially constructed; having the project leader impose a list of definitions or state his or her values with no further discussion does not mean complete buy-in.

Above all, project managers/leaders need to invite discussion and questions and promptly follow up if gaps occur between their stated values and those values elaborated on by team members.

Summary

This chapter on teams and effective team building may serve as a reminder to some and act as a source of information to others about the processes, conditions, and behaviors that influences teams. The chapter provided a snapshot of some of the key areas to consider when building and developing teams, as well as identified common team pathologies that may arise over the course of a typical project life cycle.

Following good team practices to achieve successful projects is arguably not a cause-and-effect relationship. What is more certain is that if the points brought up in this chapter are disregarded, the job of identifying, growing, and developing teams may be that much more difficult.

References

Beder, S. 2000. Working long hours. *Engineers Australia*. (March): 42.

Belgard, Fisher, Rayner, Inc. 2000. Building a collaborative team environment. Team Tools Module 14: 20811 NW Cornell Road, Suite 100, Hillsboro, OR 97124.

Blount, S., and J. Gregory. 2001. When plans change: Examining how people evaluate timing changes in work organizations. *Academy of Management Review* 26:566–585.

Brown, S. M. 2002. Changing times and changing methods of evaluating training. Lesley College, Cambridge, MA.

Brown, L. 2002b. The secret to mental toughness at work. www.lisabrown.ca/mttforwork.html (accessed November 17, 2002).

Bureau of Labor Statistics. 1999–2000. Women's share of labor force to edge higher by 2008. *Occupational Outlook Quarterly* (Winter): 33–38.

Cha, S., and A. Edmondson. 2003, How promoting shared values can backfire: Leader action and employee attributions in a young, idealistic organization. Harvard Business School. Division of Research, paper 03-013.

Chen, X., and W. Barshes. 2000. To team or not to team. *The China Business Review.*

Cleland, D. I., and W. R. King. 1988. *Handbook on project management.* New York: Wiley.

Conklin, J. E. 2001. Wicked problems and fragmentation. Cognexus Institute. This paper is Chapter 1 in the forthcoming book *Dialogue mapping: Defragmenting projects through shared understanding.* CogNexus Institute. 2003. http://cognexus.org.

Conklin, J., C. Ellis, L. Offermann, S. Poltrock, A. Selvin, and J. Grudin. 1998. Towards an ecological theory of sustainable knowledge networks. Touchstone Consulting Group White Papers. 1920 N Street NW, Suite 600, Washington, D.C. 20036.

Crisp, B. C., and S. K. Jarvenpaa. 2000. Trust over time in global virtual teams. Presented at Organizational Communication & Information Systems Division of the Academy of Management Meeting.

DeBono, E. 1977. *Lateral thinking: Creativity step by step.* New York: Harper & Row.

Delisle, C. 2001. Success and communication in virtual project teams. PhD diss. Dept. of Civil Engineering, Project Management Specialization. The University Of Calgary, Calgary, Alberta.

Delisle, C., J. Thomas, K. Jugdev, and P. Buckle. 2001. Virtual project teaming to bridge the distance: A case study. PMI seminar and symposium. Nashville, TN.

Delisle, C., and D. Olson. 2004. Would the real project management language please stand up? *International Project Management Journal.* p. 327–337.

Delisle, C., and J. Thomas. 2002. Defining success to get traction in a turbulent business climate. *Proceedings of the 2nd Annual PMI Research Conference.* Seattle, WA.

DeMarco, T., and T. Lister. 1987. *Peopleware: Productive projects and teams.* New York: Dorset House Publishing Co.

Dixon, M. 2000. *Project Management Body of Knowledge.* 4th ed. High Wycombe, UK: The Association for Project Management.

Drake, B., and Morin. 2000. *CEO turnover and job security: Research highlights from a worldwide survey.* 101 Huntington Avenue. Boston, MA 02199. Toll-free 800 DBM-2242.

Duarte, D., and N. T. Snyder, 2001. *Mastering virtual teams.* 2nd ed. San Francisco Jossey-Bass:.

Duxbury, L., and C. Higgins. 2001. Work-life balance in the new millennium: Where are we? Where do we need to go? Discussion Paper No. W|12. Canadian Policy Research Networks. 250 Albert Street, Suite 600. Ottawa, ON K1P 6M1.

Dyrenfurth, M. J. 2000. Trends in industrial skill competency demands as evidenced by business and industry. 1–18. *Proceedings from the International Conference of Scholars on Technology Education. European Commission.*

European Foundation for the Improvement of Living and Working Conditions. 2000. Third European survey on working conditions. EU0101292F. http://217.141.24.196/2001/11/study/tn0111109s.html.

Forsberg, K., H. Mooz, and H. Cotterman. 2000. *Visualizing project management.* New York: Wiley.

Frankel, C. 1999. *Emotions and emotionsand economic knowledge: Drivers of self-regulatory information processing.* Pacific Graduate School of Psychology

Howard, N F., Jr., and T. Mitra. 2001. Nov. labor force projections to 2010: Steady growth and changing composition. *Monthly Labor Review* 124(11):21–38.

Gaines, B. R., and M. Shaw. 1989. Comparing conceptual structures: Consensus, conflict, correspondence and contrast. *Knowledge Acquisition* 1(4):341–363.

Gardner, H. 1993. *Frames of mind: The theory of multiple intelligences.* New York: Basic Books.

Gersick, C., and J. Hackman. 1990. Habitual routines in task-performing groups. *Organizational Behavior and Human Decision Processes* 47(1):65–97.

Grazier, P. 2002. Work in the 21st century will recognize human potential. Telephone: (610) 358-1961.

Hartman, F. T. 2000. *Don't park your brain outside: A practical guide to improving shareholder value with SMART management.* Newtown Square, PA: Project Management Institute.

Hill, R. E., and T. L. Summers. 1988. Project teams and the human group. In *Project management handbook.* 2nd ed. D. Cleland and W. R. King. New York: Wiley.

Hoffman, E. J. 2000. Developing superior project teams: A study of the characteristics of high performance in project teams. NASA's Academy of Program and Project Leadership. *Conference Proceedings.* PMI International Research Conference, Paris.

Human Resources Development Canada (HRDC). 2001. Recent immigrants have experienced unusual economic difficulties. *Applied Research Bulletin* 7 (1, Winter/Spring).

Ingham, A. G., G. Levinger, J. Graves, and V. Peckham. 1974. The Ringlemann effect: Studies of group size and group performance. *Journal of Experimental Social Psychology* 10:371–384.

Johnson, K. L., D. S. Lero, and J. A. Rooney. 2001. Work-life compendium. 150 Canadian statistics on work, family and well-being. Centre for Families, Work and Well-Being, University of Guelph.

Kerzner, H. 2001. Project management: A systems approach to planning, scheduling, and controlling. 7th ed. New York: Wiley.

Kierstead, J. 1998. Competencies and KSAO's. Public Service Commission of Canada. www.psc-cfp.gc.ca/resaerch/personnel/comp_ksao_e.htm (accessed October 15, 2002).

Lipnack, J., and J. Stamps. 1996. *Virtual teams: Reaching across space, time and organizations with technology.* 2nd ed. New York: Wiley.

Madigan, C. O. 1998. Perfecting Project Management Skills: Part One. *Business Finance Magazine* (December): 28.

Margerison, C. J., and D. J. McCann. 1994. The types of work index: A measure of team tasks. *The Occupational Psychologist* (23):24–31.

McGannon, R. 2001. Will your project team get the job done? *ESI Horizons Newsletter.* (March).

McGrath, J. E. 1990. Time matters in groups. In *Intellectual teamwork: Social and technological foundations of cooperative work,* ed. J. Gallegher, R. E. Kraut, and C. Egido. 23–61. Hillsdale, NJ: Erlbaum Press.

Meyer, P. 1998. Trouble finding good people? Stop trying to hire them. *The Business and Economic Review.*

Miller, T. E. 1988. Teamwork: Keys to managing change. In *Project Management Handbook.* D. I. Cleland and W. R. King. 2nd ed. New York: Wiley.

Mills, T., S. Tyson, and R. Finn. 2000. The development of a generic team competency model. *Competency and Emotional Intelligence* 7(4):1–6.

Ministry of Industry. 2002. Women in Canada: Work chapter updates. Statistics Canada Housing, Family and Social Statistics Division. Catalogue no. 89F0133XIE.

NASA. 2002. *Characteristics measured by TeamMates: NASA's Project Team Development Survey.* NASA Academy of Program and Project Leadership.

Next Step. 2002. Case study: London & Manchester Assurance Company Limited. Customer Service Teams. www.nextstepltd.co.uk/case_studies/archive/london_manchester.html.

Olson, D. 2001. Is a common vocabulary lacking in project management? Information Technology Project Management. Executive MBA program at Athabasca University.

Ott, M., and L. Nastansky. 1997. Modeling organizational forms of virtual enterprises, University of Paderborn, Business Computing 2. Warburger Str. 100, D-33098 Paderborn, Germany.

Pinto, J., and D. Slevin. 1988. Project success: Definitions and measurement techniques. *Project Management Journal.* (February): 67–71.

Project Management Institute. 2000. *A guide to the Project Management Body of Knowledge.* Newtown Square, PA: Project Management Institute.

Potter, B. 1998. *Overcoming job burnout: How to renew enthusiasm for work*. Berkeley, CA: Ronin Publishing.

Sommerville, J., and S. Dalziel. 1998. Project teambuilding: The applicability of Belbin's team role self perception inventory. *International Journal of Project Management* 16(3):165–171.

Statistics Canada. 2000. Income in Canada, 1998. Labour force and participation rates. Ottawa: Statistics Canada, Catalogue 75-202XIE.

Statistics Canada. 2001. Canadian Statistics: Labour force and participation rates. (Online table). Women in Canada. Ottawa: Statistics Canada, Catalogue 89-503-XPE. www.statcan.ca (accessed November 22, 2002).

Stephenson, C. M. 2000. Innovation through people: One CEO's experience. Lucent Technologies Canada Inc. Presentation to the York Technology Association.

Strassmann, P. A. 1995. Outsourcing: A game for losers. *Computerworld* (August 21): 75.

Stark, M. 2001. Five keys to successful teams. PricewaterhouseCoopers. www.pwcglobal.com/Extweb/NewCoAtWork.nsf/docid/ 5D9D4B372EC8367485256C6100771C85 (accessed November 10, 2001).

Sugarman, K. 1999. *Winning the mental way*. Buligame, CA: Step Up Publishing.

Swann, W. B., J. T. Polzery, D. C. Seyle, and Sei Jin Ko. 2002. Finding value in diversity: Verification of personal and social self-views in diverse groups. National Institutes of Mental Health MH57455 Department of Psychology, University of Texas at Austin. Division of Research at Harvard Business School.

Thomas, Janice, Delisle Connie, and Kam Jugdeo. 2002. Selling project management to senior executives—framing moves that matter. Newtown Square, PA: Project Management Institute.

Thoms, P. Forthcoming. *Driven by time: A leader's guide to time orientation*. Portsmouth, NH: Praeger/Greenwood.

Thoms, P., and D. B. Greenberger. 1995. The relationship between leadership and time orientation. *Journal of Management Inquiry* 4(3):272–292.

Thoms. P, and J. K. Pinto, J. 1999. Project leadership: A question of timing. *Project Management Journal* 30 (1).

Tucker A. L., and A. C. Edmondson. 2002. Why hospitals don't learn from failures: Organizational and psychological dynamics that inhibit system change. Boston: Harvard Business School.

Tuchman, B. W. 1965. Developmental sequence in small groups. *Psychological Bulletin* 63(6):384–389.

Walsh, M. W. 2001. Luring the best in unsettled times. www.nytimes.com/library/financial/01working-wals.html

Weick, K. E. 1995. *Sensemaking in organizations*. Thousand Oaks, CA: Sage Publications.

Wideman, R. M. 2001, June. Project Management Glossary. www.maxwideman.com/pmglossary/PMG_E00.htm (accessed March 2002).

Wilemon, D. L., and B. N. Baker. 1988. Some major research findings Regarding the human element in project management. In *Project Management Handbook*. 2nd ed. D. I. Cleland and W. R. King. New York: Wiley.

CHAPTER FOUR

LEADERSHIP OF PROJECT TEAMS

Peg Thoms, John J. Kerwin

Leadership is difficult under any circumstances, but the leadership of project teams is particularly challenging for a number of reasons. First, projects almost always involve initiating changes. In addition, projects are typically assigned to project leaders by a person or people who have their own visions of what the end product should be. The project leader must understand and achieve those visions while creating his or her own image of the project outcomes and, particularly, of the process necessary to complete the project. With the management of large projects, there may be numerous stakeholders who have their own ideas about the end product. The project leader must please everyone, even when there are contradictory opinions. Finally, unlike the day-to-day organizational management of a particular operation, projects typically have a limited duration and utilize team members who move from project to project. Project team members must be self-directed and goal-oriented, and must take on leadership roles themselves in many situations. The challenges for effective project leaders are successfully bringing about change, satisfying their stakeholders, and leading teams of contributors who may be involved on a short-term basis.

This chapter explains several of the best-known theories of leadership that have implications for project leadership. In particular, readers will be introduced to transformational leadership theory and how it applies to project teams. Readers will learn to develop and articulate a project vision. In addition, the chapter focuses on how leaders can ensure that the project vision is being implemented. We also introduce the concept of time orientation and how leaders can utilize their own and others' time perspectives to make projects more successful. Finally, we examine two complex projects as examples of how a vision can drive project decisions and inspire project teams.

Leadership Theories

Leadership is defined in different ways by different experts. For purposes of project leadership, we will define leadership as the process of influencing others to understand what needs to be done and how it can be done, coordinating and motivating the work of various individuals and subcontractors, and delivering a successful product in the context of a project. Because project leaders often take someone else's vision and execute it, it is essential that they are able to grasp the basic concept that is being developed. In other cases, project leaders develop the vision and sell their idea to the funding sources. The "how" portion of project leadership requires technical expertise that the leader brings to the project through personal knowledge and experience or through recruiting appropriate team members who have the expertise. For example, a number of subcontractors work on large projects, which makes it necessary for the leader to organize and schedule the work of various team members while providing a motivating work environment. The leader must be constantly aware of criteria that should and will be used to measure project success. Continuous assessment and monitoring of performance are a critical component of the delivery of the finished product.

Most leadership theories fall into one of two categories: contingency or universal. Contingency leadership theories suggest that different times, tasks, and organizations may require different types of leaders or leadership behavior. These models imply that leaders can and do change their behavior as the needs change. Universal theories suggest that an effective leader is an effective leader regardless of the situation. These theories describe traits and behaviors that should work with any organization. Here are some examples of each type of theory that relate to project leadership.

The *Situational Leadership Model* (Hersey & Blanchard, 1977) suggests that leaders need to use more relationship-oriented behavior in some situations and more task-oriented behavior in others. Specifically, they tell leaders that when a team member is capable, but lacks motivation, a participative style is best. When a team member is capable and motivated, it is best to delegate authority. When a team member is inexperienced but motivated, the leader should provide guidance, explain decisions, and clarify procedures. When the team member is inexperienced and lacks motivation, the leader should dictate tasks and closely monitor the work. This theory focuses on the follower's level of maturity and how leaders should interact with them.

Path-Goal Theory (House, 1971) explains how the behavior of a leader influences the feelings of satisfaction and performance of subordinates. Essentially, this theory suggests that leaders should increase the personal payoffs to team members for work-goal attainment and make the path to these payoffs easier by clarifying work direction, eliminating roadblocks, and increasing opportunities for personal satisfaction on the way to meeting the goal. According to this theory:

1. Leaders must clearly communicate the desired outcomes and any necessary steps or requirements along the way.
2. Leaders must ensure that team members are consistently rewarded.
3. The rewards must fit the needs and interests of the individual team member.

The Normative Decision Model (Vroom & Yetton, 1973; Vroom & Jago, 1988) provides a framework to help leaders determine the optimal level of participation of team members needed for effective decision making. Levels of participation are as follows:

AI. The leader makes the decision alone using only information available at the time.

AII. The leader obtains information from team members and makes the decision alone.

CI. The leader shares the problem with team members individually, gets input, and makes the decision alone.

CII. The leader shares the problem with team members in a group setting, gets input, and makes the decision alone.

GII. The leader shares the problem with team members as a group, and the group generates alternatives, reaches agreement on a solution, and reaches a decision.

The model provides questions that help the leader determine which level of participation is best for each situation. In general, the following guidelines can be used by project leaders:

1. The more important the decision, the more participation is needed.
2. The less information the leader has, the more participation is needed.
3. The more information team members have, the more participation is needed.
4. The more important it is that team members accept the decision, the more participation is needed.
5. The fewer rules, procedures, and policies, the more participation is needed.

Leadership Substitutes Theory (Kerr & Jermier, 1978) is one of the most useful for project leaders because they are likely to be working with experienced professionals on their teams. This theory suggests that certain characteristics of a project team may actually substitute for leadership. These substitutes include team members with a professional orientation, experience, ability, and training. Other substitutes for leadership include structured and routine project tasks, intrinsically satisfying work, and feedback that comes directly from the work, like automatically generated progress reports. Finally, other substitutes include a cohesive team and established formal roles and policies. This theory helps leaders understand that choosing the right team members and setting up efficient work systems make the job of a leader much easier. Individuals who enjoy what they do and receive feedback about their work through the system do not need the level of interaction and supervision required in other situations.

In addition to the lessons we can learn from the theories described. Yukl provides some other suggestions relevant to project leaders that are based on the research done on contingent leadership theories (Yukl, 2002).

1. Spend more time planning for long complex tasks.
2. Consult with team members who have relevant knowledge.
3. Provide more direction to team members with interdependent roles.
4. Monitor critical tasks and unreliable team members or subcontractors more closely.
5. Provide more coaching to inexperienced team members and to those who have stressful tasks.

Universal theories of leadership are based on the belief that a good leader is a good leader, regardless of the situation, because of personal traits and patterns of behavior that always work. The *Great Man Theory* was one of the earliest theories of leadership. This theory tells us that great leaders are born and are born to be effective. When this theory evolved, leaders were typically male and inherited their positions of power. Although few people give this theory much credibility today, other universal theories do not differ much from the idea that some leaders "have it" and others do not.

Charismatic Leadership Theory was introduced in 1947 and refined later (Weber, 1947). Charisma is a Greek word that means "divinely inspired gift." It refers to the ability to perform miracles or predict the future. Modern theorists use the term to describe the ability of a leader to solve an immediate crisis, articulate a vision, attract followers, and achieve parts of the vision. A recent version of the theory suggests that charisma is attributed to the leader by followers (House, 1977). Followers are more likely to see a leader as charismatic if the leader:

1. Advocates a vision that is different from the status quo but socially acceptable.
2. Acts in unconventional ways to achieve the vision.
3. Makes personal sacrifices, takes risks, and pays a cost in order to achieve the vision.
4. Appears to be confident.
5. Uses persuasive appeals rather than using an authoritative approach with followers.

Charismatic Leadership Theory suggests that these types of leaders influence others because followers identify with and want to please and imitate the leader. Followers measure their own success by gaining approval from the leader. Leaders who praise and recognize followers who perform well reinforce desired behavior. In addition, charismatic leaders introduce and develop new values and beliefs. For project leaders, the implications are a bit more complicated, since these theories suggest that one is either charismatic or not. Charismatic leaders are most apt to emerge during a crisis. For example, if a project is in serious trouble, there is an opportunity for a project manager to demonstrate the behaviors typical of charismatic leaders. These behaviors may occur naturally, or it might take hard work on the part of the leader to change and inspire followers.

Transformational Leadership Theory (Bass, 1985) is the most widely studied theory in the past 20 years. This theory distinguishes between transformational leaders and transactional leaders. Transformational leaders motivate followers by making them aware of the importance of their work, convincing them to sacrifice self-interest for the sake of the group, and encouraging them to achieve higher-order needs like belonging to a group and achieving important goals that help others. Transactional leaders simply exchange rewards for task performance, but do not inspire followers or build commitment to the group or organization. Bass suggested that transformational leaders engage in specific behavior that allows them to affect followers' motivation. These behaviors are as follows:

1. *Idealized influence.* Behavior that brings about strong emotions among followers who identify with the leader
2. *Individualized consideration.* Providing support and encouragement to followers and coaching them to bring about improved performance
3. *Inspirational motivation.* Communicating a positive future vision, using symbols to focus work effort, and modeling desired behavior

4. *Intellectual stimulation.* Increasing follower awareness of problems and influencing them to see problems from a different perspective

A number of other leadership experts have explored the behaviors that seem to make transformational leaders so effective in bringing about organizational change. Kouzes and Posner have studied this behavior for many years and suggest that there are five critical behaviors common to transformational leaders (Kouzes & Posner, 1995):

1. *Challenging the process.* Continuously seeking new options and exploring others. Transformational leaders want to make things better even when the current system is broken.
2. *Inspire a shared vision.* Creating a vision that appeals to the values, goals, and interests of followers. Transformational leaders have an uplifting and rewarding image of the future that they effectively communicate to followers.
3. *Enable others to act.* Empowering followers to make decisions and act. Transformational leaders educate followers, allow them to make important contacts, and encourage them to create and try new things.
4. *Model the way.* Modeling the behavior that they value and want others to practice. Transformational leaders set an example for the level of performance they expect from others. They are dramatic, take their work personally, and tell stories which reinforce their values.
5. *Encourage the heart.* Rewarding and encouraging followers. Transformational leaders reward and recognize followers publicly. They provide feedback to followers and create winners by helping people become successful, even when it seems unlikely.

Project Vision

Vision is the common thread between the various theories of transformational leadership. According to these theories, it is necessary to be able to create and communicate a positive vision of the future in order to bring about major changes in organizations. Vision is not a new concept. There are legends and stories about visionary leaders that trace their origins to prehistoric times. Vision is a concept that cuts across cultures as well. Most peoples on this planet have known of the importance of vision for millennia. Over 1,000 articles and books have been written on vision in the academic press alone.

Despite the evidence of its importance, vision is an overused and widely misunderstood term in organizations. The problem is that few people really understand vision and fewer still have it. In most organizations and groups, the vision is nothing more than a long general statement that appears on a plaque on the wall and makes it to the second page of the annual report. These vision statements all look alike and provide little if any guidance to followers regarding the direction of the organization. A vision that will drive project and organizational performance is much different, and we define it this way: *A vision is a cognitive image of an organization or project that is positive enough to followers to provide motivation and elaborate enough to provide direction for planning and goal setting.* It may exist in the mind of the leader or, if well communicated, be understood by all constituents of the organization.

It is critical that a vision be idealistic. Visionary leaders are idealistic. Just as challenging goals lead to higher levels of performance, idealistic visions lead project teams to higher levels of accomplishment. If a project leader envisions a project where there are no cost overruns and personally behaves in a manner consistent with that, communicates that vision to team members, and is persistent whenever even a slight overrun occurs, the project will come in closer to budget. Part of the project planning will include specific strategies to ensure no cost overruns. An idealistic vision drives the planning process in ways that a "realistic" vision (or no vision at all) will never do. Visionary leaders always start with the best possible outcome in mind.

Based on recent research (Thoms & Blasko, 1999), the ability to create a vision is related to other personality characteristics. One study found that high visioning ability is related to an optimistic style toward life and situations. People who are positive and optimistic about their chances for success are more likely to imagine a positive future for themselves and their organizations. In addition, people who have higher future time perspective are more apt to have high visioning ability. People who are present- and past-oriented are less likely to think about the future and to create images of what the future could be. Nonetheless, most leaders can learn to create and communicate a vision.

Having a vision is not enough for a transformational leader, however. The vision must drive the leader's and the organization's behavior. Imaging is an important concept in management, and we use imaging training to improve sales, athletic, and task performance. We teach salespeople to picture a meeting with a customer and themselves overcoming objections. We teach athletes to imagine a ball going into a hoop. We know that creating mental images of ourselves being successful leads to successful performance. Vision works the same way, except that a vision is often broader than just one individual's performance. We create a positive image that drives our behavior, and, in turn, our behavior influences followers (Thoms & Govekar, 1997). At the project level, the behavior most likely to lead to achievement of our project vision would include the following:

1. *Planning.* Identifying the potential barriers to achieving the vision and finding ways around them.
2. *Influencing.* Convincing followers to stay committed to the vision. Particularly when problems occur, leaders must focus attention on achieving the vision and not compromising quality for an easier route.
3. *Selecting project team members.* Recruiting and hiring people who share similar values, goals, and interests who will be able to commit to the vision and have the skill to perform at the appropriate level.
4. *Providing feedback.* Executing a vision requires continuous monitoring of progress and adherence to the vision. Followers must be given accurate feedback regarding their performance and the performance of the team. When performance is inconsistent with the vision, people must know so that they can change their behavior.
5. *Choosing the appropriate use of the leader's time.* Deciding from day-to-day how to spend the time. If one crew is not on target and the vision includes meeting all timelines, the leader will have to spend part of that day finding out what is wrong and helping to find solutions. Other tasks that do not influence the success of the vision should be set aside.

Often, project leaders let tedious repetitive tasks, some of which could be delegated, take up their time and fail to focus on the aspects that will determine the success of the project.

6. *Goal setting.* Establishing the goals for each aspect of the project and for each team member. Transformational project leaders set challenging goals, communicate them clearly, and monitor team members' progress toward their goals. The goals should answer the question, "How will the vision be achieved?" The vision answers the question, "Why do we have these goals?"

7. *Communicating the vision.* Articulating the vision or parts of the vision to the team members who are responsible for the execution of each aspect of the project. Communicating vision is done in a variety of ways on a project. For example, blueprints and storyboards are clear pictures of the leader's vision. Explaining how time frames will be followed or why using a particular type of material or a specific location is important may not be so simple to understand. Effective project leaders must translate the image in their minds into words and talk with the appropriate individuals, explain their vision, influence team members to commit to it, and provide direction. Every team member does not have to understand the entire vision for a complex project, but they had better understand what the leader is trying to achieve as it relates to them and their work.

Communicating vision takes place every day in everything that the leader does. We typically think of speeches made by leaders when we consider communication of vision, but it goes well beyond that and includes modeling appropriate behavior, relating performance feedback, and sharing success stories about people working on the project. When problems occur and it appears that the vision will not be achieved, the leader must decide whether to stick with the vision. When project leaders consistently sacrifice their vision for cheaper, faster, easier solutions to problems that occur, they send a message to followers that the vision doesn't matter.

As mentioned at the beginning of the chapter, project leaders often find themselves having to execute the vision of others who hire them. For example, the project engineer of a highway construction project is paid to make sure that the highway is built according to the blueprints and plans drawn by engineers and officials. Or, the director of a film may be executing the vision of a producer or a screenwriter who came up with the original idea. In these cases, the project leader has to understand the vision of the primary stakeholder. Nonetheless, the project leader can and should create a vision of the process that will be used to achieve the vision. Visions do not just encompass end products like a well-built highway or a successful film. The vision must also include the process. This means that the project vision may include the following: the project will fall within its budget, the workers will follow all laws, the employees will feel that they were treated fairly, and the community will be satisfied with the progress made. The ideal process is the vision that the project leader, who is charged with executing someone else's idea, must create.

Creating a Project Vision

Visionary leaders create visions automatically. Images of the future appear in their minds as naturally as the rest of us eat breakfast. That doesn't mean that they can achieve their visions, only that they probably have excellent imaginations, are future-oriented, and have an optimistic outlook toward situations. The rest of us have to focus a bit more in order to create a project vision and will also have to work hard to achieve it. One approach to vision creation that has been empirically tested involves three steps and utilizes several creativity development methods (Thoms & Greenberger, 1998; Pinto et al., 1998).

Step 1: Mapping

The first step is designed to help the project leader identify aspects of the project and elements that must be considered in order for the project to be successful. To begin, put the name of the project in the center of a sheet of paper. Identify each aspect of the project. For example, for a highway project, you would list budget, subcontractors, environmental issues, location, legal requirements, and existing traffic patterns, among other things. Distribute these aspects around the center, and consider each element of the project that relates to that aspect. For example, around environmental issues, you would list local laws, flora, fauna, landscaping, noise levels, and potential chemical pollutants, among other things. You may be able to break this down even further. For example, next to fauna, you may have listed breeding areas, endangered species, and nesting locations. You would continue the mapping, creating a detailed web or cluster of associations that reveal every aspect of the project that the vision should encompass. Figure 4.1 provides an example of what this might look like. More detail would be added by an experienced project leader who understands this type of project.

Step 2: Generating "Wouldn't It Be Great If . . . ?" Statements

The second step begins by returning to each aspect of the project identified in the first step and creating as many statements that begin with the phrase, "Wouldn't it be great if . . . ?" as appropriate to capture what you would consider to be the best possible outcomes. These statements capture the ideal outcomes for the project. Do not worry about being realistic at this point. By creating challenging goals, we achieve more in the long run even if every goal is not fully achieved. The same is true for a challenging vision. Going back to the example of the highway construction project, Figure 4.2 lists some possible statements that a project leader might generate for a highway construction project. Depending on the complexity of the project, there could be hundreds of these statements.

Step 3: Creating the Move in Your Mind

The third and final step is to pull all of the statements together and visualize the project at the end. Because this step occurs in the mind of the leader, it is difficult to impossible for others to monitor it. It requires the leader to spend a significant portion of time imaging how the project will progress and how each stage of the project will look. The leader would imagine the work proceeding, how team members are interacting, how the work site sounds and smells, the condition of the equipment, the attitude of the workers, the cooperation of the various subcontractors, and so on. In essence, the leader pictures every aspect of the

FIGURE 4.1. EXAMPLE OF THE DEVELOPMENT OF A MAP.

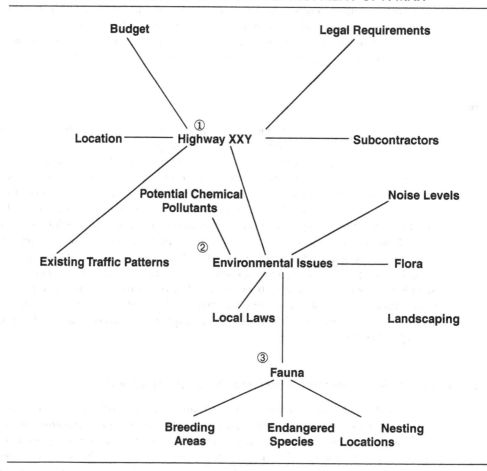

project before it occurs using the map and the statements generated in previous steps. This becomes the vision that drives the project. Visionary leaders do this automatically. Those of us who are not born visionary leaders must force ourselves to imagine successful projects and to mentally rehearse the behavior it will take to be successful. Mentally rehearsing difficult situations and specific challenges is a technique that has been used successfully for many years by the best athletes, salespeople, actors, and leaders.

Visionary leaders may not imagine every aspect of a project at the same time. It would be impossible for most humans to hold a clear image of every aspect of a complex project in one's mind at once. Most likely, leaders think about their projects in pieces and envision the best-case scenario for each aspect. At any point in time, the overall vision may be symbolized by positive feelings and the drawing, blueprint, model, or storyboards of the completed project. The images of the various pieces of the project come into focus only as each becomes important. As they emerge, the leader imagines that aspect of the project and

FIGURE 4.2. EXAMPLES OF "WOULDN'T IT BE GREAT IF . . . ?" STATEMENTS FOR A HIGHWAY CONSTRUCTION PROJECT.

Depending on the complexity of the project, there would be hundreds of these statements.

Wouldn't it be great if we had a complete list of all legal requirements with a list of the regulatory bodies and a contact person for each?

Wouldn't it be great if we had software that would track all legal issues as the project proceeds and alert us to any potential problems?

Wouldn't it be great if inspectors from each regulatory body made visits to the construction site at the appropriate time?

Wouldn't it be great if the project was completed with no labor stoppages?

Wouldn't it be great if all overtime expenses could be eliminated?

Wouldn't it be great if every worker had perfect attendance during the length of the project?

Wouldn't it be great if no OSHA inspections were requested during this project?

communicates it to team members. The leader of a construction project may move continuously around a work site, causing different images to come to mind as the focus shifts from excavating to framing to meeting with subcontractors. The key is that the leader is constantly comparing the work in progress against the vision of the ideal. It is critical to monitor progress toward the vision, looking for discrepancies between performance and the vision, and correcting the errors.

In the leadership literature, there is a lot of discussion about groups developing a vision. Although developing a vision is an activity in which many project teams engage, someone must commit to the vision in order for it to drive behavior and impact performance. Visioning or imaging only works if it is in the mind of the person who must execute it. It is highly unlikely that an entire team will imagine the same things and commit equally to a vision. Leaders do not have to use a participative style when they create a vision, but they do have to sell it to others. It can be useful, however, to involve other leaders in the process in situations where you do not have the necessary expertise or need to generate many ideas. Just don't assume that everyone on your team will see the vision the same way that you do or at the same time.

Project Leadership and Time Orientation

Recent research suggests that every person has a unique temporal alignment (Thoms & Pinto, 1999). *Temporal alignment* is made up of a variety of psychological constructs or biases that relate to time. Essentially, this means that we are all oriented primarily toward the past, the present, or the future. Each time orientation has its own strengths and weaknesses. In addition, different types of tasks may require unique time orientations. This suggests that different projects and different stages of projects require different temporal alignments. In other words, sometimes it is better to be past-oriented than future-oriented. Other times it is better to be present-oriented than future-oriented. It depends on the situation. When the leader has the temporal alignment that the situation demands, we call it *attunement.*

Effective leaders must occasionally behave in ways that fall outside of their temporal comfort zone in order to achieve goals. Most people understand that sometimes even shy people must be outgoing and friendly and that extroverted people must sometimes work alone. Because the concept of temporal alignment is relatively unknown, few people understand that sometimes we must be present-oriented when we would rather be working on an innovative project that won't be completed for ten years. Understanding temporal alignment and adapting to various situations can make leaders more effective.

The Future-Oriented Project Leader

Future-oriented project leaders are particularly effective when a project environment is dynamic and changing. They are the most likely of the three types of leaders to imagine a positive vision of the future. Envisioning the future is automatic behavior for most future-oriented leaders. Because most people are not future-oriented, visionary leaders must use their visions to motivate followers—to pull them into the future with them. Future-oriented project leaders tend to be very good at gathering information about what other organizations are doing to improve their effectiveness. This information gathering is done with the future in mind. As they read an article about another company trying something new, the future-oriented project leader is considering how the approach might work on their project. Future-oriented project leaders are constantly looking for opportunities—which may even be new projects. They are also apt to challenge current systems, seeking ways to improve. This creates tremendous frustration for followers who have done the work to make things flow smoothly and do not think about the future. Real problems can occur when future-oriented project leaders make changes without waiting for team members to catch up. One example of this would be a product development project leader who begins to market a product the company is adding to its line before the engineers on the team finish the design and production begins. Future-oriented project leaders must force themselves to slow down to accommodate the operational stages of change. This is difficult and will require effort depending on the strength of their future orientation.

The biggest problem that we see with future-oriented project leaders is their failure to recognize past accomplishments. This isn't just true for followers, but also for the leaders themselves. The past matters less to them. Most of the rest of us expect to be recognized for our performance. Some future-oriented project leaders may be less good at performance appraisals. They do not focus on the past year, may not remember what each individual reporting to them did or did not do, and tend not to value past performance. At best, feedback is usually delivered when the performance takes place. These leaders must use specific methods in order to do performance appraisals well. One example would be making notes when a follower does something out of the ordinary and putting the note in a file. Leaders who have administrative staff may ask them to keep track of specific information that can be easily accessed when an appraisal must be done. At the end of the evaluation period, the notes can be organized and used to develop feedback on performance.

Sometimes, future-oriented leaders are viewed as "off in the clouds," "daydreaming," "out there," or "in their own world." Most people are present-oriented, and they deal with life one day at a time. People who spend a great deal of time imagining the future are not mentally in the present much of the time and may even be seen as strange, depending on how well they can hide it. In meetings, future-oriented project leaders may not hear everything that is said or focus on the subject of the meeting. They are thinking about the future and are not available to the team, cognitively. This creates anxiety for team members who must get tasks done and need guidance or permission from leaders. Not hearing information provided in meetings can create critical problems, especially given that future-oriented people rarely read minutes of meetings.

Future-oriented project leaders are especially comfortable and effective in dynamic organizations. Because their orientation lends itself to taking advantage of opportunities and trying to make the future better, they work best on flexible projects with open management styles. They thrive in situations where creativity is rewarded, where they have access to decision makers, where leaders empower followers to make decisions, and where intuition or hunches are encouraged.

The Present-Oriented Project Leader

Leaders who live for the present and focus their attention on what is happening today are probably the most common of the three types. In a leadership role, present-oriented project leaders are very effective at dealing with day-to-day issues. This is true for several reasons. First, they care about the present. Problems are addressed one day at a time instead of speculating on how each aspect of the business may someday cause a problem. Their project teams tend to produce high-quality work when the leader communicates well, is accessible to constituents, and provides worthwhile feedback to team members.

Second, they are cognitively available on a daily basis. They will be able to help team members address concerns and develop methods to accomplish specific tasks. It would be unlikely to find this leader not paying attention or staring off into space when help is needed. The present-oriented project leader is likely to be circulating around the work site talking, observing, anticipating concerns, and answering questions.

Present-oriented project leaders are a good fit when leading projects similar to others they have done in the past. When they find themselves in dynamic situations, they will manage the production areas well, providing a critical anchor for those who are not future-oriented. Present-oriented project leaders are very good at solving problems. Because, they are not as reluctant to talk about the past as future-oriented leaders are and not as reluctant to talk about the future as past-oriented leaders are, they are willing to research the cause of the problems and help develop solutions for the future. When they have good communication skills, they are effective at providing information to team members and can be charged with this responsibility. They will deal better than other types of leaders with problem project workers. They will be more apt to go through the procedures in place, follow disciplinary guidelines, forget about problems that occurred five years ago that are not relevant to the current situation, and focus on solving the problem one day at a time.

The downside of present-oriented project leaders is that they may not make major contributions to the planning process. They may be too busy and may even sabotage the process if it will disrupt their operations. When they sabotage planning, they often kill it. They may also have a tendency to micromanage details and overlook the big picture. Vision is something they often lack, and they will need to make sure they have tangible things like blueprints or models to use to drive their behavior.

The Past-Oriented Project Leader

Past-oriented project leaders are very good at remembering and using the history of their project, department, and organization. This is often valuable. For example, it is important when a problem occurs and it is necessary to trace past meetings, decisions, and behavior. In this case, it is critical that information is preserved and that an oral history of the events leading to the problem can be re-created. Another contribution of past-oriented project leaders is their ability to trace patterns in their industry and to identify trends that may recur. These trends and patterns are very useful in predicting what may happen in a particular project or industry. Past-oriented project leaders with extensive experience will also remember the past behavior of specific leaders in other organizations, problems with certain subcontractors, and political issues that occurred on previous projects.

Past-oriented project leaders remember and value the contributions team members have made in the past. Future- and present-oriented project leaders tend to focus on "What have you done for me lately?" Past-oriented project leaders are more likely to forgive recent performance gaps if they know one is capable of good work based on past behavior. The past-oriented project leader believes that what followers did in the past matters. And, of course, it does. Projects with high turnover pay a price in terms of flexibility (bringing in new team members actually slows down processes and the ability to make changes), loyalty (workers without job security are also planning their next career move), and continuity of service to customers.

The downside to this orientation toward the past is that sometimes past-oriented project leaders overlook current performance problems. Previous performance can become a halo over the team member. The leader may simply think of the team member as a good employee without noting more recent problems or an inability to respond to current performance demands. The biggest problem with past-oriented project leadership is that the future is sometimes viewed as something that is going to happen to the organization, rather than something the organization will create. Past-oriented project leaders tend to respond well to trends but are less likely to set them. Maintaining the status quo is the result. Sometimes this works and sometimes it does not. These leaders are hard to engage in planning. They are often too busy, distract others involved in planning meetings, and fail to provide critical information. Once the strategic plan is complete, it is common for past-oriented project leaders to ignore the plan. An additional problem with past-oriented project leaders is a frequent lack of attention to day-to-day operations.

Applications to Project Leadership

A project that must deal with rapidly changing conditions in a dynamic industry needs future-oriented project leaders. If that is not your natural orientation, you will have to adapt to the situation. Highly future-oriented leaders must adjust when their projects are in the operational stages. Instead of spending their time thinking about the next project, they must focus on day-to-day activities, monitor progress, and provide feedback to team members. This may also mean that leaders who understand their own time orientation will select other team members who complement theirs.

In other words, if the project involves problem solving, a future-oriented project leader must recruit others who are more past- and present-oriented and use their abilities to do the research into the past and to explore the causes and solutions to the problems. A present-oriented project leader may use a future-oriented team member to attend strategy sessions, report back, and respond to issues related to the plan. Past-oriented project leaders are very good at remembering the local inspectors from previous projects and the issues they raised which can help the team anticipate problems. Attunement does not necessarily mean that a leader has to change in every situation, but it does mean that the project leader must find some way of understanding and getting the required work done so that every issue is adequately addressed. Because projects are often very complex, a variety of time orientations will be required during the course of each.

Developing Team Members into Leaders

Project leaders must rely on team members to carry out most of the tasks involved in large and complex projects. The most important part of delegating authority or empowering team members is making sure that they understand the vision—the vision of the constituents and the vision of the project leader. If the vision is understood, the team member must gain experience by managing small aspects of the project and moving up to larger and larger amounts of responsibility. This is a good opportunity to develop team members in order to complement the skills, expertise, and personality of the project leader.

For example, on a highway construction project, a promising team member may be assigned to research and develop the protocol for dealing with environmental issues. The project leader may need to monitor the team member's progress carefully the first time. After the team member has started to work successfully, the project leader can loosen the supervisory strings and allow the team member to make a few decisions. After each decision is carried out, the leader and the team member should debrief and talk about what should be done differently in the future. Then, the team member can be assigned a more complicated aspect of the project. As the team member develops more skill and confidence, the project leader can turn over more responsibility. This should be going on with several team members simultaneously.

Research tells us that most managers choose to hire, develop, and mentor people who are like them in terms of work experience, educational background, and personality. We do this because we are most comfortable with people who are like us. However, effective leaders specifically choose to work with people who are different because they bring skills, abilities, and personality traits that we do not have. They complement us and give our project teams and organizations balance. A heterogeneous project team will have more conflict and will take longer to make decisions, but it will also be more creative in developing new ideas, seek more and better solutions to problems, and find ways to make a leader's vision a reality despite barriers.

Effective project leaders also encourage and empower team members to take a leadership role on their own. In many organizations, the chain of command is strictly followed. Project teams tend to be different because we select experts to work on our teams. Expertise is a tremendous source of power, and project leaders need to give tested team members the authority and freedom they need to make decisions and follow through. This authority and freedom should be earned either through past project experience or a gradual building of levels of responsibility on the current project.

Examples of Visionary Project Leaders

To illustrate how visionary project leaders work, we have provided two examples. The first example is of a highway construction project. The second example is of a film production. Although film production is a relatively unexplored area in the project management literature, movie making shares the characteristics of typical projects utilizing a director as a project leader and crews of technical experts. Many of the problems of project teams result when various technical experts disagree on how different aspects of the project should be done. The role of a project leader is explored in these illustrations.

Example 1: Highway Construction Project. The East Side Access Highway took years to develop. After local and state officials interacted with the federal government during a ten-year planning period, ground was broken and this $130 million vision was a "done deal" by the time project engineers Reggie Jannetti and Dan Pellegrini assumed their leadership roles. They had to take the vision for this 6.2-mile highway link from Interstate 90 into downtown Erie, Pennsylvania, and convert the master plan from the blueprints of highway and bridge designers into a well-built, safe roadway. Currently, the East Side Access Highway is under construction. Although the project leaders were given the blueprints for the highway, they had to envision the construction process and make sure that the plans were executed in a safe, legal, timely, and cost-effective way.

Dick Corporation, a general contractor from Pittsburgh with extensive experience in road and bridge construction, was awarded the initial contract for $45 million to build two phases of the five-phase highway. Before ground was broken, the right-of-way officials had to buy land and properties in the highway's path so that the designer's vision could become a reality. In the initial stages, there was no one leader or engineer who could steer this huge project before the earth-movers arrived. Instead, teams of professionals, each with its own leader, had to utilize its area of expertise. To be successful, they would have to understand the vision of the constituents and the designers and the complexities of the job while sharing the overall project goal of completing a modern-day highway on budget and on schedule.

The Project Leaders: Project Engineers and Crew Chiefs. Reggie Jannetti is the project engineer in charge of a one mile phase of construction. A number of different crews and teams are involved and have to be managed during this phase of the construction. There are crews working for PennDOT, Jannetti's employer. Then there is a team of 15 inspectors who were hired by a consulting firm selected by PennDOT. Dick Corporation is the prime contractor responsible for hiring numerous subcontractors. One subcontractor paves the blacktop, another does curbs and gutters, one installs drains, and another removes and clears trees. Of course, various materials including concrete, stone, gravel, and paint come from suppliers who are critical players on the team. There are approximately 50 suppliers. Direct supervision of different crews could not come from one individual. Instead, leadership must come from the project engineers. Their directions are filtered through inspectors, then down to crew chiefs who see that the work gets done. It's a day-by-day, situation-by-situation challenge. For example, if ten trees stand in the way of the construction of an $18 million bridge essential to the highway, a project engineer must decide what to do.

Innovation and Creative Options: Key Ingredients in Leadership. Both Jannetti and Pellegrini agree that project crew chiefs are put to the test every day. There are numerous challenges in highway construction, but one of the most common is field conditions that are not on the blueprints but must be dealt with in order for the highway to be built. Jannetti faced a field condition that had the potential for stopping his phase of construction. There were ten trees whose height exceeded the maximum allowed and were located under the construction site of the bridge that was part of the highway. Chopping the trees was an easy option, except that they had been marked as untouchable for ecological concerns. It was time for an innovative solution. Jannetti and his staff, after consultation with all parties involved, topped off those trees so construction could proceed. In return, they planted over 50 new trees on surrounding Penn State University property. It was a trade-off that came out of Jannetti's experience in the field coupled with a talent for innovative solutions—a prime component of effective project leadership.

Example 2: Project Leader of a Film Production.

Pete Jones of Chicago had a vision for a movie that he converted into a film script called *The Stolen Summer*. Jones spent one year writing his story of the struggles of two adolescent boys, one Jewish and one Catholic, growing up in Chicago, and submitted his screenplay to Miramax Studios in a nationwide competition. To his surprise, Jones, a first-time writer/director, won the competition. After the shock wore off, he realized that he would now assume the leadership responsibility of numerous crews of Hollywood professionals in order to carry out his creative vision. He had won the coveted green light to begin *The Stolen Summer*.

The Vision of The Stolen Summer. *The Stolen Summer* was not like most film projects that are pitched to Hollywood studios in a steady stream—industry estimates are that there is only one acceptance given per hundred submissions that are pitched each month. Jones considered himself lucky to get the go-ahead for his film. On the other hand, Jones knew that Miramax was a bottom-line-conscious studio whose box-office successes such as *The Matrix* and *Hamlet* were as creative as they were cost-effective film productions. Jones knew that he would have to be as frugal as he was creative. In addition, Ben Affleck and Matt Damon would be looking over his shoulder as executive producers of *The Stolen Summer*. Jones had gone from selling insurance to Hollywood director. He would now test his ability to communicate his creative concept, scene-by-scene, while infusing his actors and crew chiefs alike with the enthusiasm and direction needed during a nonstop, three-month production schedule. In short, he had to become an overnight leader.

The Storyboard—A Filmmaker's Blueprint. During the first phase of preproduction, Jones had to sell and convince studio executives that his screenplay was not only creative but also doable for the allotted budget. After considerable negotiations, he passed his first test of leadership by convincing key players that his project was financially viable. Then he proceeded to the most difficult phase of production: converting his vision and script into a scene-by-scene storyboard. This visual blueprint for his movie had to translate into a clear, concise schedule. *The Stolen Summer* began as a 132-page screenplay consisting of actor's dialogue, camera shots, scene and mood descriptions, and lighting effects. This was translated into a scene-by-scene detailed storyboard that was drawn by an artist and would serve as the creative backbone of the film. Using the storyboards, Jones had to lead his entire production crew from conceptualization to the actual shooting of the film, scene by scene.

His first task was to cast the actors for his movie by working with a casting director and his staff. At the same time, he had to meet with the costume designer and her staff in order to create the wardrobe look of the seventies in Chicago. Then he had to bring the line producer on board. Collectively, they had to connect the frame-by-frame pictures on the movie's storyboard into specific filming locations around Chicago. The second week of production began with intensive meetings with his director of photography, who was responsible for the cinematic look of the film. Then the production designer, who functions like an interior designer, joined the team in order to create sets and add props. By the end of the week, the lighting director, chief sound technician, and camera operator joined the planning process as Jones was ready to leave the safe confines of his offices and lead his crews on location in Chicago. There, they would be joined by electricians, carpenters, grips, production assistants, security personnel, caterers, and other logistical personnel. Jones was now the leader of a dozen separate film production crews. The total number of personnel on location was 52 people each day. Despite the enormity of this first-time task, Jones assumed the role of a film director who must be an effective and dynamic project leader in order to be successful.

A Day on the Lake: Leadership in Action. As in the case on most film shoots, circumstances arise that cause changes or alterations of the vision on the storyboard. The final call is the director's, who must lead the crews accordingly.

As a first-time venture for Jones, *The Stolen Summer* was experiencing more than its share of production problems and resultant financial crises. His idea for shooting in Chicago was true to his vision, but his producers, actors, and crew chiefs were looking for ways to cut the location filming short and film several scenes in Hollywood in order to save time and money. Jones had to be strong when suggestions for changes were made, especially when a critical scene in the story needed to be shot on the shores of Lake Michigan.

His line producers insisted that the scene could be shot in a more controlled environment in Los Angeles. The director of photography had numerous problems conceptualizing the shots, especially with scenes where the crew would be required to shoot in the water on a makeshift floating scaffold in order to get an overhead shot. The camera crew joined in the criticism, questioning the significance of the shot versus the time required to shoot it. In addition, the water was ice cold. The result was an across-the-board questioning of Jones' insistence on this shot for just a few seconds of a usable scene in the finished film. In light of all this, Jones' challenge was formidable—he could be dogmatic and demand that the shot be done here on the shores of Lake Michigan; or he could take the time to communicate the importance of these scenes to each of the concerned crews and convince them of the visual importance of the scenes to the finished product. In his mind, where else was he going to get the gorgeous sunset over the horizon of Lake Michigan edited with cutaways of the Chicago skyline? In Jones' vision, these shots were essential to the film, and he had to lead the crew through a long, cold day on the lake in order to put these shots in the can.

First, he had to convince his three producers. If they said no, everybody would pack up their gear and head back to Hollywood. He was persuasive and won their approval. Then he had to describe the look and camera angle of the scenes so that his director of photography and the camera crew would be on the same visual page with him. The audio technician and his crew were next as they all waded into the chill of the lake to record sound. Makeup and costume personnel had their hands full trying to cope with windy conditions and freezing actors. Jones had to rally his crew. Although it was a far cry from the demands of selling insurance, which was what he did before writing the screenplay, his sense of commitment and dedication to his movie were greater than the bone-chilling temperatures of Lake Michigan.

Jones led his crew that day to the windy shores and into the cold waters of Lake Michigan. In the final cut of *The Stolen Summer*, the overhead shots from the floating scaffold were not used and his child actor got so cold that he was unable to complete the filming. However, the shots of the boy walking along the shore were brilliant and added a great deal of emotional impact to the film. That made everyone happy, and the shoot was considered a success.

The ability to lead during the production of a film requires a clear vision of the project, well-honed people skills, incredible patience, a thorough knowledge of each crew's function and responsibilities, and a burning passion for the film that translates into enthusiasm and commitment. As a first-time director, Jones had the vision and the passion. As far as the future is concerned for Jones, public acceptance at the box office will be the ultimate indicator whether or not he will have future projects that get the green light.

Summary

Clearly, there is a considerable difference between *The Stolen Summer* and the East Side Access Highway project. But despite these differences, there was a vision that had to be translated to understandable terms, teams of technical experts that had to be managed, and dozens of workers who had to be motivated. Pete Jones, Reggie Jannetti, and Dan Pellegrini share some common leadership traits even though filming on the shores of Lake Michigan in three months is a world apart from building a six-mile roadway in three years. But the same project commitment, innovative solutions to problems, and the ability to keep crews enthusiastic about their work assignments are essential to their successful leadership.

Effective project leaders share many common characteristics and exhibit similar behaviors. We can learn a great deal from leadership theories, but even more by observing effective leaders. Paying attention to how problems are solved, how innovative strategies are developed, and how great project leaders communicate and motivate are the best ways to improve our leadership ability. In addition, volunteering for challenging assignments, learning from our missteps, and using our vision to drive our day-to-day behavior will help us develop into great leaders.

References

Bass, B. M. 1985. *Leadership and performance beyond expectations.* New York: Free Press.

Conger, J. A., and R. Kanungo, R. 1998. *Charismatic leadership in organizations.* Thousand Oaks, CA: Sage Publications.

Hersey, P., and K. H. Blanchard. 1977. *The management of organizational behavior.* 3rd ed. Englewood Cliffs, NJ: Prentice Hall.

House, R. J. 1971. A path-goal theory of leader effectiveness. *Administrative Science Quarterly* 16:321–339.

———. 1977. A 1976 theory of charismatic leadership. In *Leadership: The cutting edge,* ed. J. G. Hunt and L. L. Larson. 189–207. Carbondale: Southern Illinois University Press.

Kerr, S., and J. M. Jermier. 1978. Substitutes for leadership: Their meaning and measurement. *Organizational Behavior and Human Performance* 22:375–403.

Kouzes, J. M., and B. Z. Posner. 1995. *The leadership challenge: How to keep getting extraordinary things done in organizations.* 2nd ed.. San Francisco: Jossey-Bass.

Pinto, J. K., P. Thoms, J. Trailer, T. Palmer, and M. Govekar. 1998. *Project leadership from theory to practice.* Newtown Square, PA: Project Management Institute.

Thoms, P. Forthcoming. *Driven by time: Leadership and time orientation.* New York: Praeger.

Thoms, P., and D. Blasko. 1999. Preliminary validation of a visioning ability scale. *Psychological Reports* 85:105–113.

Thoms, P., and M. A. Govekar. 1997. Vision is in the eyes of the leader: A control theory model explaining organizational vision. *OD Practitioner* 29:15–24.

Thoms, P., and D. B. Greenberger. 1998. A test of vision training and potential antecedents to leaders' visioning ability. *Human Resource Development Quarterly* 9:3–19.

Thoms, P., and J. K. Pinto. 1999. Project leadership: A question of timing. *Project Management Journal* 30:19–26.

Vroom, V. H., and A. G. Jago. 1988. *The new leadership: Managing participation in organizations.* Englewood Cliffs, NJ: Prentice Hall.

Vroom, V. H., and P. W. Yetton. 1973. *Leadership and decision making.* Pittsburgh: University of Pittsburgh Press.

Weber, M. 1947. *The theory of social and economic organizations.* Translated by T. Parsons. New York: Free Press

Yukl, G. 2002. *Leadership in organizations.* Upper Saddle River, NJ: Prentice Hall.

CHAPTER FIVE

POWER, INFLUENCE, AND NEGOTIATION IN PROJECT MANAGEMENT

John M. Magenau, Jeffrey K. Pinto

When we speak of power in project management settings, it is important to understand the meaning of the term and why it is so vital to achieving project goals. So much has been miscommunicated about the importance of power that it is easy for many of us to be mistrustful or wary of ever using power in organizations. The truth, however, is that successful project managers must become both adept and comfortable at using power all the time. Adept because they have to recognize how to use it well in order to avoid its misuse. Comfortable because it is the principal means by which major decisions are made, resources allocated, teams governed, and stakeholders satisfied. Successful project managers understand the constructive uses of power, as must each one of us.

Power is simply the ability to get activities or objectives accomplished in an organization the way one wants them to be done. Among the key activities that power enables us to achieve are the ability of affect decisions and control resources (Dubrin, 1995). Some definitions of power set it up as a confrontational issue; for example, the belief that power implies forcing someone to do something that they would not ordinarily do otherwise. Other definitions are more benign, stating that power is simply the mechanism to positively affect a project team's natural inertia. Underlying all these definitions of power is the belief that power "enables" some members of the organization to pursue objectives. Whether those objectives are for the good of the organization as a whole or are purely self-centered is another issue. Nevertheless, as part of any discussion of the project management process, it is important to understand the nature of power in our organizations, its various bases, how one gets and holds of power, and its potential effects on power holders.

Sources of Power

There are many different ways in which individuals can acquire power within organizations. The types of power individuals choose to exercise depend, in large degree, upon the types of people they are, the opportunities available to them, and the companies they work for. Sometimes, gaining power is merely the result of luck or fortunate circumstances. To understand how each of us may acquire power, it is necessary to consider some of the more common sources of power and the reasons why they exist. Table 5.1 gives a list of some of the more common types of power routinely found in organizations.

Positional Power. Positional power typically refers to the power an individual gains from occupying a position with the organization. For example, departmental managers would have a higher degree of positional power than their subordinates. Another common term for positional power is authority, stemming from the right to direct the behavior of others because of the higher position one may occupy in the organizational hierarchy. Positional power offers some important subcomponents of power, all based on positional authority. They include (1) legitimate power, (2) reward power, and (3) coercive power.

1. *Legitimate power.* Legitimate power is the power that is granted to people because of the position they occupy within the organization's chain of command. The higher up the corporate ladder an individual sits, the more legitimate power they are able to exercise in performing their duties. One benefit to a project team of enlisting the active support of senior executives is their ability to directly affect change or impact on the project's development.
2. *Reward power.* Often coupled with legitimate power is the power of distributing rewards to members of the organization as a reward for their performance or compliance with directives. This power to provide rewards often serves as the motivational "carrot" to

TABLE 5.1. TYPES OF POWER.

1. *Positional power.* Power that derives to an individual from the position they occupy in an organizational hierarchy
2. *Personal power.* Power stemming from personal qualities or personality characteristics within an individual
3. *Resource power.* Power that derives from one individual's control over critical resources needed for the project
4. *Dependency power.* The power that one individual or organizational unit acquires when others depend upon them, their output, or resources they can provide
5. *Centrality power.* The power an individual or unit receives from being closely linked to the primary activities within an organization
6. *Nonsubstitutability power.* The power that comes from the perception that an individual or group possesses a competency or skill that cannot be replicated by others within the firm
7. *Coping with uncertainty.* The power that derives from the ability to effectively cope with environmental uncertainty

induce project teams to perform to optimal levels. It is important to note, however, that reward power is only viable when the manager actually can employ it in meaningful ways. For example, in many organizations where employees are unionized, all monetary rewards are collectively bargained in advance, greatly limiting a manager's ability to provide additional rewards for performance. As a result, they may have to employ creative, nonmonetary rewards for superior performance.

3. *Coercive power.* In addition to the power to reward compliant or appropriate behavior, power that enables the leader to punish noncompliance is referred to as coercive power. Coercive power is best understood as the power that derives from the fear of punishment. Less powerful individuals are inclined to follow the direction of the leader who wields coercive power because of their wish to avoid punishment. Research suggests, however, that coercive power is only of limited usefulness as a motivator and tends to work best in the short term.

Within the arena of project management, the whole issue of positional power becomes much more problematic. Project managers in many organizations operate outside the standard functional hierarchy. While that position allows them a certain freedom of action without direct oversight, it has some important concomitant disadvantages, particularly as they pertain to positional power. First, because cross-functional relationships between the project manager and other functional departments can be ill-defined, project managers discover rather quickly that they have little or no legitimate power to simply force their decisions through the organizational system. Functional departments usually do not have to recognize the rights of project managers to interfere with functional responsibilities; consequently, novice project managers hoping to rely on positional power to implement their projects are quickly disabused.

As a second problem with the use of positional power, in many organizations, project managers have minimal authority to reward team members who, because they are temporary subordinates, maintain direct ties and loyalties to their functional departments. In fact, project managers may not even have the opportunity to complete a performance evaluation on these temporary team members. Likewise, for similar reasons, project managers may have minimal authority to punish inappropriate behavior. Therefore, they may discover that they have the ability to neither offer the carrot nor threaten the stick. As a result, in addition to positional power, it is often necessary that effective project managers seek to develop their *personal* power bases.

Personal Power. Another set of "power bases" relate to an individual's personal power, or the power that derives from characteristics or qualities within an individual. In the case of personal power, the source of power is not the result of an organizational mandate or position in the hierarchy; it is due to the traits that individuals possess. As a result, whereas positional power tends to collect within the standard authority structure of a company and is most obviously seen in higher levels, personal power can be used by people regardless of their formal position in the organization. When we think of the informal leader within a social group or the most respected members of a research and development team, we are observing that some individuals possess or use personal power to a greater degree than their peers. Just as positional power could be broken down into subcategories, so too can we look at various forms of personal power, including: (1) referent power, (2) expert power, (3) information power, and (4) connection power.

1. *Referent power.* Referent power simply refers to the situation that other organizational members like someone and want to be like them; that is, the power holder acts as a reference point for others. Referent power is an extremely significant form of power, as advertisers are well aware. Their use of star athletes to endorse products is their acknowledgment of the fact that a large percentage of people in our society will be swayed by the opinions of those they hold in high regard. Adolph Hitler's charismatic presence and oratorical abilities gained him a measure of personal power long before he became first chancellor and finally dictator of Nazi Germany. Likewise, within organizational circles, fine examples of referent power can be found at all levels of the organization. When members of a loading dock gang gravitate toward a friendly or physically large coworker, we find evidence of referent power. When a junior manager willfully conflicts with a superior and is lionized by coworkers for having the guts to stand up to the boss, he is experiencing a level of referent power. In all of these examples, the power holder is one who has the ability to sway the opinions of others through the dynamism of personal power; power that is evidenced by the regard with which the power holder is held by other members of the organization.
2. *Expert power.* A second form of personal power is called expert power and refers to the follower's belief that the power holder has some expertise or knowledge that the others need to perform their jobs. A common example of expert power can be found within the R&D departments of most organizations. Those individuals who are generally regarded by their peers as having expert knowledge will typically wield far more *real* power in the laboratory than the designated lab manager, particularly if that manager is not perceived to have an acceptable level of expertise.
3. *Information power.* Information power is another form of personal power and has strong ties to expert power. It is defined as the belief by followers that the power holder either possesses or has access to information that is necessary to perform a job. Some managers serve as the conduit for all forms of organizational information. Whether that information is conveyed in memos, gossip and hallway rumors, or direct access to upper management, these individuals hold a form of power because they possess this information. Other organizational members are willing to defer to them precisely to the degree that they perceive that this information, whether gossip or activity-based, is relevant and useful for their own work and organizational survival.
4. *Connection power.* The final type of personal power is connection power. The power of connections is well known in all organizational settings. Some individuals, regardless of their formal position, possess tremendous power solely because of their connections to other, powerful people. The classic adage, "It isn't *what* you know, it's *who* you know" succinctly illustrates the importance of connections as a source of power, not only with peers but, potentially, with superiors as well.

Note that each of the personal power bases illustrates a similar feature: The "power" found in each derives directly from human relationships. By definition, unless we are willing to explore relationships with peers and superiors in organizations, we cannot reasonably expect to develop any form of personal power. Additionally, these personal power bases offer project managers a wide range of options. Not everyone is blessed with a magnetic personality making it easy to cultivate and maintain referent power. Likewise, some people are adept at creating a network of powerful connections, while others, either through personality or external circumstances, do not have similar opportunities. On the other hand, acquiring information or developing expertise is within most managers' control and should be explored as alternative bases of power (Pinto, 1996).

Resource Power. Some individuals and organizational units have more power than others because they have the ability to control important resources. For example, when a project manager needs a critical resource, such as a lead programmer, for a project and must get the department head's approval to recruit such an individual, the department head is able to wield resource power over the project manager. Somewhat facetiously, it has been observed that the "real" golden rule states, "Whoever has the gold makes the rules." With resource power, there is a degree of truth to this adage. To the degree I can control your access to resources you require, I have power over you.

Dependency Power. Closely related to resource power, dependency power suggests that when one person or department possesses something another organizational unit needs, they have dependency power. It is very common for organizations to set up many operations so that dependencies exist between different departments. Before a construction project can begin, materials must be purchased. Prior to the start of engineering and prototype development, a corporation's design department must complete their work assignments. In both of these examples, one department operates with dependency power over another in that the follow-on unit "depends" upon the first department completing their work before the next can initiate their operations. Following this logic, we can expect that the more a department depends upon other units, the lower its relative power is within the organization.

Centrality Power. An important opportunity to acquire power within organizations occurs when individuals or departments have high visibility or are perceived as being closely linked to the primary activities within the organization. For example, a Senior Vice President for Projects within a construction organization will possess a significant amount of power because of the centrality of his or her role in the overall organization. Because project management is key to organizational success, the head of projects is likely to be very visible and central to firm operations (Daft, 2001).

Nonsubstitutability Power. A final common form of power is referred to as nonsubstitutability power. This form of power is best understood as the power an individual or department has when their activities or expertise are unique and not easily substituted. If, for example, a project organization has only one person with the important contacts or personal relationships with a potential client, that person has power because she cannot be easily substituted. In the 1970s, when computers were not as widely used as today, programmers had tremendous power within many companies because no one else could replicate their skills. As programming has become more widely disseminated throughout society, the nonsubstitutability power of computer programmers has diminished (Pettigrew, 1973).

Power versus Influence

Influence and influencers are pervasive in our society. Television and radio advertisements, televangelists, and salespeople represent examples of some of the most common types of influence we experience on a daily basis. Note that in each case, the influencer cannot *force* your compliance. Each of us has the power to simply change the channel or leave the store if we are offended or threatened by the influencer's message. How, then, do we explain the success that these people have in raising money and gaining sales? In a different context, how can we explain the success that some of our peers have in gaining compliance from other organizational members even though they have no direct authority over them? The answer to these questions typically focuses on the greater influencing ability that some of us possess relative to others.

Because influence is a key component in organizational life, it is important to distinguish between influence and power. Many managers define power in terms of influence as a convenient shorthand. Even textbooks make the same definitional link—suggesting, for example, that power is the capacity to influence others. More appropriately, we can define *influence* as one person's ability to get another to do something they want when there are no gross power differences between the two parties. That is, the influencer has no formal ability to "force" the other person to seek some goal or perform some task. From this definition, it is clear that there are some important similarities between power and influence: Both are used to change another's behavior. We will see, however, that the two concepts are very different in some important ways while demonstrating that a thorough knowledge of influence tactics can be an important tool for better managing within the organization's political climate.

Table 5.2 demonstrates some of the key differences between power and influence. We can classify these differences under three headings: scope and generality, strength of foundation, and tenure. *Scope and generality* refers to the nature of how one is able to use influence versus power. Typically, influence, in order to be successful, is situation-specific; that is, one who is adept at influencing others knows intuitively when and under what circumstances to attempt to change the other's behavior. Good influencers do not misuse or overuse their abilities, because they know that the more often they employ them, the more likely that coworkers will begin to refuse to comply with their wishes.

In addition, good influencers rely almost exclusively on face-to-face meetings, or at least opportunities in which one side can view the other. There are two primary reasons for preferring to meet head-to-head rather than using telephones or other media. First, it is

TABLE 5.2. POWER VERSUS INFLUENCE.

	Power	Influence
1. Scope and generality.	Cuts across situations and relationships.	Situation-specific and usually face-to-face.
2. Strength of foundation.	Strong base. Does not have to be done well to work.	Weak base. Must be used well or will not work.
3. Tenure.	Long-term.	Short-term.

much harder for the average person to refuse another during direct contact. Memos or telephone calls offer an impersonal approach that makes it easier to refuse an influencer's requests. The second reason is that good influencers are often adept at reading body language and other forms of nonverbal responses from their "target." When we observe good influencers in action, we see them constantly altering the "angle of attack" or promotional pitch as they perceive that one line of argument is either likely to be accepted or rejected. This sensitivity to the other individual's reactions is simply not possible while using other media.

Power does not accept the constraints of situation and approach. When some individuals have power over others, they are in the position to operate without regard to concerns about scope and generality. The "boss" or power holder is in the position to force compliance regardless of the situation and via whatever means they choose. Having power enables me to communicate through multiple means, any of which is just as effective for me as face-to-face meetings.

Another distinction between influence and power lies in *strength of foundation*. This concept refers to the fact that influence, in order to be effective, must be used well; that is, because my base (or foundation) of actual power over another may be weak, I must substitute effective planning, preparation, and role-playing in influencing another. Simply put, influence must be done "well" in order to work. Power, on the other hand, gives a manager a much stronger base from which to operate. The manager with power over another does not have to be constrained to exercising that power in clever or situation-specific ways. He or she simply tells another to do something and that subordinate is bound to comply. Further, because the power base is strong, power holders do not have to be particularly sensitive in using their authority. "Do it because I say so!" may be all the information that the power holder is required to convey.

A final difference between power and influence refers to its tenure. Power is much more long-lived than influence. As noted initially, just as influence is situation-specific, so too is it used sparingly. To overuse influence, particularly with any one individual, is often to lose influence. On the other hand, power lasts. As long as some functional managers occupy higher positions in the organizational hierarchy than others, they will have legitimate power over them.

Forms of Influence

What are some of the more common methods by which one individual can seek to influence the behavior of another? Because, as noted, influence is highly personal as well as being situation-specific, there are several tactics one can employ in trying to apply influence. Note, however, the underlying characteristics of the forms we will discuss in the following; they work best in face-to-face settings, they are situation-specific, and they implicitly assume that the influencer cannot directly force his or her will on the other person. Among the common methods are the following.

Persuasion. Persuasion is a tactic in which one person attempts to influence another simply by arguing the merits of their position. Persuasion suggests that if the target individual will simply give a fair hearing to the influencer, he or she will be won over on the strength of the argument. Persuasion is a sound influence tactic to employ when the influencer perceives that their arguments or evidence supporting them are strong. For example, in a project on which two organizations are cooperating, technical disagreements between the firms can often be resolved through persuasive argument in which detailed discussion, data analysis, and objective assessment are used as tools by one of the project partners.

Ingratiation. Ingratiation is the art of flattery, cajolery, or a search for common ground to win favor and gain another's willingness to cooperate. Ingratiation, as an influence tactic, offers the simple argument that it is easier to catch flies with honey than vinegar. For example, a project manager working for Chrysler Corporation made a point of saving all organization notices (transfers, promotions, awards of advanced degrees, and so forth). He filed them away, and prior to calling on another manager or executive for the first time, he always read up on any relevant information concerning that manager. Thus, he knew what the person's educational background was, where he or she was from, and something about the person's work history. This tactic offered the project manager a valuable source of common ground that made contacts cooperative and supportive of his needs.

Pressure. Pressure is a form of influence that applies external considerations as a supplement to the message itself; for example, time constraints or cost issues. Pressure tactics frequently seek to limit the target's freedom of choice or movement in order to gain compliance. A salesperson who claims that a product will only be on sale for a limited time is hoping that the added pressure of the time constraint will reinforce the influential message. Pressure, of course, can backfire. In fact, research in advertising has found that message content that the listener perceives to be too pressure-inducing will often have a boomerang effect, as the target individual resents the too-obvious display of pressure tactics.

Multiple Pressure. This alternative is an extension of simple pressure, in which the target is receiving the combined attention of two or more additional individuals. In employing this approach, other members of the project team "gang up" on the recalcitrant member or target for influence. Research suggests that pressure from multiple sources can be a very effective influence tactic, as it is often extremely difficult for most of us to resist the combined influence of several others.

Guilt. Guilt can be a powerful but sometimes overlooked form of influence; guilt implies a relationship based on obligation between the two parties. As a result, one party may attempt to sway the other by an appeal to his sense of duty or obligation, regardless of whether or not obligation actually exists. As with research on the effects of pressure tactics, recent research on guilt appeals in advertising demonstrate that in order to be effective, guilt must be used moderately. Too blatant examples of guilt-inducing appeals actually lower a person's willingness to cooperate and instead promote active resistance and resentment.

The preceding are just some of the more common influence tactics that project managers may use. Note the common feature of each of these tactics, regardless of how seemingly powerful or effective: They are only successful when employed correctly, used sparingly, and limited to exchanges that allow for immediate feedback (such as face-to-face meetings). Influence, though not the same thing as power, can still be an excellent method for inducing compliance when used effectively. Its effects are particularly apparent when we examine the importance of influence tactics in both political behavior and project negotiations.

Developing Influence Skills

How does a project manager succeed in establishing the sort of sustained influence throughout the organization that is useful in the pursuit of project-related goals? A recent article (Keys and Case, 1990) highlights five methods managers can use for enhancing their level of influence with superiors, clients, team members, and other stakeholders (see Table 5.3). They suggest that one powerful method for creating a base of influence is to first establish a reputation as an expert in the project that is being undertaken. This finding was borne out in research on project manager influence styles. A project manager who is widely perceived as lacking any sort of technical skill or competency cannot command the same ability to use influence as a power mechanism to secure the support of other important stakeholders or be perceived as a true "leader" of the project team. One important caveat to bear in mind about this point, however, is that the label of "expert" is typically a perceptual one. That is, it may or may not be based in actual fact. If one party perceives that another is an expert, that belief is often sufficient grounds for the leader to maintain influence over others.

A second technique for establishing greater influence is to make a distinction between the types of relationships encountered on the job. Specifically, managers need to make

TABLE 5.3. FIVE KEYS TO ESTABLISHING SUSTAINED INFLUENCE.

- Develop a reputation as an expert.
- Prioritize social relationships on the basis of work needs rather than on the basis of habit or social preference.
- Develop a network of other experts or resource persons who can be called upon for assistance.
- Choose the correct combination of influence tactics for the objective and the target to be influenced.
- Influence with sensitivity, flexibility, and solid communication.

conscious decisions to prioritize their relationships in terms of establishing close ties and contacts with those around the company who can help them accomplish their goals, rather than on the basis of social preference. Specifically, from the perspective of seeking to broaden their influence ability, project managers need to break the ties of habit and expand their social networks, particularly with regard to those who can be of future material aid to the project.

The third tactic for enhancing influence is to network. Networking consists of the practice of establishing a series of social and professional relationships with as large a set of individuals as possible. As part of creating a wider social set composed of organizational members with the power or status to aid in the project's development, canny project managers will also establish ties to acknowledged experts or those with the ability to provide scarce resources the project may need during times of crisis. It is always helpful to have experts or resource providers handy during times of munificence.

A fourth technique for expanding influence process suggests that for influence to succeed, project managers seeking to use influence on others must carefully select the tactic they intend to employ. For example, many people who consider themselves adept at influencing others prefer face-to-face settings rather than using the telephone or leaving messages to request support. They know intuitively that it is easier to gain another's compliance when the request is direct and personal. If the tactics that have been selected are not appropriate to the individual and the situation, influence will not work.

Finally, and closely related to the fourth point, successful influencers are socially sensitive, articulate, and very flexible in their tactics. For example, in attempting to influence another manager through a face-to-face meeting, a clever influencer will recognize how best to balance the alternative methods for attaining the other manager's cooperation and help. Through reading the body language and reactions of the other manager, the person attempting to use influence may instinctively shift the approach in order to find the argument or influence style that appears to have the best chance of succeeding. Whether the approach selected employs pure persuasion, flattery, and cajolery, or the use of guilt appeals, successful influencers are often those people who can articulate their arguments well, read the nonverbal signals given off by the other person, and tailor their arguments and influence style appropriately to take best advantage of the situation.

Block's Framework

Peter Block (1987) proposed a framework for developing strategies as a part of the influence process that often occurs between managers. For example, if Bob, the Project Manager, is interested in gaining support for a proposed technical change to a current project, he may have to first gain the cooperation of other significant project stakeholders, including Anne— Head of Technical Services, Nancy—V.P. for Engineering, and Tom—Lead Designer. How Bob chooses to attempt to influence these individuals to gain their cooperation with his plans is an important step and needs to be carefully considered.

Figure 5.1 shows Block's framework for developing influence strategies. There are some important points to note about the framework in general. First, the model makes clear that

FIGURE 5.1. BLOCK'S FRAMEWORK.

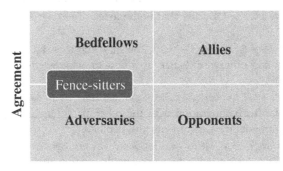

influence must be done on an individual-by-individual basis; that is, there is little reason to attempt to develop a blanket influence tactic. By definition, each individual is different, and the steps taken to attempt to influence that person have to acknowledge this fact. Each person to be influenced merits an individualized strategy. Second, the model identifies two important features that weigh heavily on our ability to influence others: agreement and trust. Agreement refers to short-term or issue-relevant assessments. In other words, do you and I agree substantially on the current issue at hand? The second dimension, trust, is based on longer-term, relationship assessment. That is, regardless of our level of agreement on the current issue, if we have a positive working relationship, we may continue to trust each other and each other's intentions, even in the face of short-term disagreements. Note that agreement and trust do not have to coincide. I can disagree with your position without sacrificing the trust relationship we have developed. Likewise, it is possible for us to work together based on an issue of mutual agreement without ever trusting each other or each other's motives.

Figure 5.1 illustrates five groups that we are likely to confront at different times as we attempt to use influence effectively: (1) bedfellows, (2) allies, (3) opponents, (4) adversaries, and (5) fence-sitters. For each of these potential groups, we can formulate an approach that is most likely to gain us their compliance or, at least, minimize their potential negative impact on our goals.

1. *Bedfellows.* Bedfellows are characterized by circumstances in which we (they and ourselves) have an immediate issue of mutual concern or interest, although there is no longer-term basis of trust between us. For example, during World War II, the Western allies, headed by Winston Churchill and Franklin Roosevelt, were able to make common cause with Joseph Stalin, dictator of the former Soviet Union in their joint desire to defeat Nazi Germany. There was little trust between the Western leaders and Stalin, but circumstances and immediate concerns enabled them to put aside differences and

collaborate in dealing with the general menace of Nazism. Block argues that the best influence strategy when you are attempting to influence bedfellows is to do the following:

- *Reaffirm agreement.* Ensure that both parties are in agreement on the key issue of mutual concern.
- *Acknowledge the caution that exists.* Because the trust factor is low, there is little point in pretending otherwise; hence, it is best to acknowledge this caution immediately and directly.
- *Be clear what we want from bedfellows and what we intend to do.* Create a joint "map" of our plan for cooperation with the other party, including our commitments and the commitments we expect from the bedfellow.
- *Try to reach agreement on how we intend to work together.* Establish a shared protocol for how we can together achieve our goals, even in the absence of mutual trust. The less the trust between partners, the more important it is that a formal agreement of some sort is established.

2. *Allies.* Allies are other parties or individuals with whom we share both immediate agreements on current issues as well as long-term trust. For example, in almost all significant foreign policy issues in recent decades, the United States and Great Britain have worked hard to present a united front based on both the long-term quality of our relationship as well as the desire to maintain agreement on how to approach the specific issue at hand. Allies are our best source of support in influence situations because they are fundamentally in favor of our views and choices of action. The best influence strategy when dealing with allies includes the following:

- *Reaffirm the agreement on the project or vision.* First, take steps to ensure that we are in substantive agreement with our allies on the issue at hand.
- *Reaffirm the quality of the relationship.* It is also key to take time to reinforce the long-term nature of the trust relationship that exists between us and our allies.
- *Acknowledge the doubts and vulnerability that we have regarding the project or other issues at hand.* With allies, it is not necessary to hide any difficulties or concerns that we may face in achieving our goals for the project. Because we have a relationship that is both trusting and affirmative, we can work with allies to seek answers to any lingering questions.
- *Ask for advice and support.* Allies should be used to strengthen our position regarding the specific project issues we face. As a source of support and information, we can depend on allies to work with us rather than at cross-purposes.

3. *Opponents.* Opponents are those within the project stakeholders who disagree with us regarding the specific issue at hand, even while the long-term, trust-based relationship remains healthy. For example, a company may have a strong, long-term relationship with its suppliers, although disagreeing with them on an issue of immediate concern. The key to working with opponents is to always remember the importance of maintaining the level of trust that exists, regardless of short-term disagreements. Among the steps to employ are the following:

- *Reaffirm the quality of the relationship.* The keystone of our relationship with opponents is trust. The first step in dealing with them is to reemphasize that trust remains the cornerstone of our interactions.

- *State our position.* We need to make a clear case for our position, recognizing that disagreements have arisen. Our goal is to influence through persuasion, not emotion.
- *State in a neutral way what we think their position is.* As objectively as possible, we need to demonstrate that we understand their principal objections with our decisions and make them aware that we appreciate their views. Because trust is key, it is vital that our attempts to persuade opponents cannot be misinterpreted into them thinking we are deliberately misleading them.
- *Engage in problem solving.* The key to resolving disagreements with opponents is to treat these issues as mutual problems requiring joint discussion and solution. Even if we are unsuccessful at persuading opponents to cooperate, the steps we take to influence them must be perceived as objective and free from deception.

4. *Adversaries.* In any circumstance involving the need to influence others, there will be adversaries who oppose our views both from an immediate "disagreement" perspective, as well as from a long-term lack of trust. Adversaries differ from other groups in that it is doubtful that we can successfully influence them; that is, turn them around to our point of view. The successful approach, therefore, is usually to find a method for minimizing their potential for negatively affecting our position and project. Among the most appropriate steps in dealing with adversaries are the following:
   - *State our position, including vision for the project.* Be clear as to our goals and directions for the project when speaking with adversaries.
   - *State in a neutral way what we think their position is.* It is important to demonstrate our own objectivity when dealing with adversaries. While this approach may not positively influence their perceptions, it ensures that we have dealt with our adversaries in a direct and honest manner. It is also helpful to demonstrate that while we disagree with adversaries, we understand their views and the means by which they came to them.
   - *End the meeting with our plans and no demands.* In adversarial relationships, it is unlikely that either side is going to effectively influence the other party or alter that person's perceptions and way of thinking. The best approach, as a result, is usually to maintain clear lines of communication, make our position unambiguous, and proceed with our plans.

5. *Fence-sitters.* Fence-sitters are usually described as wavering between agreement and disagreement with our position. Because no long-term trust has been established, fence-sitters are interested exclusively in the current issue at hand. Our strategy for influencing fence-sitters must be carefully considered because they often represent an important group in any situation in which influence is being applied. The key steps to appealing to fence-sitters include the following:
   - *State our position on the issue at hand.* Be clear and direct as to how we view the problem or issue, what the various options are that we are presented with, and why we have chosen the present course. The key is to be as comprehensive as possible to demonstrate that we have considered the issue from multiple perspectives.
   - *Ask where the fence-sitters are on the issue.* Attempt to collect as much information as possible about the views of these individuals. Find out, for example, their principal objections to your position, misunderstandings, or discomfort. It is important to gain

as much useful information as possible because it will allow us to formulate a plan for best alleviating their concerns.

- *Apply gentle pressure.* Following the process of collecting information, we are now able to develop the best approach to influencing fence-sitters. First, consider the reason for their concerns. If it was due to a misunderstanding of our position, we can now clearly explain our views and why they are appropriate. If their concern was due to some other substantive reason, we are in a position to take steps to minimize this concern. Applying gentle pressure can be coupled with a plan for correcting their misapprehensions about our approach to the problem.

- *Encourage fence-sitters to consider the issue and what it would take to gain their support.* We are attempting to influence through bargaining behavior. Rather than exerting excessive pressure or threats that can actually turn a fence-sitter into an adversary, we are looking to find the key that will make them bedfellows. In this attempt, it is important to acknowledge any concerns or issues they might have and, if possible, create a circumstance in which we can engage in a trade: their cooperation for our cooperation on some important issue of theirs.

Block's framework is a useful model for recognizing the various types of individuals we are likely to face in using influence, ranging from the trusting and cooperative all the way to the hostile and suspicious. Creating different strategies for dealing with each of these possible constituencies in a project-related problem will make our task much easier as we seek to use influence tactics most effectively.

Negotiation Skills

Project managers need good negotiation skills. Earlier in this chapter we stated that most project managers lack the positional power or authority to order people to do things. As a consequence, they must rely on less formal types of power such as personal power and use influence tactics and negotiation to accomplish their objectives.

Negotiation is ubiquitous in project management. For example, there are negotiations with department managers for the services of the members of the project team. There are negotiations with suppliers to obtain the materials and contractors to obtain the services needed to keep projects on schedule and within budget. There also are negotiations among team members and with upper management about project specifications, team member responsibilities, task completion dates, and budgets. And there are negotiations with customers about costs, project completion dates, and project change orders.

Effective negotiation requires careful preparation, the choice of a negotiation strategy appropriate to the situation, and the skillful use of power and influence to implement the chosen strategy. For example, a project manager may be very aggressive in a negotiation with a supplier if there are several sources that can provide equally good materials or services. On the other hand, if there is only one good supplier, thus giving that supplier nonsubstitutability power, a more conciliatory strategy may be required. Block's framework suggests problem solving or compromise may be the best choice if maintaining the trust of team members or customers is important to future project success or repeat business. In this section we discuss the characteristics of negotiation, the basic steps in preparation for negotiations, negotiating strategies, and considerations in choosing a negotiation strategy.

What Is Negotiation?

Negotiation is a process involving two or more people who start with apparently conflicting positions and attempt to come to an agreement by revising their original positions or by inventing new proposals that reconcile the interests underlying them. Elaboration of the parts of this definition will help explain what negotiation is and when it is used.

Negotiation is a process. Negotiation involves an exchange of proposals and counterproposals that take place over time. The time might be a few minutes, or it could be several months or even years.

Negotiation involves two or more people. The simplest negotiations involve two people. However, many negotiations involve more than two individuals. For example, a project may entail multiparty negotiations between several companies. Each company may field a team of negotiators who represent the interests of others (constituents such as higher-level management) not present at the negotiation table. When several actors are involved, the social structure of negotiation can become very complicated. Negotiations take place not only between the representatives of the various organizations but also within a negotiating team and between team members and their constituents.

Negotiators have apparently conflicting positions. There would be no need to negotiate if the people involved did not perceive differences between their initial positions. Sometimes further discussion reveals real differences between positions, but at other times further discussion may lead to the discovery of solutions that reconcile or integrate what initially appeared to be conflicting interests.

Negotiators attempt to reach agreement. Although it may not always be accomplished, usually the purpose of negotiation is to reach voluntary agreement about something. None of the negotiators have enough of a power advantage (or if they have it, they decide not to use it) vis-à-vis the other to make decisions unilaterally. Both sides believe they can improve their outcomes through negotiation. Negotiators will continue trying to reach agreement as long as they believe an agreement is possible and preferable to the alternative of not reaching agreement.

Negotiators revise their initial positions. During negotiation the parties usually change their initial proposals. Often this requires both sides to make concessions toward some middle-ground position. Consequently, negotiators have an expectation of give-and-take during the negotiation process. Sometimes negotiators are able to invent or discover options that reconcile or integrate their underlying interests in such a way that a high degree of satisfaction is achieved by all concerned. This allows both parties to satisfy their interests to a greater extent than would have been possible if they had simply conceded to some middle-ground position.

Preparation for Negotiation

Although the time and money a negotiator invests in preparation depends upon the importance of a particular negotiation, successful negotiation requires thorough preparation. Without careful preparation, negotiators are likely to find themselves reacting to events rather than influencing them. In highly competitive situations, a well-prepared opponent can put an unprepared negotiator on the defensive. The chances of concluding an agreement that satisfies the reasons for negotiating in the first place are reduced when thoughtful preparation does not occur. An opponent's confidence and commitment to a position may increase if the other side is unprepared. In this section a summary of the basic steps of preparation for negotiation is presented. A more detailed description of the steps involved in preparation for negotiation can be found in Lewicki, Saunders, and Minton (1999, pp. 29–69).

Consult with Constituents and Other Project Stakeholders. Negotiators need to consult with the various groups or individuals they represent and those who will be affected by the negotiations. For example, negotiators should learn about the requirements upper management has with regard to project costs, schedules, liability, intellectual property, and so on. In addition, negotiators need to consult with engineers, production managers, subcontractors, and suppliers, who will be critical to implementing any agreement that is reached. Consultation with constituents and others allows the negotiator to identify his or her side's interests and to frame, prioritize, and develop positions on the issues, along with arguments to support them.

Identify Interests. Next, the negotiator needs to think about the reasons for negotiating with the other side. What are the basic wants or concerns that an agreement should satisfy? A project manager negotiating a change order may need to complete the project within budget and on schedule. Although there surely will be an interest in avoiding the additional cost of making unnecessary changes, there also may be an interest in keeping clients happy in order to increase the chances of future business.

Interests are different from the specific concrete proposals that will be discussed at the negotiating table. They are the concerns underlying the proposals, and proposals are intended to satisfy them. There may be more than one proposal that satisfies an interest equally well, and alternative proposals can be evaluated in terms of how well they satisfy interests. The success of the negotiations also can be evaluated in terms of how well interests have been satisfied by any agreements reached. Interests are not limited to the obvious or tangible substantive matters such as project cost or completion date. Other interests may relate to the process of negotiation, the relationship with the other side, and preserving important principles (Lax and Sebenius, 1993). Interests also can be classified as instrumental or intrinsic for each of these three categories.

Interests are *instrumental* if favorable terms are valued because of their impact on future dealings. Interests are *intrinsic* if favorable terms are valued regardless of their effect on future dealings. A concern about possible future business is an instrumental interest, as are concerns about the impact of the completion date for a production facility on the ability of a company to honor a possible contract with one of its clients. On the other hand, staying within a budget is an intrinsic interest in a negotiation over a change order for a project.

Aside from the outcomes of negotiation, the parties may also have an interest in the *process* of negotiation. Some may prefer to conduct business in a smooth, harmonious fashion characterized by honesty and openness. Others may enjoy a more aggressive and competitive process and the feeling that they have extracted as many concessions as possible from the other side. When it comes to *relationships*, most negotiators want to be treated with respect by the other side. Having the other's respect makes the negotiator feel better while negotiating (intrinsic interest), and it also may have important implications for the negotiator's treatment in future negotiations (instrumental interest).

Negotiators also may have interests in *principles* that they wish to establish or have honored when negotiating. These principles may assume great importance because they can have precedent-setting implications in future negotiations.

Thinking about different types of interests is useful because it reminds negotiators that they should consider more than substantive interests when preparing for negotiation. There may be nonobvious but important immediate and longer-term interests at stake in a negotiation related to process, relationship, and principles.

Learn about the Other Side. Negotiators should not only think about their own interests, they also should try to learn about the other's interests, likely issues, sources of power, probable negotiation strategies and tactics, alternatives to reaching agreement, as well as any other information relevant to the negotiation. Any pertinent information they can learn about the other side prior to negotiation will help them anticipate their behavior during the negotiations.

For example, in advance of negotiation, a project manager might learn though a company sales representative that the other side may be willing to spend more money for a project if a firm completion date can be guaranteed. Or the other side may have plans for several other projects over the next several years. Alternatively, a project manager may learn from industry sources that the other company is experiencing financial difficulty and needs several future construction projects to remain profitable. These are valuable items of information for negotiators to have as they prepare for negotiation.

There are several sources of information about the other negotiator. Archival data may be used to learn more about the other party (Lewicki et al., 1999, p. 64) and may come from publicly available sources such as Dun & Bradstreet, financial statements, newspaper articles, stock reports, and legal judgments. Other ways of obtaining useful information about the other side may be visiting the other side's place of business, having preliminary discussions with them about what they would like to achieve in the forthcoming negotiations, trying to anticipate their interests through role-playing, and talking to people who know or who have previously negotiated with them. Previous negotiations with the other side may provide notes and experiences that are useful. If other information is unknown prior to negotiation, many times it can be obtained through careful observation and listening as negotiations progress.

Negotiators form perceptions of the other side based on the information they collect. These perceptions can be very useful in anticipating the other's behavior but it is important to remember that such information may be incomplete and the specific circumstances of the upcoming negotiation may lead to different behavior. Therefore, it is important not to act hastily on the basis of untested assumptions in ways that may create obstacles to reaching agreement and to remain receptive to new information as the negotiations progress.

Identify issues. An issue is a continuum on which negotiators can take different positions. The interests of the two sides define the issues to be negotiated. For example, one negotiator may have an interest in minimizing project costs and the time required to complete the project, in being treated fairly by the other side, and in establishing principles for allocating the costs of changes on this and future projects. These interests might lead to the following issues on which the two sides will take different positions: the total cost of the project, the project completion date, penalties for failure to complete the project on schedule, the ground rules for conducting the negotiations, and the criteria to be considered or procedures to be used in making changes during the project. On each of these issues each side can take different positions. For example, one side will probably favor lower project costs and an earlier completion date, while the other may propose higher cost and a later completion date.

Although having many issues to negotiate can complicate and lengthen negotiations because there are more things to negotiate, multiple issues also can facilitate negotiations by creating the possibility of trade-offs. For example, one side might agree to a penalty clause if it does not complete the project on time if the other agrees to pay 100 percent of the cost of overtime associated with any project changes.

Prioritize Issues. Negotiation very often requires making a concession on one issue in order to gain on another. To maximize their outcomes, negotiators should make larger and more frequent concessions on less important issues than they should make on more important ones. For example, if the completion date for a project is more important than the total cost of the project, a negotiator may be very inflexible about the completion date but willing to pay more to ensure that the project is completed on schedule. For these reasons, negotiators need to evaluate the relative importance of the issues.

Develop Proposals. Once issues and underlying interests have been identified, it is time to think about proposals. A proposal should usually encompass the entire set of bargaining issues that need to be resolved in order to reach agreement. It is beneficial to think about at least three alternative proposals prior to starting negotiation. The first is an *initial offer* that will be presented to the other side at the start of negotiation, the second is a *target point* or proposal that represents the negotiator's preferred outcome, and the third is the *resistance point*, or the least favorable proposal a negotiator is willing to accept for the foreseeable future.

Because negotiation involves give-and-take, initial offers should be more favorable to the negotiator than a target point and a target point should be more favorable to a negotiator than a resistance point. It can be advantageous to have the other side present their initial offer first. This allows the negotiator the opportunity to gauge the other side's position on various issues and avoids the possibility of offering the other side a proposal that is more favorable than necessary.

Sometimes a resistance point is based on what Fisher, Ury, and Patton (1991) call a BATNA. A BATNA, or *best alternative to a negotiated agreement*, represents a negotiator's best option outside the current relationship. In a case of negotiating with suppliers, a BATNA might be what another vendor would charge for a product or service. Or the best alternative might be mediation, arbitration, or a court settlement. Establishing a BATNA before negotiation can help prevent agreeing to less favorable terms than are available elsewhere or rejecting a proposal that is more favorable than the alternatives. In addition, a negotiator can strengthen a negotiation position by finding a better BATNA.

When the negotiator is establishing an initial offer, target point, and resistance point, it is very useful to consider objective standards or standards of fairness that may apply to the situation. These will help in supporting proposals and make them seem more realistic and supportable to the other side. There are a number of principles that negotiators can use to support their demands in a negotiation (Magenau and Pruitt, 1979).

Perhaps there is an industry-standard practice or custom that one or both project managers might use to support their position regarding the change orders. If there is no industry norm to rely on, perhaps the project managers might adopt an *equality* norm that would support a 50-50 division of the costs and benefits. An *equity* norm would favor allocation of project revenue according to the amount of investment or risk each of the parties have at stake in the project. Under a *needs* rule, the greatest portion of the cost would be assumed by the party with the greatest capacity to pay (least needy). In other situations, market value, scientific judgment, efficiency, costs, what a neutral third party would decide, or moral standards may be used as possible objective criteria (Fisher et al., 1991).

It is likely that a negotiator will be asked to explain how the specifics of a proposal were established. Use of objective standards will help ensure that a plausible explanation is available and that proposals appear realistic. Proposals without any supporting rationale may be dismissed without serious consideration and may lead to loss of credibility for the negotiator. Serious negotiation is unlikely to occur until a more realistic position is adopted, or the other side may become convinced that further negotiation is a waste of time.

What if there are several standards that could be applied in a negotiation? Assuming that some are more favorable to a negotiator than others, the most favorable could be used to establish an initial offer. A target point could be set according to a standard less favorable to the negotiator but that perhaps is more compelling than the standard underlying an initial offer. Other objective standards, if available, could be used to support proposals intermediate between an initial offer and target point or a resistance point.

Linking an initial offer, target point, and resistance point to principles can be a double-edged sword and should be done with caution. On the one hand, this tactic may make proposals more rational, defensible, and likely to prevail. On the other hand, adopting positions linked to principles may lead to rigidity that makes reaching agreement more difficult (Pruitt and Carnevale, 1993). That is, opposing negotiators may adopt different positions favoring their own interests, each supported by competing principles. If this occurs, there is a danger that one or both negotiators may become so committed to their respective positions that they are psychologically unable to make the concessions that are necessary to reach agreement.

To avoid the psychological trap of becoming too committed to positions based on principles, Fisher, Ury, and Patton (1991, p. 88) advise negotiators to "frame each issue as a joint search for objective criteria," to "reason and be open to reason," and to "never yield to pressure, only to principle." In other words, negotiators should be firm about reaching an agreement based on principles of fairness but at the same time be flexible enough to consider different principles that may legitimately apply to a given situation.

Negotiators should prepare for negotiation by developing a series of proposals that are supportable by objective criteria. At the same time they should be flexible enough to acknowledge the validity of other legitimate principles that may apply to the situation. This flexibility may motivate a search of innovative solutions that reconcile or integrate various principles previously thought to be contradictory. Or it may lead to a compromise solution that incorporates some elements of various principles. One side may even adopt the position of the other side if, after discussion, it proves to be fairer and easier to implement.

Team Organization

In important complex negotiations, typically each side is represented by a team instead of a single individual. Teams should be composed of members who bring the necessary negotiating skill and specialized technical expertise to the negotiating table, and team members should be briefed on their responsibilities prior to negotiations. It is difficult for a single individual to talk, listen, think, write, observe, and plan simultaneously. For this reason, Kennedy, Benson, and McMillan (1982) recommend that negotiation teams include three generic roles: leader, summarizer, and recorder.

The *leader* is usually the senior member of the bargaining team and serves as the team coordinator and spokesperson, and generally leads the team's effort. The leader usually does most of the talking at the negotiation table but may call upon others to speak when needed. Channeling communication through a leader allows a team to control the information provided to the other side and avoid having its negotiating position undermined by revealing sensitive information or within-team disagreements to the other side.

The *summarizer* listens carefully to the arguments of the other side and gives thinking time to the leader by intervening when appropriate. This often can be done by asking the other side for clarification of a point or by summarizing the other side's position or arguments on some issue. The summarizer should be someone who is trusted and who can work closely with the leader.

The *recorder* remains silent during the negotiation unless called upon to speak about a particular issue related to his or her expertise. There could be several recorders who attend all or some of the negotiations sessions depending on the requirements of a particular meeting. These might include lawyers, engineers, architects, cost estimators, environmental experts, programmers, production schedulers, or other technical specialists knowledgeable of some critical aspects of the project. The responsibilities of this position also include watching the other team for verbal and nonverbal cues that might reveal something about their interests, priorities, or points of internal disagreement. Recorders may be called upon to report their observations to the team during private meetings away from the negotiating table.

Negotiating Strategies

Negotiation strategies can be classified into five basic categories (Pruitt and Carnevale, 1993; Lewicki et al., 1994):

- Concession-making
- Contending
- Compromising
- Problem solving
- Inaction/withdrawal

Concession-making involves changing a proposal so that it provides less benefit to you but more benefit to the other side. Concession-making may be your response to the other side's successful use of power and influence.

Contending involves trying to persuade the other side to make a proposal more favorable to you but less favorable to them. Contending may involve the use of coercive power communicated in the form of threats or influence tactics such as strong persuasive arguments, pressure, multiple pressure, guilt, and making statements about your unwillingness to make any additional concessions.

Compromising represents a strategy that is intermediate between concession-making and contending. A middle ground is sought that involves an exchange of concessions and some degree of sacrifice for both sides. A compromise strategy utilizes a "carrot and stick" approach as compared with the heavier reliance on the "stick" used with a contending strategy. Compromising involves a more positive approach to the other side than contending and therefore is probably better implemented with more positive forms of power such as reward (e.g., reciprocating concessions), information, and expert power (e.g., using objective information or expert opinion to support a compromise agreement) or referent power (e.g., increasing your attractiveness or prestige in the eyes of the other side). Influence tactics such as ingratiation and moderate persuasion also would be consistent with a compromise strategy. However, the use of coercive power and the influence tactics employed in a contending strategy may be needed to motivate concessions if the other is unwilling to move to a middle-ground position.

Problem solving involves efforts to find agreements that are highly favorable to both parties. If successful, both sides can avoid the deep concessions that may be necessary under a concession-making or compromising strategy. Problem solving often involves a joint effort in which the two sides work together by openly sharing information about interests and priorities and by discussing the pros and cons of various proposals.

Problem solving is more likely to occur when there is a positive long-term relationship and where trust exists between negotiators that allows for open exchange of information and frank discussions. It is a negotiation strategy that is generally incompatible with the use of coercive power and aggressive influence tactics used with a contending strategy. On the other hand, the use of reward power, referent power, expert power, and information power are consistent with problem-solving strategy.

Inaction and withdrawal are similar. Withdrawal involves terminating negotiations for the foreseeable future without an agreement. Inaction involves attempts to delay or avoid engaging in serious negotiations. If they chose withdrawal, both parties might adopt their BATNAs. An inaction strategy might involve ignoring the other's requests to negotiate change orders, refusing to discuss changes, or diverting the conversation to other topics when the other side proposes negotiation. Inaction is likely to be favored by those with advantages they wish to preserve and who view negotiation with others as an attempt to change the status quo. Inaction and withdrawal may be effectively employed by those who have resource power, dependency power, or nonsubstitutability to convince the other side that concessions are necessary to start or resume negotiation.

Choosing a Strategy

Pruitt (1983) proposed the dual concern model to explain how negotiators choose a negotiation strategy. What the model says about how negotiators actually choose a negotiation strategy provides a good analytical framework for considering strategic alternatives. According to the dual concern model, the choice of a negotiating strategy is primarily influenced by two key concerns. First is the negotiator's level of concern for his or her own outcomes, and second is the level of concern for the outcomes of the other side. A third consideration is the feasibility of the strategy under consideration. The relationship between the first two variables, or the dual concerns, and the choice of a strategy are shown in Figure 5.2.

The model indicates that negotiators choose contending or problem solving strategies when they have a high level of concern for their own outcomes. This makes sense because most people are less willing to sacrifice their own interests when they place great value on them. Thus, we are less likely to engage in concession-making or compromise when important interests are at stake, and we are unlikely to choose inaction or withdrawal if we believe these interests can only be satisfied by the other side.

The choice between contending and problem solving depends on a negotiator's level of concern with the other party's outcomes. If the negotiator has a high level of concern for the outcomes of the other side, he or she will choose problem solving because this is a strategy designed to satisfy the needs of both sides. If there is little concern for the other side's outcomes, contending is the strategy of choice because it seeks to maximize the negotiator's own interests at the expense of the other side. If the negotiator's concern for his or her own outcomes is low but the negotiator's concern for the other's outcomes is high, the negotiator is likely to choose concession-making in order to satisfy the other side. When both concerns are low, there is little interest in negotiation and inaction and withdrawal are probable strategy choices.

The importance negotiators place on these concerns and the feasibility of various strategies can change over the course of negotiations and across various issues under discussion. As a result, the choice of a strategy can vary over the course of negotiation and with different

FIGURE 5.2. THE DUAL CONCERN MODEL.

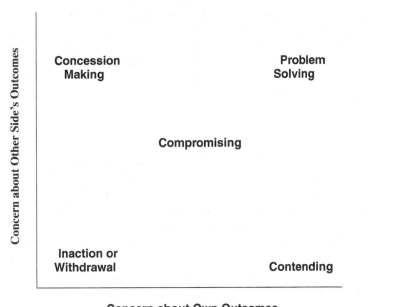

Source: Adapted from D. G. Pruitt (1983). Strategic choice in negotiation. *American Behavioral Scientist* 27:167–194.

issues. The strategies are not mutually exclusive in the sense that different strategies might be employed at different stages of the negotiation or on different issues. For example, the negotiator might begin by contending because he or she believes the other is weak and will concede in the face of our pressure tactics and persuasive arguments but shift to a compromising strategy after hearing the other side's arguments and learning that the other side is more resistant to change than anticipated. In addition, some issues in the negotiation may be perceived by the two sides as win-lose or zero sum. They may attempt to resolve these by contending, concession-making, or compromising, while other issues may be viewed as having problem-solving potential.

The dual concern model can be used as a guide to preparation for negotiation if negotiators ask themselves the following questions:

- What is my level of concern for my own outcomes?
- What is my level of concern for the other side's outcomes?
- What is the feasibility of the strategy suggested by my level of the preceding two concerns?

Concern for Own Outcomes. In thinking about their level of concern for their own outcomes, negotiators should consider the importance of the interests that will be affected by the outcomes of negotiation, the opportunity costs associated with the negotiation, and the extent to which they are willing to risk conflict with the other side in order to achieve their desired outcomes. The importance that constituents place on the outcomes the negotiator achieves also will influence the calculation, especially if constituents can hold the negotiator accountable for results. For example, I am likely to rate the importance of my side's outcomes as high if the company's profitability is significantly affected by the outcome of the negotiation, the current contract is larger than other contracts my company has, I think the other side has few alternatives to signing a contract with me, or my boss informs me that my future with the company hinges on negotiating contract terms favorable to the company.

Concern for the Other Side's Outcomes. Concern for the other side's outcomes will be greater if there are interpersonal bonds of friendship with the other side that exist or that the negotiator wishes to establish. Concern for the other side's outcomes also will be greater if the negotiator is dependent on the other side to achieve desired outcomes or if the negotiator will be dependent on them in future negotiations. It is important to keep such future dependencies in mind because negotiators sometimes become so absorbed in the current negotiation that they forget about the impact of their behavior on their long-term relationship with the other side. For example, we would expect a negotiator's concern for the other side to be greater if the other negotiator is an old and good friend, needs an agreement with the other side to complete a project before a fast approaching deadline, and wants to do business with the other again in the future.

In terms of Block's (1987) framework, in a negotiation between allies and opponents, concern for the other's outcomes would be highest where maintaining a positive long-term relationship is a key consideration. Concern for the other side's outcomes may also be a consideration when negotiating with bedfellows and fence-sitters, but it is clearly more of a short-term consideration and probably not as strong as it is for allies and opponents. There is likely to be low concern with the other's outcomes when negotiating with adversaries.

The Feasibility of Problem Solving

If the negotiator has a high level of concern with his or her own outcomes as well as a high level of concern for the other side's outcomes, the dual concern model suggests a problem-solving strategy. However, the model suggests that the choice of problem solving also depends on its feasibility.

Problem solving becomes more feasible as the perception of common ground on the part of the negotiators increases (Pruitt and Rubin, 1986). This occurs because greater common ground makes the likelihood of finding alternatives that satisfy the interests of both parties seem more promising. Common ground on an issue increases as the amount of overlap between the negotiator's resistance points becomes greater and to the extent that the parties believe that alternatives favorable to both sides exist or can be invented. The concept of common ground is obviously closely aligned with the agreement dimension in the Block (1987) model for choosing influence strategies.

Negotiators also are more likely to adopt a problem-solving strategy when they are confident in their own problem-solving ability, they have a track record of solving problems with the other negotiator, they have access to third parties who can assist them with locating or devising mutually beneficial alternatives, and they believe that the other side is inclined to problem-solve too. A negotiator's belief that the other side is ready to problem-solve depends on trust. Trust in the other in turn depends on a negotiator's belief that the other is concerned about his or her interests. Our trust in the other side is likely to increase when the other has a positive attitude toward us, is similar to us, is dependent on us, or has been helpful to us in the past, especially if the help has been voluntary.

The relationship between trust and problem solving is likely to occur as long as trust is coupled with the belief that the other side is strongly committed to their own outcomes. If the other side appears to have a weak commitment to their outcomes, trust can encourage contending rather than problem solving. This can occur because when we believe someone is concerned about us, but at the same time has a weak commitment to his or her own goals, we are likely to conclude that that person will concede easily.

There are several things a negotiator can do to encourage problem solving. One is to explore the possibility of common ground. Many negotiators have a "fixed-pie" bias (Bazerman, 1983). They assume, many times incorrectly, that there is a fixed amount of what is being negotiated and that one side can benefit only if the other side loses. Instead of making this assumption, you should approach negotiations with an open mind and try to explore areas of common ground.

Negotiators also can enhance their own confidence in their problem-solving ability by improving their communication skills. They can try to build the other side's confidence in their problem-solving ability by verbally reinforcing the other's problem-solving efforts or by encouraging them by clearly stating their desire to find a mutually beneficial agreement. If discussion of easier-to-solve issues is scheduled first on the agenda, success in solving them may lead to the belief and that more difficult issues can be solved later on in negotiations. Confidence may be increased if trusted and respected neutral third parties who can be relied upon to facilitate problem solving are identified. Finally, an indication of the other's readiness for problem solving can be gained through careful observation of clues in their behavior, by direct questioning, or by analyzing the factors relevant to their dual concerns.

The Feasibility of Contending

Contending is more feasible when (1) the other side has low concern for their own outcomes and therefore little resistance to conceding, (2) a negotiator has the power and the skill to effectively employ influence tactics and the other side lacks the power and skill to counter the tactics used, and (3) there is little risk of alienating the other side by using such tactics. Therefore, before deciding to use contentious tactics, a negotiator should attempt to evaluate the other side's level of concern for its own outcomes. The negotiator should consider the sources of power and influence tactics he or she has available as well as those available to the other side, and evaluate the risks of alienating the other side by deploying the tactics under consideration.

The Feasibility of Inaction and Withdrawing

The feasibility of inaction depends on the time pressure a negotiator is experiencing. There are two sources of time pressure: the cost of continued negotiations and the necessity of meeting deadlines. Time pressure caused by the continued cost of negotiation might result from lost business opportunities incurred while busy negotiating, the cost of supporting a negotiation team away from home, or loss in value of the object of negotiation as in the case of perishable items. Deadlines represent times in the future when significant costs will be incurred if there is no agreement.

In considering the feasibility of inaction as a strategy, negotiators should ask themselves what costs are associated with delaying the negotiation and what significant costs will result from failure to meet specific deadlines. For example, a supplier might try to use inaction for a time to postpone negotiation on price reductions but later realize that negotiation is necessary to avoid the loss of business.

The feasibility of withdrawing depends on the attractiveness of a negotiator's BATNA. Negotiators who have very attractive alternatives outside of their relationship with the other side are more likely to consider withdrawal as a feasible alternative.

The Feasibility of Concession Making

Concession making is a feasible strategy as long as one is able to make concessions. Because this strategy provides a way of reaching agreement quickly, it is favored when time pressure is high. The capacity to concede depends on how close the negotiator is to his or her resistance point. The closer the negotiator's current position is to the resistance point, the more difficult each additional concession becomes. For example, if my current proposal is for $500,000 to complete the work on a project, I am more willing to agree to a $50,000 reduction in my asking price if my resistance point is $350,000 than if it is $425,000.

The Feasibility of Compromising

Compromising can be thought of as a less intense form of concession making or problem solving (Pruitt and Rubin, 1986). The difference between compromising and pure concession making is that under a compromising strategy, negotiators attempt to coordinate the exchange of concessions with the other negotiator toward some middle ground rather than concede unilaterally. It differs from problem solving in that the compromise under consideration involves more sacrifice for the negotiators and the agreements reached tend to provide lower joint benefit.

The conditions making a compromise strategy feasible are similar to those underlying the feasibility of concession making and problem solving. That is, one must have the ability to concede to the proposed compromise alternative. And, as with problem solving, the perceived feasibility of a compromise solution is increased when there is a perception of common ground. Also, as in the case of problem solving, compromising is more likely to occur if negotiators previously have successfully exchanged concessions, believe the other side is ready to exchange concessions, and third parties are available who can help coordinate concessions if needed.

The use of conditional language is strongly recommended when attempting to coordinate concessions toward a compromise solution. For example, an negotiator might say, "If you agree to my proposed project completion date, I will agree to your proposed cost" (Kennedy, Benson, and McMillan, 1982, p. 88). Conditional proposals let the other side know precisely what you want in return for your concession. It also is clear that the concession is offered only if the other side reciprocates. If the other side does not agree to the proposed terms, the concession is withdrawn.

Summary

The nature of their work makes skill in the use power, influence, and negotiation indispensable to successful project managers. Power and influence are commonly used in negotiation, and understanding their use is valuable in preparation for and in the conduct of negotiation. In this chapter we defined power, influence, and negotiation. We discussed the sources of power and how to acquire it. Various types of influence were also described, as well as when they can be used most effectively.

Thorough preparation is essential to successful negotiation, and preparation involves several recommended steps. As with the use of power and influence, successful negotiation depends on the use of a strategy appropriate to the situation. It is important to remember that the choice of a strategy is contingent upon the dual concerns of the negotiators and the feasibility of the strategies under consideration. Different strategies are recommended under different circumstances. Further, negotiators should be prepared to change strategies over the course of a negotiation as the circumstances require.

References

Allen, R. W., D. L. Madison, L. W. Porter, P. A. Renwick, and B. Y. Moyes. 1979. Organizational politics: Tactics and characteristics of actors. *California Management Review* 22(1):78.

Bazerman, M. H., 1983. Negotiator judgment: A critical look at the rationality assumption. *American Behavioral Scientist* 27:211–228.

Beeman, D. R., and T. W. Sharkey. 1987. The use and abuse of corporate politics. *Business Horizons* 36(2):26–30.

Block, P. 1987. *The empowered manager.* San Francisco, CA: Jossey-Bass.

Daft, R. L., 1999. *Leadership: Theory and practice.* Fort Worth, TX: Dryden Press.

———. 2001. *Organization theory and design.* 7th ed. Cincinnati: South-Western.

Fisher, R., W. Ury, and B. Patton. 1991. *Getting to yes: Negotiating agreement without giving in.* New York: Penguin.

French, J. R. P., and B. Raven. 1959. The bases of social power. In *Studies in Social Power*, ed. D. Cartwright, 150–167. Ann Arbor, MI: Institute for Social Research.,

Gandz, J., and V. V. Murray. 1980. Experience of workplace politics. *Academy of Management Journal* 23:237–251.

Goodman, R. M., 1967. Ambiguous authority definition in project management. *Academy of Management Journal* 10:395–407.

Hickson, D. J., C. R. Hinings, C. A. Lee, R. E. Schneck, and J. M. Pennings 1971. A strategic contingencies theory Of intraorganizational power. *Administrative Sciences Quarterly* 16:216–229.

Kennedy, G., J. Benson, and J. McMillan. 1983. *Managing negotiations*. Rev. ed. Englewood Cliffs, NJ: Prentice Hall.

Keys, B., and T. Case. 1990. How to become an influential manager. *Academy of Management Executive* IV(4):38–51.

Lawrence, P. R., and J. W. Lorsch. 1967. Differentiation and integration in complex organizations. *Adminstrative Science Quarterly* 11:1–47.

———. 1969. *Organization and environment.* Homewood, IL: Irwin.

Lax, D. A., and J. K. Sebenius. 1993. *Interests: The measure of negotiation.* In *Negotiation: Readings, Exercises and Cases*, ed. R. J. Lewicki, J. A. Litterer, D. M. Saunders, and J. W. Minton. 2nd ed. Barr Ridge, IL: Irwin.

Lewicki, R. J., D. M. Saunders, and J. W. Minton. 1999. *Negotiation* 2nd ed. Boston: Irwin/McGraw-Hill.

Magenau, J. M., and D. G. Pruitt. 1979. The social psychology of bargaining: A theoretical synthesis. In *Industrial relations: A social psychological approach*, ed. G. Stephenson and C. Brotherton. Chichester, UK: Wiley.

March, J. G., and H. A. Simon. 1958. *Organizations.* Hoboken, NJ: Wiley.

Markus, M. L. 1983. Power, politics, and MIS implementation. *Communications of the ACM* 19:321–342.

Mayes, B. T., and R. W. Allen. 1977. Toward a definition of organizational politics. *Academy of Management Review* 2:675.

Mintzberg, H. 1983. *Power in and around organizations.* Englewood Cliffs, NJ: Prentice Hall.

Payne, H. J. 1993 Introducing formal project management into a traditionally structured organization. *International Journal of Project Management.* 11:239–243.

Pettigrew, A. M. 1973. *The politics of organizational decision-making.* London: Tavistock.

Pfeffer, J. 1981. *Power in organizations.* Marshfield, MA: Pitman.

Pinto, J. K., and O. P. Kharbanda. 1995. Project management and conflict resolution. *Project Management Journal* 26(4):45–54.

———. 1995. *Successful project managers: Leading your team to success.* New York: Van Nostrand Reinhold.

Pruitt, D. G. 1983. Strategic choice in negotiation. *American Behavioral Scientist*, 27:167–194.

Pruitt, D. G., and P. J. Carnevale. 1993. *Negotiation in social conflict.* Pacific Grove, CA: Brooks-Cole.

Pruitt, D. G., and J. Z. Rubin. 1986. *Social conflict: Escalation, stalemate and settlement.* New York: McGraw-Hill.

Slevin, D. P. 1989, *The whole manager.* New York: AMACOM.

Thamhain, H. J., and G. Gemmill. 1974. Influence styles of project managers: Some project performance correlates. *Academy of Management Journal* 17:216–224.

CHAPTER SIX

MANAGING HUMAN RESOURCES IN THE PROJECT-ORIENTED COMPANY

Martina Huemann, Rodney Turner, and Anne Keegan

In this chapter we describe the characteristics of human resource management (HRM) in the project-oriented organization. Human resource management is a specific and strategically important process in the project-oriented organization. It includes recruitment, disposition and development, leadership, retention, and release of project management personnel.

The contents of this chapter are based on recent research into the HRM in the project-oriented organization and project-oriented society. First we describe the changing nature of HRM in the project-oriented society and consider the impact on project management personnel and their careers. We then consider the different types of project personnel who need to be managed in the project-oriented organization and describe the HRM processes in the project-oriented organization. We end by briefly describing the role of the PM office in managing project management personnel.

Human Resource Management in the Context of the Project-Oriented Society

A change toward a project-oriented society is observable. Gareis and Huemann (2001) define a project-oriented society as one that does the following:

- Considers projects and programs as an important form of (temporary) organization for achieving strategic and change objectives
- Supports a relatively high number of project-oriented organizations

- Has specific competencies for managing of projects, programs, and project portfolios
- Has structures to further develop these management competencies

The fact that there are increasingly more projects performed in society is explained by the evolutionary demand for projects (Lundin and Söderholm, 1998). Not just traditional industries, but many others, including the public sector, perceive temporary organizations such as projects and programs as appropriate to perform business processes of medium to large scope. Beside traditional contracting projects, other types, such as in marketing, product development, and organizational development, have gained in importance. Projects and project management are applied in new social areas, such as local municipalities, associations, schools, and even families. "Management by projects" becomes a macroeconomic strategy of the society, to cope with complexity and dynamics and to ensure quality of the project results (Gareis, 2002). Further, project management is being established as a profession. The Project Management Institute estimates that there are about 16 million people worldwide who consider project management as their profession (Gedansky, 2002).

Individuals Work More Often in Temporary Organizations

In project-oriented societies, there is a trend for individuals to get temporary assignments as they work on successive projects and programs. Project participants move from one project to another, often from one company to another, and even from one country to another. This creates a picture in our minds of "project nomads," whom we might think of as having an adventurous life. However, the personnel manager of an international engineering company pointed out that these nomads have to move from one place to the other because the country is too poor in which to settle down permanently. Similar pictures are drawn by Drucker (1994) when he describes the knowledge workers and Handy (2002) when he describes the life of a the self-employed "flea." Handy (1988) previously described such people as being like freelancers, literally mercenaries at the time of the crusades, who were not part of the regular army. Temporary employment and self-employment is increasing. Lifetime employment and permanent careers become rare. Acquiring project management competencies, keeping them state-of-the-art, and getting them certified becomes an issue, even for those project management personnel who belong (permanently) to a project-oriented organization. An individual has to take on the responsibility for the acquisition of the competencies demanded and of his or her professional development to keep employable.

Characteristics of HRM in Project-Oriented Organizations

What are the features of project-oriented firms that influence the nature of employment within them? Projects are temporary organizations undertaken to bring about change (Turner and Müller, 2003; Lundin and Söderholm, 1998). Some, primarily functional, organizations undertake occasional projects to enact specific changes. They can adopt classical human resource management practices and assign resources to projects from within the functions as necessary. But for project-oriented organizations, projects are their business; the majority of the work they do is project-based. Turner and Keegan (2003) showed that they need a different approach to human resource management than the classical approach adopted by functionally oriented organizations.

As temporary organizations, projects are unique, often novel, and transient. Being unique, the organization has never done exactly this before. They often require novel processes and have novel resource requirements. Being unique and novel, the method of delivery can be uncertain. The consequences on human resource management requirements are as follows:

- The present and future resource requirements of the organization are uncertain.
- People follow careers other than climbing the ladder up the functional silo.
- People may not have a functional home to belong to.

Uncertain Requirements

In the classically managed, functional organization, resource requirements are assumed to be well determined. The jobs to be done are well known from past experience. A job description is written for a job, defining what is to be done, the levels of management responsibility required, and the competence required, including levels of education and training and past experience. Somebody is recruited in accordance with that specification. There is a saying, "You grade the job and not the person." The requirements of the job are defined, and the best match is found to those requirements.

That level of certainty often does not exist in the project-oriented organization:

1. Projects are unique and transient, with high uncertainty. It is often not possible to define precisely the requirements of the current job. You need to recruit people known to work well on projects and, to an extent, let them define the job around themselves. (Though this is true of many other management positions as well, of course.)

2. Contract organizations often cannot precisely predict the levels of resource requirements into the immediate future. They may have several jobs at the moment, with one coming to the end, and several bids out. For instance, consider that they have five bids out, with a normal success rate of winning one bid in five. If they achieve that, they will have one job to replace the one coming to an end. If they are successful with none, their workload will fall; if they win two, they may just cope; if they win three, they will be overloaded. Keegan and Turner (2003) report that the only way project-oriented organizations cope with this uncertainty is by employing between 20 percent and 40 percent contract staff. They report one organization employing up to 80 percent contract staff. This is essential to cope with fluctuating and uncertain workloads.

3. As for forecasting future resource requirements, if it is not possible to predict resource requirements one month out, how can anyone predict them one year out? Organizations can assume they will carry on doing the same types of projects, and they will try to use economic forecasts to predict future numbers of projects in the industry. However, it is much less certain than in a functional organization.

The Spiral Staircase Career

The consequence for people's careers is good news and bad news. The bad news is they do not have the comfortable certainty of a clear career path where they can climb the ladder up the functional silo. The good news is they have much more varied and interesting careers. Projects, being transient, cannot provide careers, but each project can be a learning opportunity in a career. Projects provide an opportunity for a broad sweep of learning experiences. Keegan and Turner (2003) coined the phrase "the spiral staircase career" to reflect that people will move through a series of varied and wide-ranging jobs. They might spend time in the design function, time as lead designers on a project, and time as project managers. Rather than each move being a whole step up the ladder, moves can be half or even a quarter of a step sideways and upwards. People can also avoid the Peter Principle—namely, being promoted to the level of their incompetence. If they find themselves in a job that does not suit them, they can take a move sideways, which does not carry any stigma, compared to taking a step down the ladder of the functional silo.

No Home Syndrome

Coupled with varied career is the "no home syndrome" (Keegan and Turner, 2003). People spend their working lives moving from one project to another. They generally do not have a permanent home, or a permanent sense of belonging. They work on one project for 9 to 18 months; then that team breaks up and they move to a new team. This creates the nomadic life mentioned previously, but it also increases the need for team building on projects to create a sense of belonging to the project (Reid, 2003). A practice adopted by many project-oriented firms is the creation of the PM office, or an expert pool of project managers. This can provide workers a "home base" between projects and a place to continue to belong to and seek support while working on projects. Sometimes the PM office may be virtual but still satisfy these needs.

Project Management Personnel

In project-oriented organizations, we can differentiate several different types of resources, including line management, technical experts, and project management personnel. Project management personnel are those human resources who need to draw on project management knowledge and experience to fulfil their roles. They include project managers but also include people in other project roles. The HRM practices we discuss in this chapter apply to project management personnel in the first instance. The project-oriented organization may apply similar processes, or conventional ones, to people working in line management or as technical experts. Project management personnel include people working in temporary structures such as projects and programs, and people working in permanent structures such as a project management office, a project portfolio group, or an expert pool. The former group includes the following:

1. Project personnel, such as:
   - The project owner, project sponsor, or project champion
   - The project manager, project leader, or project director
   - The project management assistant
   - The project controller
   - Project team members and project contributor
2. Program personnel, such as:
   - The program owner or program director
   - The program manager, project director, or program coordinator
   - The program assistant or program controller
   - Program office members

People working in permanent structures include the following:

- Project management office personnel such as the office leader and office members. They are the process owners for the project management process within the project-oriented company. Further functions of the project management office are described later in this chapter.
- Project portfolio group members who take the responsibility to manage the project portfolio from a strategic perspective. Usually these members of the project portfolio group are managers of those business units of the permanent organization, which are frequently involved in projects and programs.
- Quality management personnel such as project or project management auditors and reviewers, project or project management coaches, and project or project management consultants.
- Expert pool personnel such as the leader of the project expert pool and the members of the project expert pool. From these expert pools the project personnel is drawn.

The project portfolio office leader, project portfolio group members, and project expert pool leaders are often labeled as "project executives." Employees in the project-oriented company often have more than one role and can therefore belong to different groupings of project management personnel. For example, one person can be a program manager for one project and at the same time work as project coach for a different project.

Competences of Project Management Personnel

As part of their HRM policies and practices, project-oriented firms need to define competence requirements for all these project management personnel. (Competence development is described in Gale's chapter). Competence is the knowledge, skills, and behaviors (experience) a person needs to fulfill his or her role (Huemann, 2002). Project management personnel need a set of several competencies covering not just the management of projects but also the following:

- *Project management.* Knowledge and experience about project and program management including methods and processes
- *Organization.* Knowledge and experience about the project-oriented organization at its specific processes like portfolio management, assignment of projects and programs, and so on
- *Business.* Social networks, product, industry, and so on
- *Technical.* Technical, marketing, engineering, and so on
- *Cultural and ethical awareness.* As in the case of international projects.

How these competencies are described is specific to the company and the project management approach used. It may be traditional, emphasizing scope, cost, and time, as in PMI's PMBOK (2002); it may be more holistic, emphasizing process orientation, as in PRINCE2 (OGC 2002); or it may emphasize also project context and organization, as proposed by Gareis (2002) and Morris (1997) or the Association for Project Management (APM, 2000). There is always a lot of discussion on how much technical competencies the project managers need to manage a project. The range goes from nontechnical competencies to being a technical expert as well as a project manager. The more project management is considered a profession in the organization the less technical competencies may be asked for. Figure 6.1 illustrates minimum competence requirements of a senior project manager in an engineering company. The competence profile required very much depends on the size of the project, its type, and the industry. The competencies will be developed through the individual's career, which we discuss next. However, there is a different emphasis in the project-

FIGURE 6.1. MINIMUM COMPETENCE REQUIREMENTS OF A SENIOR PROJECT MANAGER.

Competences	Knowledge					Experience				
	5 very much	4 much	3 average	2 low	1 none	1 none	2 low	3 average	4 much	5 very much
Project and Program Management	▓								▓	
Management of the Project-Oriented Company		▓						▓		
Business		▓						▓		
Project Contents				▓			▓			

oriented firm. In the traditional organization, the individual gains his or her knowledge and experience working within one function, in the stable organization. In the project-oriented firm, the person gains knowledge and experience through a series of projects, fulfilling different roles on those projects, following the spiral staircase career. Such careers need careful management.

Career Development

We have already seen how project-based ways of working fundamentally change the careers of individuals. Rather than climbing the ladder up the silo, they follow a spiral staircase career, with wide and varied career experiences. There are several practices project-based organizations adopt to support project management careers, including the following:

- A defined project management career
- Measuring "up" in novel ways
- Career committees
- Project management communities
- Individual responsibility

A Defined Project Management Career. Many project-based organizations from both the engineering and high-technology industries recognize project management as a defined career path. Table 6.1 shows a typical seven-step career for many high-technology companies. Many organizations support the career path with professional certification for the Project Management Institute or International Project Management Association, and with formal education programs. IBM, for instance, requires its personnel to take PMI PMP certification followed by a master's degree.

Some organizations have parallel career paths. One high-technology company profiled by Keegan and Turner has a career structure like an upside-down table, with four recognized careers in the company:

- Project management
- Line management
- Technology management
- Sales and marketing

The four careers followed a common structure up to stage 3 in Table 6.1. This enabled people to follow a spiral staircase, sampling different possible careers, until they reached the start of stage 4. Then they were expected to specialize, climbing the ladder up one of the four legs for the remainder of their career.

Measuring "Up". Discussing career movement in terms of climbing a ladder or a spiral staircase implies there is some measure of "up"—some measure of increasing seniority and responsibility. Traditionally, the way to measure up was by the number of people managed

TABLE 6.1. SEVEN-STEP CAREER MODEL.

Stage	1	2	3	4	5	6	7
Name	New start	Team member	Team leader	Junior project manager	Project manager	Senior project manager	Program director
Responsibility			Single function	Several functions	Several companies	Complex projects	Many complex projects
IPMA certification			Level D		Level C	Level B	Level A
PMI certification				PMP			
Education				Certificate	Diploma	MBA, MPM MSc (PM)	

and the budget of the individual's department. This idea was widely discredited early in the history of HRM but is still applied by many organizations.

In spite of this, many organizations continue to reward people according to the number of their direct subordinates right up to the present day. Recently one of the authors was interviewing a man from a company that made electronic equipment. The company wanted to projectize their business and wanted to know what might stand in the way of that. The reward structure was seen as a potential barrier: The company still rewarded people according to the number of direct subordinates. That meant the manager of an engineering department with 1,000 engineers would be scaled as very senior. On the other hand, the manager of a project of £5 million, with a profit margin of £500,000 and of critical importance to the UK's defense, would have very few direct subordinates and so would be scaled as very junior.

Many project-oriented organizations measure "up" in other ways. A practice common in high-technology and engineering companies is to measure "up" by control of risk (which is related to impact on profit). In organizations in both industries, the head of a function or department may not be the most senior person in that department. For example, as part of the spiral staircase career, a potential project manager may return to manage the design function for a while and while doing this might not be the most senior member of the team. Management of the design function carries a certain level of risk, while being a senior designer may carry more.

We interviewed somebody from the engineering industry who had gone from being projects director on the company's board to director of a $1.5 billion project for a major client. The project was considered to be of such high risk that the director of that project was a more critical role than a company board member.

Career Committees. The career development process in Table 6.1 does not happen on its own. Many organizations have committees of senior project or program managers, managing the development of project management professionals. In the engineering industry these committees tend to be fairly ad hoc. The process is managed, but in an informal way. In high-technology firms, the process tends to be more formally managed, linked to career development process of the organization as a whole. This may also be the role of the PM office, as discussed later in the chapter.

Project Management Communities. Project management communities are often used to aid organizational learning. These are networks of project managers to support the development of individual project managers, through mentoring or via events where project management professionals can meet and exchange experiences. Developing the competence or maturity of an organization and its people are closely linked. Project management communities can be company-internal or company-external. Project management associations, such as the Project Management Institute and the International Project Management Association, fulfill the role of external communities. The European Construction Institute in Europe and the Construction Industry Institute in the United States provide an external community for people from the engineering construction industry. Internal communities are maintained in large, high-technology companies, such as IBM, Telekom Austria, and the information services (IS) department of the Dutch bank ABN AMRO. In large companies the project management community also helps project managers to meet, which would otherwise be difficult. It can also help promote the profession and facilitate knowledge management.

Individual Responsibility. Although companies maintain committees to manage careers and communities to support development, individual project management professionals are expected to take responsibility for their own career development. It is easy to get lost on the spiral staircase, both to lose your way and for people to stop noticing you. Turner, Keegan, and Crawford (2003) report that in the engineering industry personal ambition is a key criterion for identifying someone as a potential project manager. Many organizations through their career committees and project management communities provide people with guidance on setting annual objectives and development plans, including training and certification. But individuals must take personal responsibility for achieving their plan.

A key issue that frequently occurs when a person has a development objective and a new project comes along that provides that opportunity. In this case, is the person made to finish his or her current project and be denied the opportunity, or can they be switched mid-project? Enlightened companies switch people mid-project, as long as the current project is not in start-up or closeout. It is better for the company that the individual gets the development opportunity, and individuals will be more loyal to companies that provide them with appropriate development opportunities.

Project Management Profession

The establishment of a project management career path contributed to the development of the project management profession in the project-oriented society. Many project-oriented organizations require potential project management personnel to seek certification. Such certificates prove that the person has a certain level of project management knowledge and experience. Project management certification is offered by global project management associations such as the International Project Management Association (www.ipma.ch), the Project Management Institute (www.pmi.org), or the many national project management associations. Figure 6.2 shows the IPMA project management certification as offered by Project Management Austria (PMA). This also illustrates the changing competence requirements of project management personnel at different levels. This structure and the associated levels have been adopted by many project-oriented companies in Austria, such as Unisys and Telekom Austria. The certification structures are associated to the project management career structure in these companies, and there is a link to the reward system. Certification is perceived as an external quality check for the project management personnel, and in many cases the customers ask for certified project managers.

Processes of Human Resource Management

In this section we describe the processes of human resource management, their specific characteristics in the project-oriented organization, and the methods applied. These processes include the following (Schein, 1987; Keegan, 2002):

FIGURE 6.2. PROJECT MANAGEMENT AUSTRIA 4 LEVEL CERTIFICATION BASED ON IPMA CERTIFICATION.

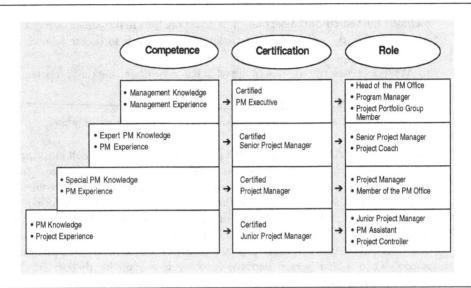

- Recruitment
- Disposition
- Development
- Leadership
- Retention
- Release

Recruitment

Recruitment comprises the search for competent personnel to meet current or future resource requirements. Search and selection can be done for the company in general or for a specific project or program. One can differentiate between company-internal and company-external recruiting. Company-internal recruiting for a specific project or program draws on expert pools within the company. Another possibility is to recruit project management personnel from outside the company from project management networks. Freelance personnel often appreciate the advantages of being part of a network. For example, the platform www.myfreelancer.at offers a marketplace for freelancers in the IT sector. Networks increase the flexibility of the company by enabling it to build relationships with cooperationpartners and experts and to have access to them in case of demand. We saw previously that the only way many project-oriented organizations cope with fluctuating demand is through the use of contract staff, with levels typically ranging between 20 percent and 40 percent. Often the recruiting is performed by a single project or program, the project portfolio office, the project management pool, or expert pool and not by the human resource management department.

Methods for Search and Selection. The traditional, formal manner of recruiting people to a post, whether recruiting them internally or externally, is to do the following:

1. Write a job description, describing what has to be done, and the competence (knowledge, skills and experience) required.
2. Identify candidates for the job, often through advertising.
3. Find the best match between the job description and the candidates.

The person is matched to the job. That process does not work so well in the project-oriented organization, especially for the people working in temporary structures. Jobs cannot be defined with precision. You need to find people with competencies required to work on projects and let them define the jobs that need to be done in the circumstances. In project organizations much less formal recruiting practices are adopted. People are often initially recruited to work on an individual project, perhaps on a freelance or contract basis. Because of the large number of contract workers, it is easy to take someone on initially as a peripheral worker. Then if that person performs well, or is a good fit, the person can be offered permanent employment and even be placed on the project management career development track.

Maintaining Networks. To recruit people in this way, it is essential to maintain networks in the industry, with clients, competitors, and suppliers, and with universities and professional associations. We mentioned previously that some organizations maintain a project management community external to the organization, belonging, for instance, to a professional association or network. As well as providing development opportunities for existing project management personnel, they can also be a source for new personnel. Companies also often maintain strong links with universities, offering students temporary employment during the summer vacation, to see if they are a good fit, and offer them employment if they are.

Assessment Centers. Crawford (2003) describes the use of assessment centers for competence assessment and development. Assessment centers (Woodruffe, 1990) use a process lasting from two to five days, during which candidates are put into a simulated project environment to see how they perform. An assessment center consists of a standardized evaluation of behavior based on multiple inputs. Judgments are made from specifically developed assessment simulations. These judgments are pooled in discussions among the assessors, and the participants themselves have an opportunity to give and receive feedback on the instruments and measures used. The discussion results in evaluations of the performance of the candidates on the competencies or other variables that the assessment center is designed to measure. The effectiveness of assessment centers depends on the involvement of senior personnel from the project management community to observe the candidates and give informed feedback. Their dedication and willingness to be involved often depends on their being convinced that the center is as much a development opportunity for them as for the organization. Assessment centers are highly resource-intensive. It therefore makes sense to assess the threshold competencies of the majority of project personnel through other means and to reserve assessment centers for the assessment of the performance of candidates for promotion to senior project management positions. The exercises used in project management assessment centers have to be specific to reflect the typical processes of the project-oriented organization. Table 6.2 shows some examples.

Disposition

As project management personnel work on several projects and programs at the same time, the disposition of project management personnel to the different projects and programs is a critical issue in the project-oriented organization (Eskerod, 1998). Coordination and disposition through organizational unit, which is of permanent nature, is required. This function may be carried through the PM office. Disposition comprises the following:

- Allocation of PM personnel to projects and programs
- Optimization of allocation of resources in case of multiproject engagement
- Organization and support of the transition of PM personnel from one project to another

In many organizations disposition is closely linked to the coordination of the project portfolio. The project portfolio database is often linked to the management of personnel resources.

Development

The objective of personnel development is to improve the competence of project management personnel by offering the possibility of gaining knowledge and experience. Development activities are carried out either on the job, in a project or program assignment, or in general outside project assignments. Development can be limited to project management personnel employed or extended to include freelancers in the network. Responsibility for

TABLE 6.2. PROJECT-MANAGEMENT-RELATED EXERCISES IN A PROJECT MANAGEMENT ASSESSMENT CENTER.

AC Method	PM-Related Exercises
Presentation	Project start, project controlling, project closedown situation
One on one	Project controlling, feedback
Group discussion	Nearly all PM topics
Role-playing	Nearly all PM situations, e.g., meeting of project manager and project owner in a crisis situation
Analysis	Interpretation of a portfolio report, PM audit result interpretation, planning of the project start workshop

the development of individuals is generally taken on by the career committee and the project management community as described previously, although as we saw, the individual is also often required to take responsibility for his or her own development.

Methods adopted for the development of project management personnel include the following:

- Education and training
- PM competence assessment
- Assessment centers for development
- Coaching
- Feedback
- Training on the project
- Job rotation (within the project or program, between projects, and to other organizations)
- Support networks and communities
- Career development committees

Turner and Huemann (2000, 2001) describe education programs offered in several project-oriented societies globally. Project management training is often organized by the project office, in cooperation with the HR department, using either internal or external trainers. What often happens is that firms train their junior project management personnel internally. They are coached in the organization's ways of working. With middle- to senior-level personnel, levels 5, 6, and 7 in Table 6.1, training is conducted externally. There are several reasons for this:

- There are fewer people to train.
- They need general management skills in addition to project management skills.
- They can network in the industry on the courses.

The project management community also plays an essential role in developing individuals, providing coaching and mentoring, as well as the opportunity for people to network internally and develop personal competence through assimilation. Coaching and mentoring in the project management community is also part of leadership in the project context.

Project management competence assessments are either used in combination with the project management training or included in assessment centers (for development as well as for selection). Figure 6.3 shows some results of a project management competence assessment of project managers, which was used to get an overview on the status of the project management personnel in a company. Based on the single results, tailored activities for the further development of the project management knowledge and competence were taken.

Feedback is a method that is not very often used. Feedback methods are introduced to project managers in project management courses at the University of Economics and Business Administration Vienna. The course participants, who are experienced project managers, give feedback to each other after they have been working together on some training projects. The discussion that springs up from this is often that project managers are lacking feedback from the project owners as well as from project team members. A special form of

FIGURE 6.3. RESULTS OF A PROJECT MANAGEMENT COMPETENCE ASSESSMENT.

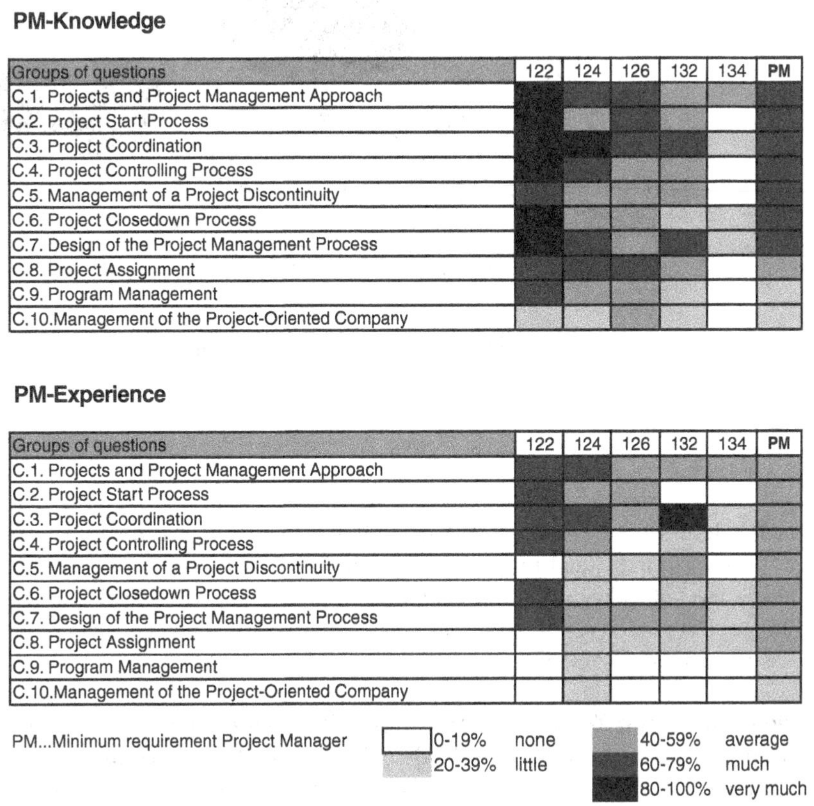

PM-Knowledge

Groups of questions	122	124	126	132	134	PM
C.1. Projects and Project Management Approach						
C.2. Project Start Process						
C.3. Project Coordination						
C.4. Project Controlling Process						
C.5. Management of a Project Discontinuity						
C.6. Project Closedown Process						
C.7. Design of the Project Management Process						
C.8. Project Assignment						
C.9. Program Management						
C.10.Management of the Project-Oriented Company						

PM-Experience

Groups of questions	122	124	126	132	134	PM
C.1. Projects and Project Management Approach						
C.2. Project Start Process						
C.3. Project Coordination						
C.4. Project Controlling Process						
C.5. Management of a Project Discontinuity						
C.6. Project Closedown Process						
C.7. Design of the Project Management Process						
C.8. Project Assignment						
C.9. Program Management						
C.10.Management of the Project-Oriented Company						

PM...Minimum requirement Project Manager

0-19%	none	40-59%	average
20-39%	little	60-79%	much
		80-100%	very much

feedback is the 360-degree review, shown in Figure 6.4. (Philips is one company that uses it for the project managers.) Here the project manager does a self-assessment based on a questionnaire. Other environments like project owner, project team members, suppliers, and customers give feedback to the project manager based on the same questionnaire.

Leadership

Leadership and team building are topics to which individual chapters are devoted elsewhere in this book (Thoms & Kerwin on leadership, and DeLisle on teams). However, they are both essential HRM processes, and for completeness, we discuss some core and new topics here.

In general, leadership is needed

FIGURE 6.4. CONCEPT OF THE 360-DEGREE FEEDBACK.

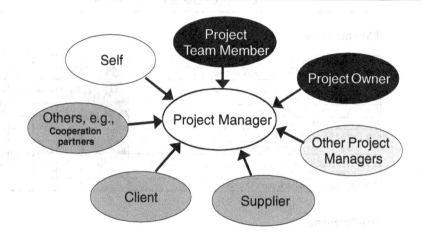

- at the level of the project-oriented company, with (inter alia) the company providing leadership to individual projects, programs, and project participants; and
- at the level of individual projects and programs, with project leaders (the project owner, the project manager, and the subteam leaders) providing leadership to other project participants (see Table 6.3).

Empowerment as a Key Value in the Project-Oriented Company. In high-performance organizations, people are *enabled* to do their best work. They have the adequate tools, standards, policies, and procedures. They are well trained and trusted. Empowerment is about goal setting, providing frameworks and limits within which subordinates can operate but

TABLE 6.3. LEADERSHIP IN THE PROJECT-ORIENTED COMPANY.

Level	Leader	Others
Project	Project sponsor Project manager Project subteam leader	Project manager Project team Project subteam Project team members
Program	Program sponsors Program manager	Project managers Program team Program team members
Project-oriented company	Portfolio group PM office Line manager	Program managers Project managers Project members PM office members

allowing freedom within those limits. Both van Fenema (2002) and Müller (2003) deal with the issue of empowerment and show that it leads to better project performance. It is needed by the following:

- The project itself
- The project team
- The single project manager
- The single project team member

Projects, as we have seen, are temporary organizations used to deliver change in organizations (Turner and Müller, 2003). It is a less efficient form of organization than the line organization, but it is more effective at delivering change, as it is more responsive and has lower inertia. However, if the change is to be achieved, the project must be removed from the line organization, and so empowerment of the project is essential. Empowerment here means to reduce the interventions of the line organization to a minimum and let the project work. The quality is ensured by providing adequate project management tools, standards, and guidelines.

Empowerment of the project creates the issue of the principal-agent relationship between the project sponsor and the project team, and so effective communication mechanisms must be put in place for empowerment to work (Müller, 2003). The project manager and project team must be made aware of the client's requirements, and the client needs to be made aware of progress. With effective communication, empowerment is possible; and with empowerment, effective project management is possible. An example of a symbolic act to empower the project team is to let them all sign the project charter. It is standard practice in some project-oriented companies that at the end of the project start workshop, the project owner and all the project team members (including the project manager) sign the project charter on a flip chart.

In researching communication between project sponsors and project managers, Müller (2003) found that high-performing projects were correlated with high collaboration between project managers and sponsors, and medium levels of structure. Collaboration was related to clearness of objectives and relational norms, and structure to clearness of work methods and "mechanicity". In Müller's sample, medium levels of structure and high levels of collaboration (empowerment) were necessary conditions for project success. Empowerment means that the project sponsor should set clear objectives, relational norms (high collaboration), and defined boundaries, but leave the project manager freedom to find the best solution within those boundaries (medium structure). Empowerment is not tight structure (no freedom), but equally it is not laissez-faire management (no objectives and boundaries).

Figure 6.5 illustrates three paths to falling collaboration. On failing projects, or projects with unclear objectives, tight structure and control tends to be adopted. Where there is remote working or infrequent reporting (van Fenema 2002), there is low collaboration with medium levels of structure. Where the clients and project managers objectives are misaligned, or reporting is informal, collaboration falls and anarchy reigns.

Empowerment of project managers includes clear agreements on the role and agreements on frequent communication structures and decision making with the project sponsor.

**FIGURE 6.5. COLLABORATION AND STRUCTURE ON HIGH
PERFORMING PROJECTS.**

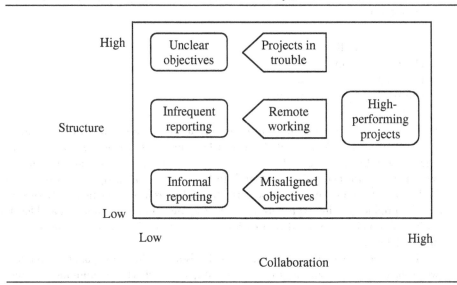

The role of the project manager should be described in relation to the role of the project sponsor. While the project sponsor has to take care of the company's interest and is responsible for strategic decisions and providing context information to the project, the project manager takes care of the project interest, is responsible for operative decisions and contributes to strategic decisions, and provides project information to the project sponsor.

Just as the project manager needs to be empowered by the project sponsor, so too does the project participant need to be empowered by the project manager, which leads us to the leadership role of the project manager.

Project Management Includes Project Leadership. Project management involves both people leadership and task management. Turner and Müller (2003) liken the role of the project manager to that of the chief executive of the temporary organization that is the project, and quote the classic text by Barnard (1938), who said the role of the chief executive is

> to formulate purposes, objectives and ends of the organization . . . This function of formulating grand purposes and providing for their redefinition is one which needs systems of communication, experience, imagination, interpretation and delegation of responsibility.

That sounds like the role of the project manager, who has to delegate, guide project team members, motivate, set goals, provide information, make decisions, and give feedback. But the project manager is not the only one taking on a leadership function. Within the project, leadership has to be provided by the project sponsor, the project subteam managers, and the project manager. The project organization chart in Figure 6.6 shows an empowered project organization. In the figure a differentiation is made between the leading of single individuals and the leading of teams.

Moderation functions have to be distinguished from leadership functions. While facilitation of workshops and meetings includes preparation of meeting, moderation of decision processes, structuring communication processes, and such is normally directionally neutral, while leadership sets interventions to steer into a direction. Recognizing this difference can allow the project manager to understand that in a workshop situations like project start-up, project crisis, and project closedown, he or she has to take over two roles: the role of the leader and the role of the neutral process facilitator. That might be difficult. Sometimes project and program managers support each other in such situations by bringing in someone else, for example, a manager from another project, to facilitate in the workshop. This helps the project manager to concentrate on the leadership function.

Leadership functions can be further described by looking at the subprocesses of project management and the subprocesses of team development

Leadership in the Project Start. Project start is the most important subprocess of project management. From a leadership point of view, during the project start the project team has to be established and the "big project picture" has to be commonly defined. In a project start workshop, not only do the project plans need to be created but the rules and values

FIGURE 6.6. LEADERSHIP IN THE EMPOWERED PROJECT ORGANIZATION.

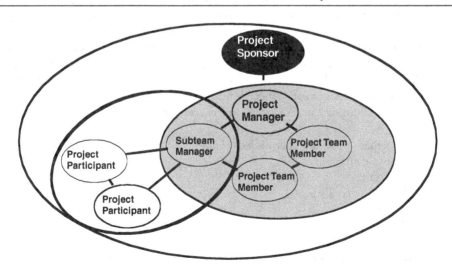

for working together have to be established and agreed on. Benchmarking of project management processes (Gareis and Huemann, 2003) suggested that traditional phases of team formation and maintenance (forming, storming, norming, performing) should be reconsidered, and instead the team actively work on the establishment of a project and team culture. That includes agreeing project rules and understanding roles in the project. Doing the forming and norming of the project team together in the project start situation reduces the storming and leads earlier to the performing of the project team. Further elements in establishing a project culture and identity are as follows:

- Project name
- Project motif and logo
- Newsletters

The name is important and agreeing on a project name might be quite sensitive as "nomen est omen." Guess what happened to a huge IT project by the name of Atlantis? It sank like Atlantis.

As part of the start-up process, it is important to build psychological attachment among project team members, especially for teams working remotely. Flying people to a central location for the start-up meeting may seem a great expense, but it can be invaluable for team formation. Once established, the team newsletter will help maintain psychological attachment.

Leadership in the Project Coordination and Project Controlling. In the project coordination, leadership functions of the project manager comprise delegating work packages, setting objectives, and giving feedback to project participants. As a communication structure, regular team meetings and meetings of single individuals are appropriate.

What leadership means in the project controlling process very much depends on the project management approach used. Project organization, project culture, team, and communication structures have to be questioned regarding whether they are still appropriate. If the project roles and rules are not adequate, they have to be adapted and agreed on within the team. Methods used here are reflection and feedback. Gareis and Huemann (2003) describe the project management competence of project team, which has to be built up and maintained during the project. A project team has to be able do the following:

- Together create the "big project picture."
- Create commitment.
- Use synergies.
- Managing conflicts.
- Learn.
- Together create the PM process.

One method to foster these competencies is to do have the team do a self-assessment of their team project management competence. An example question of such a self-assessment is given in Figure 6.7. Thus, one might go so far and say that the project team as such has a leadership function.

Leadership in the Project Closedown. While during project start the team is established, in project closedown the leadership function is to support the resolution of the project team, which includes making agreements for the final work, but it should also involve an emotional closedown by reflecting the process of working together and giving each other feedback. Very often, because of time pressure caused by another project waiting in the wings, this closedown is neglected.

Retention

Turner, Keegan, and Crawford (2003) reported that project managers tend to stay longer with one organization than other project participants. They feel their commitment to one firm as part of their career development. But commitment is a two-way street. Project management is a core competency to project-oriented organizations, and so they should make an effort to retain their project managers. One way to do that is make a commitment to their development as project managers, which means helping to manage their careers through the spiral staircase, giving them development opportunities as they arise. Sponsoring them through certification and master's programs demonstrates a clear commitment to their development. Building a psychological contract in this way can engender the commitment of individuals to the company.

Incentive systems and motivation are closely linked. Different incentive systems are possible in the project-oriented organization. We can look at incentives for project managers

FIGURE 6.7. SELF-ASSESSMENT OF THE PROJECT TEAM COMPETENCE.

Creation of "Big Project Picture" in Team		
1= none, 2= little , 3= average , 4= much, 5= very much	Knowledge	Experience
Common performance of workshops and meetings		
Use of project plans for communication		
Context orientation		
Holistic view		
Interpretation		

or for the whole project. Most incentives are monetary. Often there are difficulties in distributing the reward at the end of the project. The incentive system of the company is very much linked to the culture of the company. Again, size of budget may distort real priorities. Rewards linked to the budget of the project can lead to competition amongst the project managers to manage projects with the highest budget, which are often not the ones that are the most complex and the most strategically important. Creative incentive systems are more personalized. For example, after a very busy period in a project, a project manager gets as an incentive a week-end trip with his family. Recognition through little things can have a huge impact on motivation.

Release

Finally we consider release, which encompasses both release of the project and release of the temporary workers from the organization. There are two key elements of the release process, applicable to both: organizational learning and individual review and feedback. Organizational learning is covered in the chapters by Bredillet and by Morris. Feedback and review were discussed earlier in the chapter. With the release of freelance workers from the project and from the organization, it is also important to remain in contact to maintain the organization's network and to make future cooperation possible.

We have discussed several processes of human resource management in the project-oriented organization, including recruitment, disposition, development, leadership, retention, and release. These are summarized in Table 6.4.

The Role of the PM Office in HRM

So far we have discussed the practices and processes of human resource management in the project-oriented company. In this section we consider the role of the PM office as the unit that in cooperation with the central HR department is responsible for managing project management personnel. The PM office is a permanent function within the project-oriented company (Knutson, 2001; Rad and Levin, 2002).

An organization chart of a PM-office, which could be virtual, is shown in Figure 6.8.

The objectives of a PM-office, which in some companies is called the project management center of excellence, are to

- ensure a ready supply of professional project and program managers;
- provide management support to projects and programs, often by providing project managers and program managers to projects and programs;
- develop individual and organizational competencies in the project-oriented company; and
- manage project portfolios and related services, which will not be discussed further here.

Home for Project Management and Project Management Personnel

In many cases the PM office provides a pool of project and program managers. The PM office is therefore often seen as the home for the project management personnel. In addition,

TABLE 6.4. OVERVIEW: METHODS USED IN THE DIFFERENT HR SUBPROCESSES.

Recruiting	Disposition	Development	Leadership	Retention	Release
Assessment center	Portfolio database	Assessment center	Decisions	Incentive system	Documentation of learning
Informal section	Resource database	Feedback	Feedback	Reward system	Feedback
Liaison with universities	Reflection	Reflection	Reflection		
	Education and training	Information providing Delegation Empowerment			

FIGURE 6.8. ORGANIZATION CHART OF A PM OFFICE.

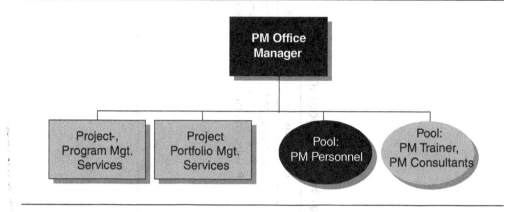

the PM office may provide HR services to the project management personnel, including the following:

- Provision of internal project management training and/or the organization of project management training with external training providers in accordance with the project management approach of the company
- Provision of coaching and mentoring of project management personnel, supporting the career committee
- Establishment and maintenance of the project management community as described previously
- Maintenance of the link with project management freelancers, by maintaining a database and inviting them to in-house networking activities and conferences organized by the project management community

Services to Empower Projects and Programs

As discussed, empowerment of projects and programs is one of the key values of the project-oriented company. To make empowerment possible and to ensure quality of the project management processes, the PM office provides and further develops the following:

- Project management guidelines and procedures, standard project plans for repetitive projects, and standard project management forms
- Project management infrastructure such as project management software software, project management portals, and collaboration platforms
- Management consulting services to projects and programs
- Management audits and (peer) reviews for projects and programs

Promoter of the PM Profession

The PM office acts as promoter of the project management profession within the company. In many project-oriented companies, the PM office is responsible for the following:

- The establishment of the project management career path
- The running of incentive and reward system suitable for project-oriented companies
- The holding of in-house PM conferences and support for project managers to attend project management conferences
- Cooperation with universities to have access to new theories and to well-educated project personnel for recruitment
- Cooperation with external project management communities and professional institutions to have access to best practices

Summary

In this chapter we showed human resourse processes and practices applied in the project-oriented organization. We argued that as individuals work more often in temporary organizations, such as projects and programs, and move from one project to the other like modern nomads. We identified specifics like the spiral staircase career and the no-home syndrome of project workers which leads to a specific view on human resource management in the project-oriented organization. We especially concentrated on project management personnel and defined the term and described the competencies which are developed through the individual's career.

We showed examples of career paths and how these are linked to certification offered by IPMA and PMI. We further discussed career committees and project management communities which support the establishment of project management as a profession.

We described specific human resource processes like recruitment disposition, development, leadership, retention and release, and the practices applied in these processes. Table 6.4 provides a compact overview on the methods used in different human resource processes in the project-oriented organization. We finally ended with the role of the PM office, which takes on human resouce management functions.

References and Further Reading

Association for Project Management (2000) *Body of Knowledge 4th ed.* www.apm.org.uk.

Barnard, C. I. 1938. *The functions of the executive.* Cambridge, MA: Harvard University Press.

Crawford, L. 2003. Assessing and developing the project management competence of individuals. In *People in project management*, ed. J. R. Turner., Aldershot, UK: Gower.

Drucker, P. F. 1994. *Post capitalist society.* New York: Harper Business.

Eskerod, P. 1998. The human resource allocation process when organizing by projects. In *Projects as arenas for renewal and learning processes*, ed. R. A. Lundin and C. Midler. Boston: Kluwer Academic Publishers.

Gareis, R. 2002. Project management for everybody: A visionary dimension of the project-oriented society. In *Proceedings of IRNOP V, the Fifth Biennial Conference of the International Research Network on Organizing by Projects*, ed. J. R. Turner. Renesse, Netherlands, May 29–31.

Gareis, R., and Huemann, M. 2001. Assessing and benchmarking project-oriented societies. *Project Management: International Project Management Journal, Finland* 7 (1, Summer): 14–25.

———. 2003. Project management competences in the project-oriented company. In *People in project management*, ed. J. R. Turner. Aldershot, UK: Gower.

Gedansky, L. 2002. Inspiring the direction of the profession. *Project Management Journal* 33(1).

Handy, C. B. 1988. *The Future of Work: a guide to changing society*. Oxford, UK: Blackwell.

———, 2002. *The Elephant and the flea: Reflections of a reluctant capitalist*. Cambridge, MA: Harvard Business Press.

Huemann, M. 2002. *Individuelle projektmanagement Kompetenzen in projektorientierten Unternehmen.*. Europäische Hochschulschriften. Frankfurt-am-Main: Peter Lang.

Keegan, A. E. 2002. Human resource management. In *Project management pathways*, ed. M. Stevens. High Wycombe, UK: Association for Project Management.

Keegan, A. E., and J. R. Turner. 2003. Managing human resources in the project-based organization. In *People in project management*, ed. J. R. Turner. Aldershot, UK: Gower.

Knutson, J. 2001. *Succeeding in project-driven organizations: People, processes and politics*. New York: Wiley.

Lundin, R. A., and Söderholm, A. 1998. Conceptualizing a projectified society: Discussion of an eco-institutional approach to a theory on temporary organizations. In *Projects as arenas for renewal and learning processes*, ed. R. A. Lundin and C. Midler. Boston: Kluwer Academic Publishers..

Morris, P. W. G. 1997. *The management of projects*, London: Thomas Telford.

Müller, R. (2003). *Communication of information technology sponsors and managers in a buyer-seller relationship*. DBA thesis, Henley Management College, Henley-on-Thames.

PMI, (2001). The PMI Project Management Fact Book, Secons Edition, Project Management Institute: Pennsylvania.

Rad, P. F., and G. Levin. 2002. *The advanced project management office: A comprehensive look at function and implementation*. Boca Raton, FL: St. Lucie Press.

Reid, A. 2003. Managing teams: The reality of life. In *People in project management*, ed. J. R. Turner. Aldershot, UK: Gower.

Schein, E. H. 1987. Increasing organizational effectiveness through better human resource planning and development. In *The art of human resources*, ed. E. H. Schein. Oxford, UK: Oxford University Press.

Turner, J. R. 1999. *The Handbook of Project Based Management*, 2nd edition, McGraw-Hill, London.

Turner, J. R., and M. Huemann. 2000. Current and future trends in the education of project managers. *Project Management: International Project Management Journal, Finland* 6 (1, Summer): 20–26.

———. 2001. The maturity of project management education in the project oriented society. *Project Management: International Project Management Journal, Finland* 7 (1, Summer): 7–13.

Turner, J. R., and R. Müller, R. 2003. On the nature of the project as a temporary organization. *International Journal of Project Management* 21(1).

Turner, J. R., A. E. Keegan, and L. Crawford. 2003. Delivering improved project management maturity through experiential learning. In *People in project management*, ed. J. R. Turner. Aldershot, UK: Gower.

van Fenema, P. C. 2002. *Coordination and control of globally distributed software projects*. PhD. thesis, Erasmus Research Institute of Management, Erasmus University Rotterdam. ISBN: 90-5892-030-5.

Woodruffe, C. 1990. *Assessment centers*. London: Institute of Personnel and Development.

CHAPTER SEVEN

COMPETENCIES: ORGANIZATIONAL AND PERSONAL

Andrew Gale

M orris (1999) argues that the "rapidly changing climate of management enablement" has three sources of change: organization, IT, and people. He said recently an important way in which the organizational context of managing projects has changed is a new focus on competence at both organizational and personal levels.

Competence

The simple meaning of the word "competence" is the ability to do something well or successfully. However, the concept of competencies, as known and understood today, developed because of dissatisfaction with the so-called intelligence tests used during the 1970s. The cause of dissatisfaction, as argued by McClelland (1973), was that the intelligence tests were discriminatory, and this not only favored certain ethnic and socioeconomic groups but also rendered the tests largely invalid. Based on his research, McClelland suggested the adoption of criterion sampling, where a sample job and required skills are tested against the performance on that job. This concept of testing performance forms the basis of the majority of competency approaches. Competence related to both occupation and social and interpersonal skills.

Boyatzis (1982, p. 21) defines competency as an "underlying characteristic of a person in that it may be a motive, trait, skill, aspect of one's self-image or social role or a body of knowledge which he or she uses." He developed this definition from that of Klemp (1980), who stated competency was "an underlying characteristic of a person, which results in effective and/or superior performance in a job." After Boyatzis, many definitions were proposed. Some have suggested that competency could be defined as the knowledge, skills, and qualities of effective managers, and pointed to the ability to perform effectively the functions associated with management in the work situation. Hogg (1993) states that competencies are the characteristics of a manager that lead to the demonstration of skills and abilities and result in effective performance within an occupational area.

Competence is linked then with individual behavior and job performance. Regarding the effective performance in a job, Boyatzis (1982, p. 2) states that the "effective performance of a job is the attainment of specific results (i.e. outcomes) required by the job through specific actions while maintaining or being consistent with policies, procedures and conditions of the organizational environment".

Figure 7.1 illustrates a model for effective job performance showing the relationship between competence of an individual, task requirements, and the environment.

Particular importance in this model goes to the competence of the individual, because each individual must demonstrate the ability or characteristics to perform specific actions for a particular job to produce the desired results. The desired results are considered as the requirements of the task, and the environment is the culture and tradition, physical, economic and technical resources available, plus organizational constraints. The likelihood of effective performance increases when two of these components are congruent. However, the effective performance will occur only when all three components fit in this model.

Since the capability of performing the job effectively may or may not be known to the individual, these capabilities can be characterized as conscious as well as unconscious aspects

FIGURE 7.1. EFFECTIVE JOB PERFORMANCE.

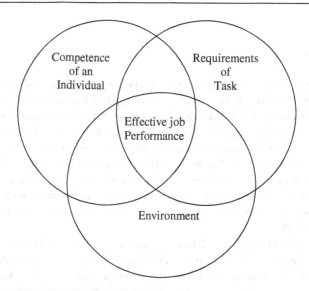

of the person (Boyatzis,1982). Describing the relationship between competence and effective performance, Boyatzis argues that to define competence, one must determine the required actions, their place in the system, and their sequence of behavior. One must also determine the results or effects and the margins or intent of the actions and results.

The definition of the concept of competence continues to cause confusion, and it may be that is it has become overdefined. Competence is a normative concept rather than a descriptive one. Competence can be delineated from performance in that competence is the ability to perform effectively in different ways in various contexts. Competence implies the integration of many aspects of practice and as such can be regarded as a psychological construct. However, performance may be measured for competence in relation to specific behaviors. Competence certainly includes the possession of particular knowledge and skills. Competence is concerned with the capacity to undertake specific types of action, and it can be considered a holistic concept involving the integration of: attitudes, skills, knowledge, performance, and quality of application.

Projects, their environments, and contexts require project managers and team members to be competent to address the challenges of their roles. Projects management bodies—for instance, Association for Project Management (APM), International Project Management Association (IPMA), Project Management Institute (PMI)—have for some time been concerned with the question of competencies, and through the development of bodies of knowledge (BOKs), they have begun to establish a frame of reference within which to relate the implications of competence so that they can be understood.

Learning Skills

Eraut (2002) argues that skills (or how things are done) must be viewed in the context of knowledge and learning, in which he distinguishes between three types of knowledge. These help to differentiate what is being learned and determine how learning occurs.

Eraut's three knowledge types are (1) codified, (2) cultural, and (3) personal. The BOKs are a form of codified knowledge. Cultural knowledge, argues Eraut (2002, p. 1), covers "much more than codified knowledge and introducing a totally different perspective. Its emphasis is on knowledge created, shared and used by groups of people working together, networking or socially interacting with each other." Finally, personal knowledge relates to the totality of a person's knowledge and background knowledge (Eraut, 2002).

Codified knowledge and cultural knowledge can often appear contradictory. Also, personal knowledge tends to be augmented by cultural knowledge. Therefore, there is much to consider in relation to the fragility of competence frameworks that do not take these issues into account.

Competence Frameworks

Whiddett and Hollyforde (2000) explain the structural properties of a useful competence framework. They say it should be clear and easy to understand, relevant, take account of expected changes, contain discrete elements (e.g., behavioral indicators) that do not overlap, and appear fair and balanced to all those affected by the use of the framework. Typically, a framework constitutes competency clusters that are broken down into different competencies with different levels. Behavioral indicators are stated for each competence listed by level (see Figure 7.2).

Frame (1999) has produced a major work on project management competence. He discussed the significance of competence and argued that in the case of project management it must be addressed at the level of (1) the individual, (2) the teams, and (3) the organization. He argues that there is a "competence dilemma" because of the conflict between the the-

FIGURE 7.2. TYPICAL COMPETENCE FRAME MODEL HIERARCHY.

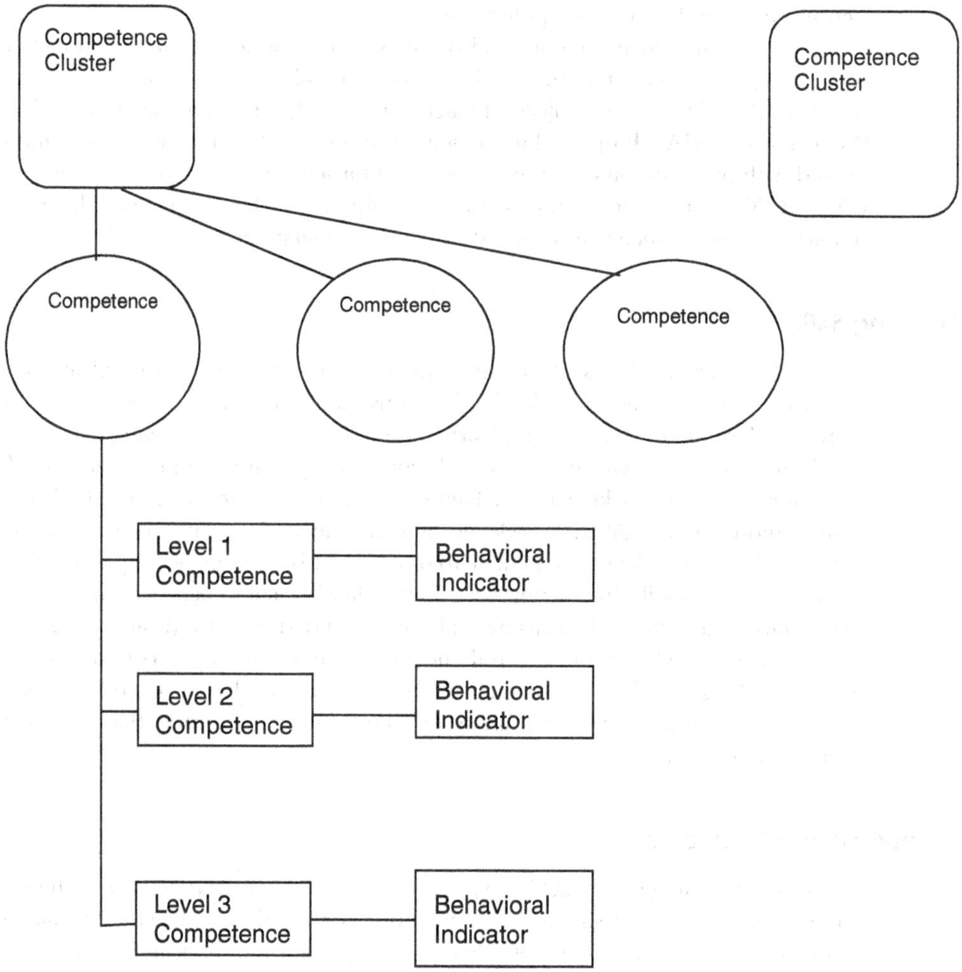

Source: Whiddett and Hollyforde, Figure 2:14.

oretical notion that all people are competent if given proper support and the working situation involving major differences in individual capabilities. He is also concerned with the microeconomic aspects of competence and the benefits brought to a team, project, and organization by a competent individual. Finally, he recognizes that an important connection must exist between a viable and healthy organization and the importance that should be attached to organizational culture and governance.

Some Recent Research

What Project Managers Do

Blackburn (2002) undertook a qualitative study into what project managers do and how project managers understand and talk about what they do. This study did not use a competence framework; rather, it focused on the network of relationships between project managers, along with environmental and technological factors. Actor-network theory (ANT) was used to interpret the project managers' stories about their work and project management techniques in the context of a variety of projects.

ANT is a school of sociological scientific knowledge in which networks constitute human actors, technological and natural agents, organisms, and human inventions. Project organizations can be viewed as generated in diverse, patterned networks but not structures with elements. The project network is more like a play with a script in which all the actors have their agendas and who seek to direct the play toward their own goals and interests. A dynamic, interactive process occurs in which both network and actors influence each other (Blackburn, 2002). The research found that project management processes enable project managers to interest and enroll team members and stakeholders and to mobilize the support of sponsors and other powerful players in a quasi-political process.

Blackburn (2002) argues there are two different prevailing schools of thought on the subject of project management: (1) methods, tools, and techniques, and (2) the heroic position, regarding the project manager as the key person responsible for bringing about project success. However, there is a third way in the form of ANT that is concerned with process not structures. Project management methodologies focus on "the right structures" and related processes linked with appropriate competencies on the part of project managers and project team members. Thus, these structures and processes form the agenda for the competence frameworks adopted and applied by project management experts. ANT is concerned with human and nonhuman actors and alliances used by people to make project organizations function. Rather than adding a "competent" project manager to a project team to ensure success, based on the prevalent perspectives on project management involving structures and processes, the ANT position is to understand project managers as competent through their effect on the actor-network of all the human and nonhuman elements that constitute the project.

Professional Competence. Cheetham and Chivers (1998) went in search of a comprehensive holistic model to address the concept of professional competence. Their research involved interviews with 80 practitioners in 20 professions. They drew on the theoretical position of the reflective practitioner, citing Schon (1983 and 1987) and challenging the techno-rational perspective on professional practice. They were concerned with what can be called "tacit knowledge" or "knowing in action," referred to also by Eraut (2002). Professional workers are able to draw on a set of solutions that fit the needs of a particular situation. Professional workers use their "artistry" to identify the need for a particular approach and make the judgment of fit, thus taking a professional approach to a situation all based on reflection— the crucial professional competence. This reflection can occur during an action or afterward and is the basis of continuous improvement. Cheetham and Chivers (1998) discuss the relationship between functional competence and personal competence. In the United States personal (or behavioral) competencies are a common focus for researchers, covering elements such as (1) confidence, (2) emotional control, and (3) interpersonal skills. Arguably, these personal competencies form a good basis upon which to predict a person's potential. However, there is no way of telling whether or not a person will be able to apply his or her personal competence to meet the demands of particular occupational requirements. Therefore, personal competence should be considered as complementary to functional competence.

Cheetham and Chivers (1998) also discuss the importance of competence in acquiring skills such as communication, problem solving, and critical analysis. These researchers further turned their attention to ethical considerations, arguing that ethics and values are usually ignored but should form an important element in professional competence. The Cheetham and Chivers (1998) model of professional competence argues that there is an over-arching set of meta-competencies:

- Problem solving
- Learning/self-development
- Mental agility
- Analysis
- Reflection

Under this are four interrelated components:

- Knowledge and cognitive competence
- Functional competence
- Personal and behavioral competence
- Values and ethics competence

The preceding all feed into professional competence, leading to outcomes that can be reflected upon in action and subsequent to the outcomes and feeding back into knowledge and cognitive competence, and values and ethics competence.

Organizational Competence

The literature reports some diverse recent findings with respect to organizational competence. Although none of these directly relate to project management, elements of these findings could be useful in the context of the project management environment, and project organization.

Stuart et al. (1995) reported the results of a research study on the importance and difficulties involved in translating existing competence frameworks into workable frameworks of specific contexts or organizations. While his work reports on small and medium enterprises (SMEs), it has general relevance to those interested in competence frameworks. A ten-step process was developed out of this research for translating a competence framework. The process enables the identification of individual team and organization development issues. From this management development, programs can be planned based on the competence-based management development initiatives. This is relevant to the interpretation and practical application of project management.

Research on organizational competence undertaken in Austria (Fischer and Schuch, 1994) identified four types of subcontractor organization. It was found that subcontractors unable to satisfy the demands of the client and who were poor in providing flexibility in production, just-in-time delivery, and quality control were unprofitable and ranked low in the hierarchy of subcontracting organizations. This has implications for the selection of suppliers and subcontractors, organization and specification of work packages, and supply chain integration in the context of complex projects.

Very (1993) reported a study in which the relationship between operational and strategic "relatedness" and the success or failure of seven large French industrial companies was investigated. No direct relationship was found between operational relatedness and success. However, the ability to exploit or strengthen competitive advantage through operational relatedness is linked with the success or failure of diversifications. These findings contribute to an understanding of the importance of "relatedness" in strategic diversification. This has a bearing on approaches to strategy.

Henderson and Cockburn (1994) discuss resource-based theory of the firm and focus attention on the role of heterogeneous organizational "competence" in competition. They seek to measure the importance of organizational competence. They distinguish between "component" and "architectural" competence by using data from ten firms, showing that the two forms of competence appear to explain variations in productivity across the firm. This has relevance to resource management in projects.

Sanchez (2002) proposes a five-mode competence-based management model. He argues that the whole field of competence has added benefits to modern management thinking. Confusion in the literature concerning the essential aspects of organizations' competence is attributable to the three reasons in the list that follows. Terminology is frequently differentially defined by different authors. The most obvious of these situations is the definition of competence itself. Different writers refer to different levels of activity within organizations, and there is often a tendency to consider competence as in some way static and unchanging in organizations. According to Sanchez (2002), different forms of competence arise from different levels of activity in organizations. He lists three levels of activity:

1. An organization's capability to create and produce products and services
2. The ability to organize, coordinate, and innovate in effective ways
3. The capability of senior management to imagine strategies for creating value in the market

The five organizational competence modes postulated by Sanchez (2002) are as follows:

1. *Cognitive flexibility to image alternative strategic logics.* The source of this mode is the "collective corporate imagination" of an organization's managers in perceiving feasible market opportunities in which the organization can create value.
2. *Cognitive flexibility to image alternative management processes.* This cognitive flexibility is concerned with managers' ability to conceive of alternative processes for implementing strategies.
3. *Coordinative flexibility to identify, configure, and deploy resources.* This depends on managers' ability to acquire or access, configure, and implement resources to achieve strategic goals.
4. *Resource flexibility to be used in alternative operations.* This relates to the organization's inherent level of flexibility. For example, it covers the extent to which an organization's resource can be described by a range of uses and the time frame within which this typically happens.
5. *Operating flexibility in applying skills and capabilities to available resources.* This relates to an organization's ability to apply resource flexibility to a range of operating conditions.

Sanchez argues that these five modes need careful interrelated management. Also, the type of competitive environment dictates the critical competence mode associated with success in that context. In stable competitive environments, Mode 5 is critical; for evolving environments, Modes 1 to 4 are critical, but led by Modes 3 and 4; and for dynamic environments, all five modes are critical, but led by Modes 1 and 2. Regarding success, Sanchez cites Bove et al., (2000): ". . . organizational competence does not depend simply on achieving excellence in one or two key success factors . . ."

The applicability of these five modes of competence to project management is easy to see. Some relate directly to key roles. This is particularly the case with respect to resource acquisition. Project managers, as change managers, need to be strategic as well as able to envisage and create linkage between many components in the project team and dynamic context.

Project Management Competencies, Standards, and Bodies of Knowledge

As practitioners of project management and the broader management community have increasingly focused on the importance of competence and standards, the leading professional bodies associated with project management have begun to address these issue of competence in a systematic manner. I have chosen, in spite of several national and regional competency models extant, to concentrate discussion around the two principal ones: in Europe and in North America.

European Bodies of Knowledge

Project management competencies have become the subject of much literature and debate. The International Project Management Association (IPMA) has produced an "International Competence Baseline" (IPMA, 1999). The baseline was produced from a study of the Association for Project Management (APM) UK Body of Knowledge (UK BoK), Swiss Assessment Structure (VZPM), German Projecktmanagement-Kanon (PM-ZERT), and the French assessment Criteria (AFITEP). The Baseline contains 42 elements on knowledge and experience (28 core and 14 additional) and a total of 18 attitudinal aspects.

The argument for improving the competence of employees is strong. The IPMA undertook a benchmarking study through all the members of the IPMA to produce the IPMA International Competency Baseline (ICB) and National Competency Baselines (NCBs) for each member country. These show the fields of project management qualification and competence. The IPMA defines competence as:

Knowledge + Experience + Personal Attitude

The IPMA classifies project management under four levels: A, B, C, and D. Knowledge and experience relate to function, and attitude relates to behaviors. Project management competence relate to the capability to manage projects professionally, by applying so-called best practices regarding the design of the project management process and the application of project management methods. Project management competencies require knowledge and experience in the subject, which enable objectives and deadlines to be met (Gareis and Huemann, 1999).

A. *Projects director.* Competent to "direct all projects of a company or branch or all projects of a program."
B. *Project Manager.* "able to manage complex projects". IPMA defines complex projects as having several interrelated subprojects and involving several companies or organizations across different disciplines and phases. These projects typically utilize many project management methods, techniques, and tools.
C. *Project management professional.* Defined as "able to manage noncomplex projects and/or assist the manager of a complex project in all elements and aspects of project management."
D. *Project management practitioner.* Said to "have project management knowledge in all elements and aspects."

Standards relating to project management competence fall into two main areas: Those relating to what project managers are expected to know and those relating to what project managers are expected to be able to do. The latter primarily takes the form of performance-based or occupational competence standards. For example, the principal competencies of a project manager in project management, as stated by Morris (1994), are as follows:

- Skills in project management methods and tools
- Team and people skills
- Basic business and management skills
- Knowledge of project sponsor role
- Knowledge and awareness of project environment
- Technical knowledge (specialized discipline skills)
- Integrative abilities of the preceding skills and knowledge.

Project management professionals working in projects where technical issues are important must have the competency to deal with them. Project managers must be able to recognize issues and be confident that appropriate action has been taken to deal with them. Professional project management competencies are achieved by the combination of education and the knowledge acquired during training, the skills developed through experience, and application of such acquired knowledge and experience. McCaffer et al. (2000, p. 113) lists the related field in which project manager should have a good working knowledge (see Figure 7.3).

To provide formal recognition that a project manager has reached a level of higher project management competence, the Association for Project Management, Body of Knowledge (1996) identified eight principal characteristics:

1. Attitude
2. Common sense
3. Open-mindedness
4. Adaptability
5. Inventiveness
6. Prudent risk taking
7. Fairness
8. Commitment

Key Project Management Competencies. The Association for Project Management Body of Knowledge (1996) identified 40 key competencies which are appropriate for the project management and are divided into four parts. The key competencies are as follows:

FIGURE 7.3. FIELDS IN WHICH PROJECT MANAGERS KNOWLEDGE IS CLASSIFIED.

Integration	Human resource	Quality
Time	Scope	Risk
Cost	Procurement	Communications

Source: McCaffer and Edum-Fotwe (2000, p. 113).

Part one: Project management

Part two: Organization and people

Part three: Process and procedures

Part four: General management

The preceding competencies are addressed fully by Crawford in her chapter.

UK Competence Standards. Occupational standards (e.g., UK National Vocational Quali-fications) are based on functional competence in which job-specific outcomes are recognized. It has been recognized that learning is a continuous process that is not limited only to classrooms. In the world of competition, in order to survive and achieve success, everyone needs to raise and maintain his or her skills. The requirement of good-quality qualifications that are recognized and valued by individuals and employers is increasing. As the demand for skills and knowledge increases, individuals and industries are bound to improve com-petitiveness and productivity.

The UK government created the work-related National Vocational Qualification (NVQ). It is flexible and widely recognized by industry as evidence of performance stan-dards, describing competence for a particular job (occupation). NVQs are classified from level one (routine and predictable activities) to level five (professional and managerial activ-ities).

The assumption behind the development and use of project management standards is that the standards describe the requirement for effective performance of project management in the workplace. This is a controversial and deterministic approach and may have only limited value in the field of project management. However, it is acknowledged that for functional aspects of project management application, the use of standards may have a very useful role to play.

American Body of Knowledge. The Project Management Institute (PMI) in the United States has developed a Body of Knowledge (PMBOK) that has become their basis for knowledge testing. The PMI has recently introduced "The Project Manager Competency Development Framework" (PMI, 2002) called the PMCD Framework. This document claims to clearly identify the interdependencies between job knowledge, skills, and behavior. The PMCD Framework is founded on the following sources: *A Guide to the Project Management Body of Knowledge* (PMBOK, 2000), *Project Management Experience and Knowledge Self-Assessment Manual,* and *Project Management Professional* (PMP Role Delineation Study). The PMCD takes a performance-based approach that presumes a causal relationship between skills, attitudes, and behaviors, and job performance (Crawford (1997). According to the PMCD Framework, there are three dimensions of project management competency:

1. Knowledge competence
2. Performance competence
3. Personal competency

These dimensions are broken down into units of competence at various levels, as is typical of the framework approach: clusters, element, and performance criteria (Figure 7.4).

The PMCD Framework dedicates a page of definition for each "unit of competence," incorporating examples of assessment guidelines. An example of such a unit of competence is "Project Cost Management" (PMI, 2002, pp. 30–31). Under this unit of competence are "competency clusters" containing processes (e.g. "Planning"). The "Planning" cluster contains three "elements," which are cross-referenced to the PMBOK Guide:

- Conduct Resources Planning (PMBOK Guide 7.1) with 12 performance criteria (e.g., "No.6. Develop resource histograms.")
- Conduct Cost Planning (PMBOK GUIDE 7.2) with ten performance criteria (e.g., "No. 5. Evaluate inputs to the cost baseline development process.")
- Conduct Cost Budgeting (PMBOK Guide 7.3) with 3 performance criteria (e.g., "No. 3. Develop a cost baseline to determine cost performance.")

Under "Examples of Assessment Guidelines" for "Knowledge Competencies" is listed "Demonstrate a knowledge and understanding of the tools and techniques utilized for planning of resources and the compilation of cost estimates and budgets," and for "Performance Competencies," "Demonstrate an ability to develop/use Cost Baseline."

The PMCD Framework states that there are five steps in the competence development methodology:

1. *Determination of applicable elements and performance criteria.* This step involves the individual or organization identifying elements and performance criteria contained in the PMCD Framework.
2. *Determination of the desired level of proficiency.* This step is concerned with determining the desired level of proficiency under each performance criterion section in the PMCD Framework.
3. *Assessment.* This stage is an assessment of the project managers to assess their strengths and weaknesses against the elements and performance criteria in order to determine any "gaps."

FIGURE 7.4. PMCD FRAMEWORK.

Project Management Competency		
Knowledge Competency	**Performance Competency**	**Personal Competency**
9 units	9 units	6 units
5 clusters per unit	5 clusters per unit	2–4 clusters per unit
Elements for each cluster	Elements for each cluster	Elements for each cluster
Performance criteria for elements	Performance criteria for elements	Performance criteria for elements

Source: PMI, (2002, p. 12).

4. *Dealing with gaps in competence identified.* Once gaps have been identified, these should be addressed to enhance performance in specific areas. Reassessment must occur after action to establish if more action is needed.

5. *Progression towards competence.* By progressively addressing gaps, individual project managers can achieve competence in each dimension included in the assessment.

The whole thrust of the American approach is performance-based with a strong component of personal competencies. It appears reductionist.

Measuring Competence

There are a number of indicators commonly used in the measurement of competence: continuous education, examinations, portfolios, self-assessment, interviews, outcome and performance measures, direct observations, and peer review (assessment). Performance management, a distinct and large subject, is inextricably related to the question of measurement of competence.

According to Gratton (1989), there are three broad techniques for measuring competence: the checklist approach, observational method, and framework approach. Checklists and observational methods are fairly low-level techniques and appear to be appropriately engaged in the assessment of performance. Performance underlies competence (Hager, 1993). The framework approach is likely to be more related to an integrated perspective of competence—a relatively holistic perspective.

The generic skills approach, while not having the large number of specific competencies of the behaviorist approach, but rather a smaller set of generic competencies, seems to lack a plausible rationale.

The integrated or task attribute approach (Hager, 1993) argues that competence may be inferred from performance. Typically a manageable number of key competencies are developed and used in a framework. Arguably the Body of Knowledge philosophy of the project management professional bodies such as the Association for Project Management and the Project Management Institute are rather reductionist, tending toward the atomistic end of the continuum.

Olney (1999), in a short paper entitled "Measuring Project Management Competence," acknowledges that interrelatedness is axiomatic of projects. She describes project management as an "integrative function." The processes required in the management of projects are complex, and coaching and mentoring are useful ways of improving outcomes. She argues that a core competency model is "individual statements—measurable and observable skill, behaviours traits as applied to best in class performers." This concept of "best in class" is related to the concept of so-called best practice. While experts can perhaps judge what constitutes "best practice," there is no escaping that "practice" has sociocultural determinants. There is then cultural bias woven into the fabric of the measurement process.

Many methods are adopted to measure competence. These include specific techniques such as so-called 360-degree feedback. This is relevant when dealing with the concept of competence, as the type of information included in 360-degree feedback can certainly be regarded as constituting the competence of an individual—namely, knowledge (of the job, organization and industry), skills (task proficiency and efficiency), and behaviors (energy and general approach).

The concept of 360-degrees is based on the practice of obtaining data from an individual's peers, direct reports, internal customers, and line manager. The technique usually involves using questionnaires (sometimes software-based) to obtain the evaluation of an individual from many sources.

Olney (1999) goes on to explain that an organization can assess the efficacy of project management coaching and professional development, in terms of improved performance, through the application of 360-degree assessment. The 360-degree technique is said to have the advantage of obtaining more objective, triangulated data, and the results are thus better received by the individual as being helpful in determining personal and professional development needs. In other words, through this technique, the individual is able to understand how others see him or her. Teams are said to benefit from increased communication among members and enhanced involvement in the development process. Company benefits are argued to be enhanced development of employees, increased internal promotion, inclusion of customers through training involvement, and a powerful driver for training and professional development.

A Case Study of Project Management Competence Measurement

Rolls-Royce plc uses an electronic questionnaire instrument as an aid to employees, their managers, and business units to assess competencies, knowledge, skills, and experience for project management roles. The questionnaire is designed to take a person no more than 20 minutes to complete. If there is discussion time with a manager, this will obviously take somewhat longer.

The questionnaire instrument is an aid to appraising a person's knowledge and experience in the topics that make up the Association for Project Management's (APM) Body of Knowledge (BoK). It is used in conjunction with a list of "Core Competencies, Knowledge and Experience for Programme Management Roles," which gives guidance on the level of knowledge and experience for each of the four job titles: A: Project Director, B: Project Manager, C: Project Management Professional, and D: Project Management Practitioner; based on IPMA certification levels. Figure 7.5 shows the program management core competencies identified by Rolls-Royce for Project Directors: Level A and for Project Managers: Level B.

Individuals and/or their managers complete a questionnaire and are given feedback on the level of knowledge and experience against topics that are considered to be central to the project management task. As well as recording an individual's level of knowledge and experience against each of the APM BoK topics, the levels of knowledge or experience can be averaged across the questionnaire and overall average of knowledge and experience can be reported across all topics. To enhance feedback, an individual's average at Level B can be broken down into B− (slightly above C), B, and B+ (almost an A).

This aid may be used by a manager in a 180-degree appraisal, by an individual in self-appraisal, or a combination of these. Details at the personal level remain confidential between a manager and/or individual and are not retained centrally. To provide useful

FIGURE 7.5. CORE COMPETENCIES OR PROJECT DIRECTORS AND PROJECT MANAGERS.

Project Director: Level A	Project Manager: Level B
Managing vision and purpose	
Business acumen	
Customer focus	Customer focus
Priority setting	Priority setting
Directing others	Directing others
Leading from the front	Leading from the front
Drive from the results	
Dealing with ambiguity	
Composure	
Comfort around higher management	Comfort around higher management
Negotiating	Negotiating
Building effective teams	Building effective teams
	Conflict management
	Timely decision making
	Motivating others
	Organizing

Source: Brown (2003).

information at a central level for training need analysis, business provide feedback to a central database.

Project complexity is difficult to define in terms applicable to all businesses. Nevertheless, project complexity is often referred to in the preceding aid when a person's experience is assessed. Following are some indicators of project complexity, on a scale of 1 (Low) to 5 (High):

1. Multidiscipline
2. Multidiscipline, multicompany
3. Multidiscipline, multicompany, multinational up to £30m in value
4. Multidiscipline, multicompany, multinational £30 to £75m in value, complex funding
5. Multidiscipline, multicompany, multinational £75m plus, complex funding, BOT/ BOOT or TCP included in scope

These five descriptions are only general indicators of project complexity, and each business unit considers what factors contribute to project complexity in its business area.

Return on Investment from Education and Training

There are some theoretical considerations relating to the content, delivery, and pedagogical strategy adopted for the program. These are all linked through a common theme of return on investment (ROI). Commercial organizations spend resources on educating and training their staff in order that the following intangible benefits occur: increased job satisfaction and increased organizational commitment. Brown (2003) prefers the concept of benefits metrics.

Some would argue that there is no valid and reliable relationship between indirect measures of the intangible benefits, mentioned previously, and the quantitative bottom-line indicators. There may be arguments based on anecdotal evidence, but these are not statistically valid associations. Rowe (1994) argues that it is not possible to make measurements that enable return on investment to be assessed or evaluated with respect to improved teamwork, improved customer service, reduced complaints, reduced conflict, or improved communication. These are measured indirectly using a number of methods. On the other hand, direct measures evaluating the success of an organization are quantitative and internationally understood (e.g., return on capital employed, or ROCE, profitability, turnover, and market share). Organizations invest in people through education and training to improve their bottom-line performance. Figure 7.6 makes linkages between investment in education and training and ROCE, relating competence, change, and the measurements necessary to close this loop.

Quantitative Approaches

Phillips (1997), on the other hand, writes convincingly on the relationship between competence and ROI. He is a devotee of developing algorithms for the calculation of real financial benefit in order to be compared with training, education, and development investment costs. He argues that organizations have moved from training for activity to training with a focus on bottom-line results. He insists that ROI methodologies must demonstrate simplicity, credibility, and soundness. The three most common measures are (1) cost-benefit ratio, (2) return on investment, and (3) payback period.

COST-BENEFIT RATIO

CBR = Program benefits/Program costs;
expressed as a ratio x:1.

RETURN ON INVESTMENT FORMULAE

ROI= (Net program benefit/Program cost) × 100;
expressed as a percentage.

PAYBACK PERIOD

PP = Total investment/Amount saved;
expressed as x years.

FIGURE 7.6. RETURN ON INVESTMENT CYCLE.

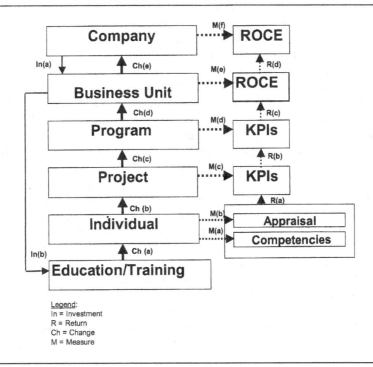

Legend:
In = Investment
R = Return
Ch = Change
M = Measure

Other methods are (1) discounted cash flow (DCF), (2) internal rate of return (IRR), and (3) utility analysis.

UTILITY ANALYSIS

$$DU = T \times N \times dt \times Sd - N \times C$$

Where: DU = Monetary value of the program
T = Duration of the program
N = Number of employees on the program
dt = True difference in performance: trained cf. untrained
Sd = Standard deviation in performance of untrained
C = Cost of training per employee

Phillips acknowledges the importance and difficulty of quantifying benefits of training and development programs. There is no doubt that some types of program lend themselves to this quantitative approach. However, it is very difficult to see how project management development programs with complex inputs and slow-to-develop intangible benefits can be fitted into this style of approach.

When one considers the emerging body of research and debate on competence in project management, the strongest argument for an association between measurable bottom-line indicators and intangible benefits from the activities of education, training, management development programs, and continuing professional development may have to do with competence (Humphreys, 2001; Rowe, 1994; Seppänen, 2002; Skulmoski, 2000; Hartman 1999).

Rowe (1994) reported the experience of BAE SYSTEMS (formerly British Aerospace) in evaluation and effectiveness of open training. Trainees were asked to keep learning logbooks to record their experiences. This included how they learned, insights gained, benefits obtained, and other relevant matters. The reality was that very few people actually made entries in their logbooks. Regarding measurement of learning experience through end-of-course questionnaires; Rowe (1994) states that ". . . issuing assessment questionnaires at the end of a course neither measures nor evaluates a training event; it simply monitors it. But it does not tell us how effective the training was in terms of meeting the business needs."

Evidence for return on investment and individual learning gain remain unrecorded. Rowe points out that in his opinion it is not possible to effectively link the effects of training to improvements in organizational performance.

Rowe (1994) stresses that there is a contradiction between what he argues to be the two fundamental goals of training for an organization. One is the need to develop and maintain a *learning organization*, and the other is to *meet business needs*. The former is associated with the development and encouragement of open-ended stimulation, while the latter is concerned with things that we need to know, involving a relatively convergent perspective.

Summary

There seems little point in yet more adventures intended to define competence as such. It seems far more worthwhile addressing the linkage between learning and acquisition of knowledge and experience. An important issue to consider is the fact that the field of organizational competence continues to grow in importance. The relationship between project objectives and organizational objectives represents another area in this field of research. Cultural and diversity dimensions have been neglected by researchers and should be a major consideration for the future. Western values underpin the bodies of knowledge, and their use in developing competence frameworks in many parts of the world may be inappropriate because of differences in cultural contexts.

The field of return on investment or benefits metrics is certainly an important area to research. The focus should be on the relationships depicted in Figure 7.6. However, a serious effort is needed to move forward on the qualitative front and to get away from the evermore statistically based quantitative obsessions so often associated with this subject. The long-term view should be developed in which continuing professional development and communities of practice are identified as both outcomes in terms of professional competence development and sources of rich qualitative data for return on investment research.

Websites

Individuals

www.hrscope.com/project_management_competencies.htm

www.pmforum.org/pmwt01/duncomp.htm

www.pmforum.org/library/papers/cbwhitepaper.htm

Books and Publications

www.pm-prepare.com/BIBLIOGRAPHY.htm

www.cbponline.com/bookstore/project_management.htm

www.majorprojects.org/cgi-bin/pub_cont.cgi?range=az

Organizations

www.apm.org.uk/Default.htm

www.ipma.ch/index.htm

www.pmi.org/info/default.asp

www.aipm.com.au/html/

www.birminghamnow.com

References and Bibliography

Alic, J. A. 1995. Organizational competence: Know-how and skills in economic development. *Technology in Society* 17(4):429–436.

Andersen, P. H., N. Cook, and J. Marceauet. Forthcoming. Dynamic innovation strategies and stable networks in the construction industry: Implanting solar energy projects in the Sydney Olympic Village. *Journal of Business Research.*

Association for Project Management, 1996. *Body of knowledge*, 3rd ed. High Wycombe, UK: Association for Project Management.

Atkinson, S. 1999. Reflections: Personal development for managers: Getting the process right. *Journal of Managerial Psychology* 14(6):502.

Australian Institute of Project Management. 1996. *National competency standards for project management.* 1st approved ed. Sydney: Australian Institute of Project Management (www.aipm.com.au/html/ncspm.cfm).

Barnes, J. K. 1999. UK education's support for its aerospace industry. *Aircraft Engineering and Aerospace Technology* 71(2):136–142.

Baxendale, T., and O. Jones. 2000. Construction design and management safety regulations in practice: Progress on implementation. *International Journal of Project Management* 18(1):33–40.

Bergenhenegouwen, G. J. 1996. Competence development: A challenge for HRM professionals: Core competences of organizations as guidelines for the development of employees. *Journal of European Industrial Training* 20(9):29.

Birchall, D. 1993. Case study B: Senior managers and competence. *Management Development Review* 6(3): 13.

Binnersley, S., and C. Rowe. 1992. British Aerospace takes off with higher education in management development. *Executive Development* 5(1):10–13.

Blackburn, S. 2002. The project manager and the project-network. *International Journal of Project Management* 20(3):199–204.

Boam, R., and P. Sparrow. 1992. *Designing and achieving competency: A competency-based approach to developing people and organization.* London: McGraw-Hill.

Bove, K., H. Harmsen., and K. G. Grunert. 2000. The link between competencies and company success in Danish manufacturing companies. VoI. 6 in *Advances in applied business strategy: Implementing competence-based strategies,* ed. R. Sanchez and A. Heene. 287–312. Greenwich, CT: Jai Press.

Boyatzis, R. E. 1982. *The competent manager: A model for effective performance.* New York: Wiley.

———. 1993. Beyond competence: The choice to be a leader. *Human Resource Management Review* 3(1): 1–14.

Brown, M. R. 2003. Rolls-Royce plc. Personal communications.

Burnes, B. 1996. No such thing as . . . a "one best way" to manage organizational change." *Management Decision* 34(10):11–18.

Cannon, M. D., and A. C. Edmondson. 2001. Confronting failure: Antecedents and consequences of shared beliefs about failure in organizational work groups. *Journal of Organizational Behavior* 22(2): 161–177.

Cassells, E. 1999. Building a learning organization in the offshore oil industry. *Long Range Planning* 32(2):245–252.

Chandler, G. N., and E. Jansen. 1992. The founder's self-assessed competence and venture performance. *Journal of Business Venturing* 7(3):223–236.

Chastain, T., and A. Elliott. 2000. Cultivating design competence: Online support for beginning design studio. *Automation in Construction* 9(1):83–91.

Chaston, I., B. Badger, et al. 1999. Organizational learning: research issues and application in SME sector firms. *International Journal of Entrepreneurial Behaviour and Research* 5(4):191–203.

Chaston, I., and T. Mangles. 2000. Business networks: Assisting knowledge management and competence acquisition within UK manufacturing firms. *Journal of Small Business and Enterprise Development* 7(2):160–170.

Cheetham, G., and G. Chivers. 1998. The reflective (and competent) practitioner: A model of professional competence which seeks to harmonise the reflective practitioner and competence-based approaches. *Journal of European Industrial Training* 22(7):267–276.

Cleland, D. I. 1999. *Project management: Strategic design and implementation.* 3rd ed. New York: McGraw-Hill.

Crawford, L. H. 1997. A global approach to project management competence. *Proceedings of the 1997 AIPM National Conference,* 220–228. Gold Coast, Brisbane: AIPM.

———. 2000. Profiling the competent project manager. *Project Management Research at the Turn of the Millennium: Proceedings of PMI Research Conference.* 3–15. Newtown Square, PA: Project Management Institute.

CRMP 1999. *Guide to the project management, Body of knowledge.* Manchester: CRMP, UMIST.

Curtis, B., W. E. Hefley, et al. 1997. Developing organizational competence. *Computer* 30(3):122–124.

Dunphy, D., D. Turner, et al. 1997. Organizational learning as the creation of corporate competencies. *The Journal of Management Development* 16(4):232–244.

Eliasson, G. 1996. Spillovers, integrated production and the theory of the firm. *Journal of Evolutionary Economics* 6(2):125–140.

Ellström, P. E. 2001. Integrating learning and work: Problems and prospects. *Human Resource Development Quarterly* 12(4):421–435.

———. 2002. *Learning challenges for knowledge-based organizations.* Sussex: University of Sussex Institute of Education.

Fairtlough, G. 1994. Organizing for innovation: Compartments, competences, and networks. *Long Range Planning* 27(3):88–97.

Fischer, M. M., and K. Schuch 1994. Technological and organizational competence of Austrian subcontractors: An empirical study based on the machine construction, steel, electrical and electronic industries. *Mitteilungen Der Osterreichischen Geographischen Gesellschaft* 136:179–202.

Frame, J. D. 1999. Project management competence: Building key skills for individuals, teams, and organizations. San Francisco: Jossey-Bass.

Fraser, C. 1999. A non-results-based effectiveness index for construction site managers. *Construction Management and Economics* 17(6):789–798.

———. 2000. The influence of personal characteristics on effectiveness of construction site managers. *Construction Management and Economics* 18(1):29–36.

Frey, R. S. 2001. Knowledge management, proposal development, and small businesses. *The Journal of Management Development* 20(1):38–54.

Gareis, R., and M. Huemman. 1999. Specific competencies in the project-oriented society. Project Management Days '99: Projects and Competencies, Vienna: Austria, November 18–19.

General, M., and S. G. Genega 1997. Leadership: Essential to managing success. *Journal of Management in Engineering* 13(4):22–23.

Godbout, A. J. 2000. Managing core competencies: The impact of knowledge management on human resources practices in leading-edge organizations. *Knowledge and Process Management* 7(2):76–86.

Gongal, K. 2000. An investigation into the relationship between project management competencies and project success. MS, diss, UMIST, Manchester.

Gottshall, W. L. 2000. Competence or collapse. *Civil Engineering—ASCE* 70(10):74–75.

Gratton, L 1998. Work of the manager. In *Assessment and selection of organizations: Methods and practices for recruiting and appraisal,* ed. P. Herriot, 511–528. New York: Wiley.

Gray, C. J. 1998. A strategic investment in training and development by the UK steel construction industry. *Journal of Constructional Steel Research* 46(1–3):281.

Gronhaug, K., and O. Nordhaug 1992. Strategy and competence in firms. *European Management Journal* 10(4):438–444.

Hager, P. 1993. Conceptions of competence. *Proceedings of the Forty-Ninth Annual Meeting of the Philosophy of Education Society,* University of Technology Sydney, Philosophy of Education, www.ed.uiuc.edu/EPS/PES-yearbook/93_docs/HAGER.HTM.

Hartman, F, and G. Skulmoski, G. 1999. Quest for team competence. *Project Management* 5(1):10–15.

Haydock, W. 1995. Management development: A personal competency approach. *Training & Management Development Methods* 9(4):7–13.

Henderson, R., and I. Cockburn. 1994. Measuring competence: Exploring firm effects in pharmaceutical research. *Strategic Management Journal* 15:63–84.

Henriksen, L. B. 2001. Knowledge management and engineering practices: The case of knowledge management, problem solving and engineering practices. *Technovation* 21(9):595–603.

Hodgson, G. M. 1998. Competence and contract in the theory of the firm. *Journal of Economic Behavior & Organization* 35(2):179–201.

Hogg, B. A 1993. European managerial competences. *European Business Review* 93(2):21–26.

Holmberg, S. C. 2001. Systemic research on competence and competence development in SMEs. *Systems Research and Behavioral Science* 18(2):101–102.

Humphreys, P. 2001. Designing a management development programme for procurement executives. *Journal of Management Development* 20(7):604–623.

IPMA. 1999. www.ipma.ch/document /+CB20DL.pdf (accessed March 18, 2002).

Jang, Y., and J. Lee 1998. Factors influencing the success of management consulting projects. *International Journal of Project Management* 16(2):67–72.

Javidan, M. 1998. *Core competence: What does it mean in practice?* Long Range Planning 31(1):60–71.

Johnston, N. M. 1991. How to create a competitive workforce. *Industrial and Commercial Training* 23(2): 4–7.

Jones, N., and N. Fear. 1994. Continuing professional development: Perspectives from human resource professionals. *Personnel Review* 23(8):49–60.

Jones, N., and G. Robinson. 1997. Do organizations manage continuing professional development. *Journal of Management Development* 16(3):197–207.

Jurie, J. D. 2000. Building capacity: Organizational competence and critical theory. *Journal of Organizational Change Management.* 13(3):264–274.

Kersten, A. 2000. Diversity management. *Journal of Organizational Change Management* 13(3):235–248.

King, A. W., and C. P. Zeithaml. 2001. Competencies and firm performance: Examining the causal ambiguity paradox. *Strategic Management Journal* 22(1):75–99.

Klemp, G.O. 1980. *The assessment of occupational competence.* Report to the National Institute of Education, Washington, D.C.

Kræmmergaard, P., and J. Rose 2002. Managerial competences for ERP Journeys. *Information Systems Frontiers* 4(2):199–211.

Lampel, J. 2001. The core competencies of effective project execution: The challenge of diversity. *International Journal of Project Management* 19:471–483.

Larsen, H. H. 1997. Do high-flyer programmes facilitate organizational learning? *Journal of Managerial Psychology* 12(1):48–59.

———. 1997. Do high-flyer programmes facilitate organizational learning? From individual skills building to development of organizational competence. *Journal of European Industrial Training.* 21(9):310–317.

Lewis, D. 1998. Competence-based management and corporate culture: Two theories with common flaws? *Long Range Planning* 31(6):937–943.

Lewis, M. A. 2001. Success, failure and organizational competence: A case study of the new product development process. Journal of Engineering and Technology Management 18(2):185–206.

Lindsay, P. R., and R. Stuart 1997. Reconstructing competence. *Journal of European Industrial Training* 21(9):326–332.

Löfstedt, U. 2001. Competence development and learning organizations: a critical analysis of practical guidelines and methods. *Systems Research and Behavioural Science* 18(2):115–125.

Lloyd, C., and A. Cook. 1993. *Implementing standards of competence: Practical strategies for industry.* London: Kogan Page.

Lysaght, R. M., and J. W. Altschuld. 2000. Beyond initial certification: The assessment and maintenance of competency in professions. *Evaluation and Program Planning* 23(1):95–104.

Maister, D. 1982. Balancing the professional service firm. *Sloan Business Review.* 24(1):15–29.

Margerison, C. 2001. Team competencies. *Team Performance Management* 7(7/8):117–122.

Martin, G., and H. Staines. 1994. Managerial competences in small firms. *The Journal of Management Development* 13(7):23.

Mathiassen, L., and P. A. Nielsen. 1990. *Surfacing organizational competence. Soft systems and hard Contradictions.* Bjerknes, G. (Ed) Organizational competence in system development. Lund: Studentlitteratur.

McCaffer, R., and F. T. Edum-Fotwe. 2000. Development project management competency: Perspectives from the construction industry. *International Journal of Project Management* 18(2):111–124.

McCain, B. 1996. Multicultural team learning: An approach towards communication competency. *Management Decision* 34(6):65–68.

McClelland, D. C. 1973. Testing for competence rather than "intelligence." *American Psychologist* 28(1): 1–14.

McCreery, J. K. Forthcoming. Assessing the value of a project management simulation training exercise. Uncorrected proof. *International Journal of Project Management.*

McGrath, R. G., I. C. MacMillan, and S. Venkataraman. 1995. Defining and developing competence: A strategic process paradigm. *Strategic Management Journal* 16(4):251–275.

Merali, Y. 2000. Individual and collective congruence in the knowledge management process. *Journal of Strategic Information Systems.* 9(2–3):213–234.

Meyer, T., and P. Semark. 1996. A framework for the use of competencies for achieving competitive advantage. *South African Journal of Business Management* 17(4):96–103.

Morden, T. 1997a. Leadership as competence. *Management Decision.* 35(7):519–526.

———. 1997b. Leadership as vision. *Management Decision* 35(9):668–676.

Morris, P. W. G. 1994. The management of projects. London: Thomas Telford.

———. 1999. Project management in the twenty-first century: Trends across the millenium. IPMA/ SOVNET International Project Management Congress, June 28.

———. 2001. "Updating the Project Management Bodies of Knowledge", *Project Management Journal.* 32(3):21-30.

Mothe, C., and B. Quelin 2000. Creating competencies through collaboration: The case of EUREKA R&D consortia. *European Management Journal* 18(6):590–604.

Mulder, L. 1997. The importance of a common project management method in the corporate environment. *R&D Management* 27(3):189–196.

Murphy, L. 1995. A qualitative approach to researching management competences. *Executive Development* 8(6):32–34.

Murphy, S. E. 1988. Organizational competence: It depends on your staff. *Training and Development Journal* 42(1):34–36.

National Vocational Qualification. www.dfee.gov.uk/nvq/what.html.

The APM Body of Knowledge: The 40 key competencies. www.apmgroup.co.uk/the 40 key.html (accessed March 2002).

Nair, K. U. 2001. Adaptation to creation: progress of organizational learning and increasing complexity of learning systems. *Systems Research and Behavioral Science* 18(6):505–521.

Newman, V. 1997. Redefining knowledge management to deliver competitive advantage. *Journal of Knowledge Management* 1(2):123–128.

Nkado, R, and T. Meyer. 2001. Competencies of professional quantity surveyors: A South African perspective. *Construction Management and Economics* 19:481–491.

Olney, J. 1999. Measuring project manager competence. *PM Network.* (October).

Österlund, J. 2001. The forgotten revenue of product development: learning new competence. *Systems Research and Behavioral Science* 18(2):159–170.

Otala, L. 1994. Industry-university partnership: Implementing lifelong learning. *Journal of European Industrial Training* 18(8):13–18.

Overmeer, W. 1997. Business integration in a learning organization: The role of management development. *The Journal of Management Development* 16(4):245–261.

Owen, K., R. Mundy, W. Guild, and R. Guild. 2001. Creating and sustaining the high performance organization. *Managing Service Quality* 11(1):10–21.

Phillips, J. J. 1997. *Return on investment in training and performance improvement programs: Improving human performance series.* Houston: Gulf Publishing Co.

Project Management Association 2000. *Project Management Institute Body of Knowledge.* Newtown Square, PA: Project Management Institute.

———. 2002. *Project manager competency development framework,* Newtown Square, PA: Project Management Institute

Quality Assurance Agency for Higher Education, *Handbook for academic review,* Annex D, www.qaa. org.uk (accessed March 2002).

Raynaud, D. 2001. Competences et expertise professionnelle de l'architecte dans le travail de conception: The competence and expertise of architects during the phase of design. *Sociologie du Travail* 43(4):451–469.

Ritter, T., and H. G. Gemunden 2004. The impact of a company's business strategy on its technological competence, network competence and innovation success. *Journal of Business Research* 57(5): 548–556.

Rolls-Royce 2000. Personal communications with Change Management Department.

Rowe, C. 1994. Assessing the effectiveness of open learning: The British Aerospace experience. *Industrial and Commercial Training* 26(4):22–27.

Seppanen, V. 2002. Evolution of competence in software subcontracting projects. *International Journal of Project Management* 20:155–164.

Sanchez, R. 2001. *Knowledge management and organizational competence.* Oxford: Oxford University Press.

———. 2002. Understanding competence-based management; Identifying and managing five modes of competence. *Journal of Business Research* 57(5):518–532.

Schön, D. 1983. *The reflective practitioner: How professionals think in action.* London: Maraca Temple Smith.

———. 1987. *Educating the effective practitioner.* San Fransisco: Jossey-Bass.

Seppänen, V. 2002. Evolution of competence in software subcontracting projects. *International Journal of Project Management* 20(2):155–164.

Simkoko, E. 1992. Managing international construction projects for competence development within local firms. *International Journal of Project Management* 10(1):12–22.

Skulmoski, G., F. Hartman, and R. DeMaere. 2000. Superior and threshold project competencies. *Project Management* 6(1):10–15.

Stuart, R., J. E. Thompson, and J. Harrison. 1995. Translation: From generalizable to organization-specific competence frameworks. *The Journal of Management Development* 14(1):67–80.

Stuart, R., and P. Lindsay 1997. Beyond the frame of management competenc(i)es: Towards a contextually embedded framework of managerial competence in organizations. *Journal of European Industrial Training* 21(1):26–33.

Sundberg, L. 2001. A holistic approach to competence development. *Systems Research and Behavioral Science* 18(2):103–114.

Tovey, L. 1993. Competency assessment: A strategic approach—Part I. *Executive Development* 6(5):26–28.

———. 1994. Competency assessment : A strategic approach—Part II. *Executive Development* 7(1):16–19.

Very, P. 1993. Success in diversification: Building on core competences. Long *Range Planning* 26(5):80–92.

von Zedtwitz, M. 2002. Organizational learning through post-project reviews in R&D. *R&D Management* 32(3):255–268.

Wang, C. K. 2001. Organizational competence analysis: Experience of a Japanese multinational competitive intelligence review 12(3):3–9.

Whiddett, S., and S. Hollyforde. 2000. *The competencies handbook*. London: CIPD.

Williamson, O. E. 1999. Strategy research: governance and competence perspectives. *Strategic Management Journal* 20(12):1087–1108.

Yan, Y., T. Kuphal, and J. Bode. 2000. Application of multiagent systems in project management. *International Journal of Production Economics* 68(2):185–197.

CHAPTER EIGHT

PROJECTS: LEARNING AT THE EDGE
OF ORGANIZATION

Christophe N. Bredillet

For the past 40 years project management has become a well-accepted way to manage organizations. The field of project management has evolved from operational research techniques and tools to a discipline of management (Cleland, 1994; Bredillet, 1999).

Introduction: Some Conceptual Issues to Knowledge and Learning in Project Management

Management of/by Projects for Implementing Strategy

Many authors emphasize this evolution in the way of managing projects. Referring to his book *The Management of Projects*, Morris writes, "This book traces the development of the discipline of project management" (Morris, 1997). Project management becomes the way to implement corporate strategy (Turner, 1993; Frame, 1994) and to manage a company; ". . . value is added by systematically implementing new projects—projects of all types, across the organization" (Dinsmore, 1999, p. ix). *Management of Projects*, which discusses the way to manage projects within the same organization (Morris, 1997), and *Management by Projects*, which describes projects as a way to organize the whole organization (Gareis, 1990; Dinsmore, 1999), are both good examples of that tendency. Projects are a form of organization that positions a company in relation to its environment. As projects are the vectors of the strategy (Grundy, 1998), project management is a way to deal with the characteristics of the whole environment: complexity (Arcade, 1998), change (Voropajev, 1998), globalization, time, and competitiveness (Hauc, 1998). Thus, with the help of project management, strategic management becomes really the management of irreversibility (Declerck et al.,

1997), concentrating on the ecosystem's project/organization/context, operation/organization/context and their integrative management (Declerck et al., 1983).

Competencies, Sources of Competitive Advantage and the Creation of Value

Projects, as strategic processes, modify the conditions of the firm in its environment. Through them, resources and competencies are mobilized to create competitive advantage and other sources of value. As resources are easily shared by many organizations, the organization's competencies are the most important relevant driver. Thus, through the organization's processes or projects, past action is actualized as experience; present action reveals and proves competencies; future action generates and tries out new competencies (Lorino and Tarondeau, 1998). Competencies (both individual and organizational) are at the source of competitive advantage and the creation of value.

The Link with Performance

Recent research is being done on the assumption that the more competent the project managers, teams, or organizations (maturity), the more efficiently they will perform, the more effective will be the performance of the projects, and the more successful will be the organization (Crawford, 1998; PMI Project Manager Competency Development Framework, 2002). Such research, and indeed the development of professional certification programs in general, seems to contradict former findings. For example, Pinto and Prescott (1988) concluded that the "personnel factor," even if designated in theoretical literature as a crucial factor in project efficiency, is a marginal variable for project success at any of the four project life cycle phases considered (for a criticism of their findings, see Belout, 1998). A working paper (Turner, 1998) shows the influence of the project managers' competencies on value of shares of a company. But performance also comes from the maturity of the organization's ability to deal with projects. And in respect of maturity, learning is especially significant. The OPM3 research program (PMI Standards Committee) and others papers (for example, Remy, 1997; Saures, 1998; and Fincher et al., 1997) explore the relations between maturity of the organizations and success of the projects.

Knowledge and Competence

To develop competencies, an individual needs knowledge. Two main views of competence development may be considered. One traditional view is that it involves applying a body of knowledge to known situations in order to produce rational solutions to problems (what I call the "have" or "quantitative" perspective). However, in a rapidly changing world and information-based society, practitioners and organizations increasingly need to respond intelligently to unknown situations and go beyond established knowledge to create unique interpretations and outcomes (Schön, 1971; Ackoff, 1974; Toffler, 1980, 1990; Reich, 1991) (what I call the "be" or "quality" perspective). As a result, it is no longer adequate to base professional development just on transmitting existing knowledge and developing a predefined range of competences on the basis that one problem equals one solution. Instead, practitioners need to be able to construct and reconstruct the knowledge they need and continually evolve their practice (Schön, 1987:35–6), thereby leading to a systemic and dynamic development of their competencies. (For a review of the link between knowledge, personal, and performance-based dimensions of competence see Crawford, 1998). These alternative approaches of going beyond traditional models of production and knowledge use while recognizing its validity in some areas are based more on the reflecting, questioning, and creating processes.

An Epistemological Perspective

The term "epistemology" refers to the study of the nature and grounds of knowledge, including how we define or recognize it. Most of the works on organizational learning, learning organizations, knowledge management, knowledge-creating organizations, and so on are based on a traditional understanding of the nature of knowledge. We could name this understanding the "positivist epistemology" perspective, since it treats knowledge as something people have. But this perspective does not reflect the knowing found in individual and team practice, knowing as an "intelligent" action, "ingenium,": This mental faculty that makes possible connecting in a fast, suitable, and satisfying way of "the separate things" (as stated by Lemoigne (1995), quoting Giambattista Vico (1708) in calling for a "constructivist epistemology" perspective). The "positivist epistemology" tends to promote explicit over tacit knowledge (see *Tacit vs. Explicit Knowledge* coming up in the chapter), and individual knowledge over team or organizational knowledge.

This integrative epistemological approach for project management suggests that organizations will be better understood if explicit, tacit, individual, and team/organizational knowledge are treated as four distinct forms of knowledge (each doing work the others cannot), and if knowledge and knowing (intelligent action) are seen as inseparable and mutually enabling. Thus, knowledge may be seen as an input of knowing, and knowing as an aspect of our interaction with the social and physical world, and therefore the dynamic interaction of knowledge and knowing can generate new knowledge and new ways of knowing.

Knowledge Management: Overview, Key Issues

A Brief History of Knowledge Management (KM): Different Cultural Perspectives

KM offers a unique concept considered by many in the industry as simultaneously progressive yet soft and difficult in application., One may suggest that this is primarily because of the intangible elements of knowledge. However, the increased topicality—if not to say pervasiveness—of the term through the writings of such well-known and recognized authors as Drucker (1993), Wheatley (2001), De Geus (1997), and Senge (1999) strongly suggest that KM is becoming accepted as a credible concept.

Although the study of knowledge dates back at least to Plato and Aristotle, entertaining the management of knowledge throughout a corporation first gained visibility by Polanyi in

1958. O'Dell and Grayson (1998) state "Polanyi's work served as a basis for the much-acclaimed knowledge management theories and books by the Japanese organizational learning guru, Ikujiro Nonaka" (p. 3).

Polanyi (1958) presented knowledge as something that can have intrinsic value placed on it; he also outlined two types of knowledge: tacit and explicit (defined in the *Tacit vs. Explicit Knowledge* section, coming up). Nonaka (1991) confirmed Polanyi's two knowledge-level concepts and introduced what he named "the knowledge-creating company" (p. 22). Nonaka proposed that organizations were not so much like machines as living organisms. That insightful biological analogy created a logical link between knowledge and organizations and started a paradigm shift in the need to pay attention to the collective thoughts of the people within the organization as knowledge contributors. The focus was on knowledge sharing and knowledge transfer. The average life span of major corporations is 40 to 50 years, roughly half that of humans. The need for development of KM is potentially the answer to expand the life expectancy of organizations and in doing so improve the overall health of the organization within the process.

Prior to Nonaka's (1991) research, Westerners (predominantly in the United States) viewed organizations as "a machine for information processing. According to this view, the only useful knowledge is formal and systematic hard data, codified procedures, universal principles" (p. 23). Wheatley (2001) maintains that even today the Japanese approach the field of KM differently than Westerners. She explains that we in the West still focus on explicit knowledge, while our Japanese counterparts find most gains in the areas of tacit knowledge ("be" side). For instance, Davenport and Prusak (1997) focus on knowledge acquisition, providing a market perspective on organizational knowledge creation (the "have" side). Peng and Akutsu (2001) propose that "there are two fundamentally different mentalities for dealing with new knowledge: linear thinking and dialectical thinking" (p. 107). Linear thinking is defined as "distaste for ambiguity and contradiction and preference for consistency and certainty" (p. 108). They differentiate dialectical thinking in "synthesizing dialectical thinking," aiming at identifying contradiction and resolving it by means of synthesis or integration, from "compromising dialectical thinking," focusing on tolerating contradiction. They come to the conclusion that mentality is a major factor in understanding how people are behaving in front of new knowledge. They found in their research that the Japanese are more dialectical than Americans.

What Is Knowledge? What Is Knowledge Management?

Part of the interest in the possible value of corporate knowledge comes from the information age. The advent of sophisticated information systems, the World Wide Web, networks, e-mail, and instantaneous sharing of information led to the realization that knowledge (and its sharing) was the fundamental element behind an organization's activities. This is not to say that information is knowledge. Deming (1993) accurately states that "information, no matter how complete and speedy, is not knowledge" (p. 106). Once information is embodied in time and gets a temporal value, it then becomes knowledge. To differentiate knowledge from data and information, I would say that the bad thing about knowledge is that many people, experts and lay people alike, treat it as some sort of higher-level information: extended, synthetic, advanced, tacit, complex, and so on, but still as information. Although information is an enhanced form of data, knowledge is not an enhanced form of information. It is quite clear, even on an intuitive level, that knowledge is not and cannot be the same thing as information, not even a form of information. It cannot be handled as information, does not have the same uses, and will resist any simplistic and expedient methodological transfers from information systems to "knowledge systems." Having information is not the same as knowing: Not every reader of a cookbook is a great chef. It is therefore very important to define knowledge in a distinct, appealing, and operational way. Simply calling more complex forms of information "knowledge" will not make them knowledge, even if repeated for years. "Knowledge" of information can be demonstrated through a statement, recall, or display. Knowledge itself can only be demonstrated through action. What is knowledge? Knowledge is purposeful coordination of action (intelligent action, or "ingenium").

Knowledge management is "the art of creating value from an organization's Intangible Assets" (Sveiby, 1999). With Sveiby (2001), we can define knowledge management by looking at what people in this field are doing: "Both among KM-researchers and consultants and KM-users there seem to be two tracks of activities, and two levels."

The two tracks of *activities* are as follows:

1. *Management of information.* Researchers and practitioners in this field tend to have their education in computer and/or information science (Hayes-Roth et al., 1983). They are involved in the construction of information management systems, Artificial Intelligence (AI), reengineering, groupware, and so on. To them, knowledge equals objects that can be identified and handled in information systems. This track is new and is growing very fast at the moment, assisted by new developments in IT.
2. *Management of people.* Researchers and practitioners in this field tend to have their education in epistemology, philosophy (Kuhn, 1970), psychology, sociology (Polanyi, 1958, 1966), or business/management/economics (Silberston, 1967). They are primarily involved in assessing, changing, and improving human individual skills and/or behavior. To them knowledge equals processes, a complex set of dynamic skills, know-how, and so on that is constantly changing. They are traditionally involved in learning and in managing these competencies individually—like psychologists—or on an organizational level—like philosophers, sociologists, or organizational theorists. This track is very old and is not growing so fast.

The two *levels* defined by Sveiby are as follows:

1. *Individual perspective.* The focus in research and practice is on the individual (AI specialists, psychologists).
2. *Organizational perspective.* The focus in research and practice is on the organization (reengineering, organization theorists, etc.) (Sveiby, 2001).

Crossing these two dimensions, we can capture one essential issue: "There are paradigmatic differences in our understanding of what knowledge is" (Sveiby, 2001). "The researchers and practitioners in the "Knowledge = Object" column tend to rely on concepts from Information Theory in their understanding of Knowledge. The researchers and practitioners in the column "Knowledge = Process" tend to take their concepts from philosophy or psychology or sociology.

Another approach of KM schools can be founded in Earl (2001). Seven knowledge management schools grouped into categories are introduced. For each of them, attributes (focus, aim, unit critical success factors, principal IT contribution, and "philosophy") are proposed. See Table 8.1.

Useful Concepts in Knowledge Management

Some key concepts emerging from the KM movement and from other disciplines can be summarized as follows.

Tacit vs. Explicit Knowledge. This idea finds its origin in Polanyi (1966) but has been applied to business and knowledge management by Nonaka (1995). It suggests that there are two types of knowledge: tacit, which is embedded in the human brain and cannot be expressed easily, and explicit (Brooking, 1999), which can be easily codified. Both types of knowledge are important, but Western organizations have focused largely on managing explicit knowledge.

Codification vs. Personalization. This distinction is related to the tacit vs. explicit concept. It involves an organization's primary approach to knowledge transfer (Hansen and Ali, 1999). Organizations using codification approaches rely primarily on repositories of explicit knowledge ("have"). Personalization approaches imply that the primary mode of knowledge transfer is direct interaction among people ("be"). Both are necessary in most organizations, but an increased focus on one approach or the other at any given time within a specific organization may be appropriate.

Knowledge Processes. Knowledge processing may be seen as a social system positioned in the value chain of an organization. Two sides may be considered as shown in the Knowledge Life Cycle exposed by McElroy (2002, p. 6):

TABLE 8.1. KM SCHOOLS ACCORDING TO EARL (2001).

Groups	Schools	"Focus"
TECHNOCRATIC	Systems	Technology
	Cartographic	Maps
	Engineering	Processes
ECONOMIC	Commercial	Income
BEHAVIORAL	Organizational	Networks
	Spatial	Space
	Strategic	Mind-set

1. Demand side with knowledge production; also called first-generation KM
2. Supply side with knowledge integration; called second-generation KM.

The overall process can be generically represented as three subprocesses:

- Knowledge production and capture
- Knowledge integration/codification
- Knowledge transfer/use

Knowledge production and capture includes all the processes involved in the acquisition and development of knowledge. Knowledge integration/codification involves the conversion of knowledge into accessible and applicable formats. Knowledge transfer/use includes the movement of knowledge from its point of generation or codified form to the point of use. One of the reasons that knowledge is such a difficult concept is that this process is systemic and often discontinuous. Many cycles are concurrently occurring in businesses. These cycles feed on each other. Knowledge interacts with information to increase the space of possibilities and provide new information, which can then facilitate generation of new knowledge.

Knowledge Markets. This concept focuses on the interest that individuals have in holding onto the knowledge they possess. To part with it, they need to receive something in exchange (Davenport and Prusak, 1997). Any organization is a knowledge market in which knowledge is exchanged for other things of value—money, respect, promotions, or other knowledge.

Communities of Practice. This idea, developed in the "organizational learning" movement, states that knowledge flows best through networks of people who may not be in the same part of the organization but do have the same work interests (Brown and Duguid, 1991; Wenger, 1998, 2002). Some organizations have attempted to formalize these communities, although theorists argue that they should emerge in a self-organizing fashion without any relationship to formal organizational structures: "learning happens, design or not design" (Wenger, 1998, p. 225).

Intangible Assets. Many observers have recently pointed out that formal accounting systems do not measure the valuable knowledge, intellectual capital, and other "intangible" assets of a corporation (Sveiby, 1997). Some analysts have even argued that accounting systems should change to incorporate intangible assets and that knowledge capital should be reflected on the balance sheet. However, the esoteric and subjective nature of knowledge makes it impossible to assign a fixed and permanent value to knowledge. Intangible assets have, however, always been integrated in strategic analysis as a source of competitive advantage.

Knowledge Management in action

Since KM is a relatively new philosophy, Davenport (1999) argues that it is "not yet tied to strategy and performance in practice" (p. 2-1). The use of the term "knowledge" in business strategy is rampant; however, Davenport (1999) believes that only a very small number of companies use a knowledge strategy as part of their business strategy.

Some authors have provided principles that facilitate the implementation of KM. Among them some are offering a humanistic perspective: Wheatley's (2001) six principles that facilitate KM are more a testimonial of Nonaka's original principle of an organization being analogous to an organism. The six principles are (1) knowledge is created by human beings; (2) it is natural for people to create and share knowledge; (3) everybody is a knowledge worker; (4) people choose to share their knowledge; (5) knowledge management is not about technology; and (6) knowledge is born in chaotic processes that take time. De Geus (1997) comes within the scope of the humanistic side of an organization by not using the term "learning organization" and provided the exploration of the possibilities of a "living company." If one subscribes to the six elements that have been provided by Wheatley (2001) to facilitate KM, the next step would be one of capture, integration/codification, and transfer/use of knowledge. These functions would be limited to explicit knowledge and not applicable for tacit knowledge at this time. It should be noted, however, that as organizations create processes to facilitate the conversion from tacit to explicit knowledge, applicability increases. Although Wheatley (2001) states "knowledge management is not about technology" in her fifth of six elements, capture, integration/codification, and transfer/use of knowledge are accomplished through technology.

With a more "information technology" perspective, Zack (1999) discusses the management of codified (explicit) knowledge and the use of four primary resources to manage knowledge. They include "repositories of explicit knowledge, refineries for accumulating, refining, managing and distributing knowledge, organization roles to execute and manage the refining process, and information technologies to support the repositories and processes" (p. 47).

As stated previously, it is not my intent to say that information technology creates knowledge. Augier and Morten (1999) state "technologies manifest themselves as representers of knowledge" (p. 253). Huang, Lee, and Wang (2001) show that "technology and systems, however, are used as facilitators in the production, storage, and use of organizational knowledge" (p. 4). Wiig (1999) outlines 16 building blocks that should be considered for introduction of KM: "1. Obtain management buy-in; 2. Survey and map the knowledge landscape; 3. Plan the knowledge strategy; 4. Create and define knowledge-related alternatives and potential initiatives; 5. Portray benefit expectations for knowledge management initiatives; 6. Set knowledge management priorities; 7. Determine key knowledge requirements; 8. Acquire key knowledge; 9. Create integrated knowledge transfer programs; 10. Transform, distribute, and apply knowledge assets; 11. Establish and update KM infrastructure; 12. Manage knowledge assets; 13. Construct incentive programs; 14. Coordinate KM activities and functions enterprise-wide; 15. Facilitate knowledge-focused management; 16. Monitor knowledge management: Provide feedback on progress and performance of KM program and activities." (p. 3–6)

Thus, Wheatley (2001) provides the principles in facilitation of KM, Zack (1999) provides the primary resources to manage codified knowledge, and Wiig (1999) provides the thought process necessary to build upon (in sequential order) to have a successful KM program.

It was mentioned earlier that according to the Japanese perspective, tacit knowledge represents the fundamental element to enable knowledge. Nonaka has stated that to facilitate true "learning organizations," tacit knowledge must be the cornerstone of future investigation. Krogh, Ichijo, and Nonaka (2000) provide five knowledge enablers as a means to develop the power of tacit knowledge: "1—instill a knowledge vision, 2—manage conversations, 3—mobilize knowledge activists, 4—create the right context, and 5—globalize local knowledge" (p. 5). They state that it is through these "knowledge enablers" that sharing of knowledge can occur and that true organizational improvement will happen. They also identify KM in its current form as a "constricting paradigm" that has three pitfalls: (1) knowledge management relies upon easily detectable, quantifiable information; (2) knowledge management is devoted to the manufacture of tools; and (3) knowledge management depends on a knowledge officer (pp. 26–28). It is through "knowledge enabling" that they propose to avoid these pitfalls.

Clearly, the field of KM is extensive, complex, and still in its preparadigmatic stage (Kuhn, 1970), though with significant application.

Organizational Learning: Mapping the Domain, Considerations

Origins and Definitions

The first publications on organizational learning (OL) appeared in the 1960s (Cangelosi and Dill, 1965), but research on learning organizations principally gained impetus with the publication *of The Fifth Discipline* by Peter Senge (1990a) and the special edition on organizational learning in *Organization Science* (1991). An overview of the development of the field can be found in Dierkes et al. *Handbook for Organizational Learning & Knowledge* (2001, pp. 926–927).

The concept of a "learning organization" or "learning by organizations" has actually been taken from the psychological concept of "individual learning" (Weick, 1991). Almost all definitions of organizational learning are based on this analogy.

One can differentiate normative and descriptive definitions. The normative definition refers to some requirements that an organization must satisfy in order to be known as a learning organization (Garvin, 1993; Hayes et al., 1988; Bomers, 1989; Senge, 1990a). Various other authors propose a more descriptive definition (Kim, 1993; Levinthal and March, 1993). Kim argues that all organizations learn, whether consciously or not. Some organizations try to encourage learning; others abandon such efforts and in doing so, obtain habits that finally reduce their learning capability. However, in both situations there are, in one way or another, learning processes taking place.

Individual and Organizational Learning

Many authors emphasize the paradoxical nature of the relationship between individual and organizational learning (e.g., Argyris and Schön, 1978; Huber, 1991; Bomers, 1989). One can observe that an organization consists of individuals, and individual learning is consequently a necessary condition of organizational learning. In contrast, the organization is capable of learning independently of each single individual but not independently of all individuals (Argyris and Schön, 1978).

An organization learns through its individual members and is thus directly or indirectly influenced by individual learning. Therefore, it is not surprising that most theories about learning organizations are based primarily on observations of learning individuals, particularly in experimental situations (Sterman, 1989; Huber, 1991; Kim, 1993).

Hedberg (1981) makes a comparison between the brains of individuals and organizations as information processing systems. Organizations have cognitive systems and memories, through which certain modes of behavior, mental models, norms, and values are retained. For that reason, organizations are not only influenced by individual learning processes, but organizations influence the learning of individual members and store that which has been learned. This may take the form of manuals, procedures, symbols, rituals, and myths. Though the individual is the only entity able of learning, he or she must be seen as being part of a larger learning system in which individual knowledge is exchanged and transformed.

Single-Loop and Double-Loop Learning

Most authors refer to two kinds of learning processes: single-loop and double-loop learning (Argyris and Schön, 1978; Bomers, 1989; Duncan and Weiss, 1979; Fiol and Lyles, 1985; Pedler et al., 1991). Single-loop learning involves processes in which errors are tracked down and corrected within the existing set of rules and norms. According to Fiol and Lyles (1985), single-loop learning is the result of repetition and routine. Examples of the result of these sorts of learning processes are successful programs and decision-making rules (Cyert and March, 1963). These are particularly important in situations in which the organization controls its environment (Duncan and Weiss, 1979). The main characteristics of single-loop learning can be described as based on repetition and routine, and within existing structures. It mainly concentrates on a specific activity or direct effect, within a simple context. The expected results may be change of behavior or performance level, and problem-solving capacity.

Double-loop learning, in contrast, involves changes in the fundamental rules and norms underlying action and behavior (Argyris and Schön, 1978). Double-loop learning generally has long-term effects with consequences for the whole organization. Crisis situations often provide opportunities for double-loop learning. Argyris and Schön (1978) defined double-loop learning as a process in which errors are tracked down and corrected with the result that underlying norms, ideas, and objectives become the objects of discussion and, where necessary, change. To summarize, the main characteristics of double-loop learning can be described as based on cognitive processes and understanding the nonroutine, and aimed at changing rules and structures. It occurs within a complex context. The results are changes of mental frameworks, development of frames of reference, and interpretation on the basis of which decisions can be made, as well as the development of new myths, stories, and cultures.

Different Approaches to Organizational Learning

Organizational learning may be seen under many different perspectives. For instance, the *Handbook of Organizational Learning and Knowledge* (Dierkes et al. 2001) proposes in Part I, entitled "insights from major social disciplines", seven perspectives: psychological, sociological, management science, economic theories, anthropology, political science and historic. Easterby-Smith (1997) describes six academic perspectives that have made significant contributions to understanding about organizational learning: psychology and organization development, management science, strategy, production management, sociology, and cultural anthropology. Argyris (1999), introducing the evolving field of organizational learning (p. 1), suggests the following "subfields": sociotechnical systems, organizational strategy, production, economic development, systems dynamic, human resources, and organizational culture.

Here I can propose four different approaches to organizational learning: contingency theory, psychology, information theory, and system dynamics (Romme and Dillen, 1997). As such, these approaches seem to constitute the main alternative frameworks for thinking about organizational learning. Each approach is presented as an ideal type, although in practice there is, of course, some overlap.

Contingency Theory. The classic interpretation of the concept of organizational learning is based on contingency theory, which views organizations as open systems, which continually adapt themselves to their environment. Thus, the learning process in organizations is seen primarily as an adaptation process (Cangelosi and Dill, 1965; Cyert and March, 1963; Meyer, 1982; Hutchins, 1991).

Psychology. Organizational learning can also be considered from a psychological perspective. Here the fundamental assumption is that organizations translate their internal and external environment in terms of their own frames of reference. One of the best-known ideas produced by this perspective is Weick's enactment principle (Weick, 1979), which implies that members of organizations develop collective perceptions of the organizational environment. These sets of beliefs are, to a large extent, unique for an organization and lead to a collective language through which agreement on experiences and insights can be reached (see also Argyris and Schön, 1978; Argyris, 1982, 1990, 1991, 1992).

Information Theory. The two previous approaches do not indicate how learning processes take place and where frames of reference come from. The approach based on information theory gives some attention to these kinds of questions. Here, organizations are primarily considered as processes of acquisition, distribution, interpretation, and storage of information. Organizational learning is a continually evolving process that results in the expansion and improvement of knowledge. This knowledge can only be labeled as organizational knowledge if it is exchanged and accepted among the participants. To shape the structure and dissemination of organizational knowledge, format and informal learning systems can be institutionalized: examples of these systems include strategic planning systems, management information systems, informal information channels, and communication networks

(Duncan and Weiss, 1979; Walsh and Ungson, 1991; Ulrich et al., 1993; Huber, 1991; Nonaka, 1991; Boisot, 1998).

System Dynamics. The system dynamics approach uses principles and concepts from system dynamics and systems thinking to understand learning processes in organizations (Morgan, 1986; Senge, 1990a). Here, the main assumption is that human organizations are characterized by "dynamic complexity," which makes models with simple cause-effect relationships no longer applicable (Senge, 1990a). Consequently, principles of system dynamics, such as positive and negative feedback, are used to show that social reality consists of circles of causality. Organizational learning must first be understood as a cohesive, holistic process before more detailed theorizing can take place (Morgan, 1986). Authors from the Sloan School of Management (MIT) have contributed significantly to the development and dissemination of concepts of system dynamics (e.g., De Geus, 1988; Kim, 1993; Stata, 1989; Senge, 1990a).

Contingency and psychological theories approaches have to some extent been integrated in more recent publications from the perspective of information theory (Walsh and Ungson, 1991; Huber, 1991) and system dynamics (Senge, 1990a; Kim, 1993).

Learning from a System Dynamics Perspective

Daft and Weick (1984), March and Olsen (1975), and Kolb (1984) have tried to formulate integrated models of learning processes in organizations. These attempts included the crucial link between individual and organizational learning. However, the model developed by Kim (1993) incorporates the key concepts and dimensions of earlier models, and in addition, describes the interactions between individual and collective mental models as a transfer mechanism between individual and organizational learning.

At the level of individual learning, Kim distinguishes between two levels: operational learning, what is learned ("know-how") and conceptual learning, the understanding and use of this knowledge ("know what"). Operational and conceptual learning are linked to two aspects of the individual mental model. Operational learning is learning at a procedural level through which the individual learns the steps required for accomplishing certain tasks. This knowledge is rooted in routines. Routines and operational learning influence each other. Conceptual learning involves thinking about the underlying causes of required actions, through which conditions, procedures, and concepts are discussed and new frames of reference are created. Then, the individual learning model consists of a cycle of conceptual and operational learning that is fed by individual mental models. Individual learning cycles—that is, the processes through which individual learning and its results are stored in the mental models of individuals—influence the learning process at the team or organizational level through their influence on collective mental models. So, the organization can only learn through its members, but in doing so, it does not depend on every single member. On the other hand, individuals are able of learning without the organization.

Changes in the frames of reference of one or more individuals can lead to conceptual learning at the organizational level, in the form of changes in collective frames of reference, or *Weltanschauung*, which in turn can lead to changes in the frames of reference of other individuals in the organization. In general, collective frames of reference evolve very slowly. Changes in individual routines can lead to operational learning at the organizational level, in the form of adapted or new organizational routines, such as standard procedures. These organizational routines in turn influence the development of (other) individual routines. The extent to which individual mental models can influence collective mental models depends on the influence that certain individuals or groups can exert. In general, top management tends to be one of the most influential groups.

Kim's model also incorporates the concept of single- and double-loop learning. This characteristic applies at the individual level as well as the organizational level. Individual single-loop learning takes place if the learning cycle leads to changing behavior of the individual. Individual double-loop learning concerns the process by which the individual learning cycle influences the individual mental model and vice versa. Single-loop learning at the collective level takes place when individual actions lead to the intended changes in collective actions. Double-loop learning at a collective level happens when individual mental models are transformed into collective mental models.

The literature describes seven types of learning disturbances possibly occurring in the learning process, which can be mapped with Kim's model:

1. *Role-constrained learning.* This occurs if individual learning processes do not have an effect on individual actions (March and Olsen, 1975).
2. *Audience learning.* The problem arises from the link between individual and organization actions (March and Olsen, 1975).
3. *Superstitious learning.* The problem in this disturbance is the link between individual and collective actions and the response of the environment to these actions (March and Olsen, 1975).
4. *Learning under ambiguity.* In the case of learning under ambiguous conditions, the causal links between events and the environment are no longer obvious (March and Olsen, 1975; Levinthal and March, 1993).
5. *Situational learning.* This takes place if the individual does not secure his knowledge or forgets to code it for later use (Kim, 1993).
6. *Fragmented learning.* The link between individual mental models and the collective model is poorly maintained (Cunningham, 1994; Kim, 1993).
7. *Opportunistic learning.* This happens when collective actions are based on the initiative and vision of only one person or a small group of individuals. According to Kim, this disturbance occurs in situations in which individuals consciously try to by-pass the prevailing *Weltanschauung* and organizational routines, because the old way of doing things is considered inadequate. The link between collective mental models and collective actions is deliberately broken in order to create new opportunities without the whole organization having to change.

The identification of learning disturbances is important. Methods and tools, mainly resulting from training and consultancy work (e.g., Argyris, 1991, 1992; Cunningham, 1994; Senge, 1990a; Swieringa and Wierdsma, 1990), have been developed to overcome learning disturbances: system archetypes and learning laboratories, team learning and the role of dialogue, learning sets and contracts, and circular organizational structures (Romme and Dillen, 1997).

In general, the literature on this subject still has a strongly theoretical nature. The information-theoretical perspective principally appears to provide a broad analytical framework for describing and understanding organizational learning. An integrative framework with interesting practical implications has been developed in the field of system dynamics. Tools and methods with practical relevance have been developed mainly as part of training and consultancy work.

From Knowledge Management and Organizational Learning to Learning Organization

The focus now shifts to look at the applied area of organizational learning, which is normally associated with the label of the learning organization. In this area KM and OL are intimately and inextricably linked.

Finding a Common Ground

"Knowledge development constitutes learning"

WEICK, *1991, p. 122*

Similarities between knowledge management (KM) and organizational learning (OL) begin with ambiguity of definition. Garvin (2000) provides no less than seven definitions of learning organizations. He synthesizes various definitions to form one that appears broad enough to be considered as a template of what a learning organization is: "A learning organization is an organization skilled at creating, acquiring, interpreting, transferring, and retaining knowledge, and at purposefully modifying its behaviour to reflect new knowledge and insights" (p. 11).

Peter Senge (1990b, 1999) outlines the mechanisms to achieve a learning organization working on the basis of systems theory. In fact, the open environment needed to facilitate KM is the same environment that facilitates organizational learning. "Knowledge management is not a stand alone process. It is closely bound up with the inputs of organizational learning and strategy that govern its nature and scope" (Rostogi, 2000).

The information era has generated the need for focus on knowledge workers versus blue-collar labor. The new knowledge workers are only successful if they operate in particular environments. The environment proposed in learning organizations appears to facilitate KM. KM and OL reflect the collective focus of minds toward meeting common interactive organizational objectives. It can be plausibly deduced that, given the research provided on KM, they become inextricably linked to learning organizational development. The title, and the subjects, of Dierkes et al., *Handbook for Organizational Learning & Knowledge*, (2001) provides a good example. To further illustrate the interconnections that exist between KM and learning organizations, Kofman, Senge, Kanter, and Handy (1995) state "the learning organization is built upon an assumption of competence that is supported by four characteristics: curiosity, forgiveness, trust, and togetherness" (p. 47). These characteristics are also those providing the proper environment for the process to share explicit and tacit knowledge.

Two Learning Organization Models.

A typology of learning organization models has been proposed in Easterby-Smith (1997), as we have seen. Dierkes et al. (2001, p. 930) identify three main learning models: ". . . based on feedback loops between the organization and its environment," ". . . portrayed learning in term of steps or phases," and the spiral model (Nonaka and Takeuchi, 1995).

Among this diversity I will introduce two major models. My choice is driven by the fact these models are widely recognized (Nonaka et al.) and/or are bringing new perspectives (Boisot). Nonaka, Toyama, and Byosiere (in Dierkes et al. 2001, pp. 491-517), after a critical review of existing studies of OL, state that "creating knowledge is a continuous . . . not a special kind of learning at one point in time." They propose the basic concepts of knowledge creation process, management of the process, and the organizational structure for knowledge creation. The knowledge creation process is composed of three layers: SECI (four modes of knowledge conversion: socialization-externalization-combination-internalization processes), *Ba* (platforms for the knowledge creation process), and knowledge assets (inputs, outputs of SECI, and moderator of the knowledge creation process between *Ba* and SECI). The foundation of SECI process is *Ba*, four types of *Ba* being considered: originating *Ba* (socialization/face-to-face), dialoguing *Ba* (externalization/peer-to-peer), systemizing *Ba* (combination/collaboration), and exercising *Ba* (internalization/on site). Knowledge assets, inputs, and outputs of the knowledge creation process form the basis of organizational knowledge creation. They also influence how *Ba* works. The four types of knowledge assets (KA) are as follows: experiential KA (tacit knowledge shared through common experience), conceptual KA (explicit knowledge articulated through images, symbols, and language), systemic KA (systemized and packaged explicit knowledge), and routine KA (tacit knowledge routinized and embedded in actions and practices). An organization, building on its existing knowledge assets, creates new knowledge through the SECI process that takes place in *Ba*, the knowledge created becoming then part of the knowledge assets of the organization and the basis of a new cycle of knowledge creation.

The proper management of the knowledge creation process requires a new model called a "middle-up-down" model. Top and middle management have a leadership role by working on the three layers of the knowledge creation process, providing knowledge vision, developing and promoting the sharing of knowledge assets, create and energizing *Ba*, and enabling and promoting the continuous spiral of knowledge creation. But for effective knowledge creation a supportive organizational structure is needed. The proposed new organizational structure is called "hypertext" organization, articulating a knowledge-base layer (accumulation and sharing of knowledge) bottommost layer, a business-system layer (utilization of knowledge), and a project-team layer topmost layer (creation of knowledge) through a dynamic knowledge cycle, continuously creating, exploiting, and accumulating organizational knowledge.

FIGURE 8.1. KNOWLEDGE CREATION PROCESS.

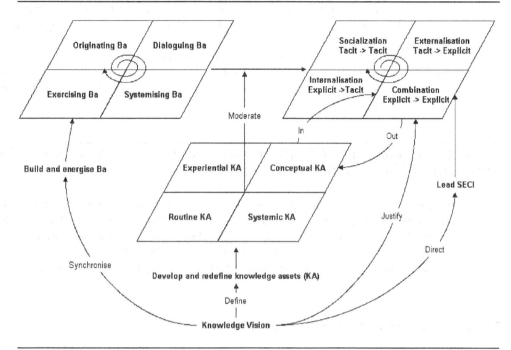

Source: Adapted from Nonaka, Toyama, and Byosiere (in Dierkes et al., 2001, p. 507). By permission of Oxford University Press.

Boisot (1998) proposes a different model grounded on an information perspective and complexity science, a set of theories describing how complex adaptive systems work. For him (p. 34), knowledge assets emerge as a result of a two-step process, constituting the two distinct phases of the evolutionary production function: creating knowledge ("process of extracting information from data") and applying knowledge ("testing the insights created in a variety of situations that allow for the gradual accumulation of experiential data"). He defines an information space (I-Space) according to three dimensions: codification (information codified/uncodified), abstraction (abstract/concrete), and diffusion (diffused/undiffused). The creation and diffusion of new knowledge occurs in a particular sequence (social learning cycle, or SLC, p. 59): scanning, problem solving, abstraction, diffusion, absorption, and impacting. Two distinct theories of learning, although not mutually exclusive, are introduced as part of identification of two distinct strategic orientations for dealing with the paradox of value (i.e., "maximising the utility of knowledge assets compromises their scarcity, and maximising their scarcity make it difficult to develop and exploit their utility," p. 90). In neoclassical learning (N-Learning), knowledge is considered cumulative. Learning becomes a stabilizing process. This approach may lead to excessive inertia and fossilization of the knowledge assets. In Schumpeterian learning (S-Learning), change is the natural order of things. Abstraction and codification are incomplete. "Knowledge may be progressive in the sense that successive approximation may give a better grasp of the underlying structures of reality, but it is not necessarily cumulative" (p. 99). S-Learning is more complex than N-Learning because it integrates both certainties and uncertainties, and it requires an "edge of chaos" culture (p. 116).

Knowledge management and organizational learning are thus linked in a systemic way through action and the quest of developing learning organizations. The diversity of episte-mologies, scientific fields, theories, perspectives, and resulting models, not mutually exclusive, exemplify the plurality of approaches.

Project Management, Knowledge Management, and Organizational Learning: A Systemic Relationship

My purpose here is to introduce and illustrate the specificity of the project environment with respect to a learning organization perspective. Most of these developments are the results of research undertaken as part of the CIMAP Research Centre–Groupe ESC Lille and are grounded on the former works of the founders Decleck and Debourse (1983, 1997). (See also "Projects as Arenas for Renewal and Learning Processes," Lundin and Midler, 1998).

Projects vs. Operations: The Nature of Projects

Every organization acts according to two fundamentals modes: an operational mode, aiming at the exploitation of competitive advantage and current position on the market and pro-viding profits and renewal or increase of resources; and an entrepreneurial mode, or project mode, focusing on the research of new position and new competitive advantage, consuming money and resources. To ensure their sustainability and development, all organizations need to combine both modes. (Declerck in Ansoff, Declerck and Hayes, 1976).

Thus, we have to face two types of activities, and I wish to propose the dichotomy operations/projects. Table 8.2 emphasizes the main characteristics of these activities. I focus here on these two types, although in reality, activities may be a blend of these two pure types.

With this differentiation in mind, you can now look at the characteristics of a project team not only in charge of project activities but, to some extent, of "operations" activities as well, and you can view the project team as a learning organization.

Project Team and Learning Dynamic

I would like firstly to compare some characteristics of groups and teams. Wenger and Snyder (2000, p. 142) draw a comparison between several forms of team organizations: community of practice, formal work group, informal network, and project team. To this it seems im-portant to add the concept of *Ba*, as described previously. There are some fundamentals differences between project team, community of practice, and *Ba*, as summarized Table 8.3.

To understand the specificity created by the project environment and project team as far as learning is concerned, let us synthesize some of the key perspectives (see Table 8.4).

TABLE 8.2. OPERATIONS VS. PROJECTS.

Operations	Projects*
Ongoing and repetitive activities, being prone to influence of numerous factors.	Nonrepetitive activities, one-shot.
The factors of influence are mainly internal (endogenous) rather than environmental, and they can be manipulated by the operation manager.	Decisions are irreversible.
	Projects are subjects to multiple influences.
	The main influences come from environment (exogenous) and may vary considerably.
The environmental factors explain only a low part of the fluctuation of outputs.	The decision maker cannot usually handle an important number of variables (exogenous variables).
The inputs present random variations.	
It is possible to measure and to estimate the probabilities associated to these variations.	It is very tough to measure the effects or these influences.
The variation of inputs can be made statically stable.	Project is generally not in statistical stability and it is not possible to associate probabilities to the effects one try to measure.
Future effects can be predicted with a specified margin of error.	
Nonusual variations coming from perturbations external to the operation lead to slight penalizing and never to disaster.	A "bad" decision and/or a non controllable influence of a major event may lead to catastrophic result.
Operations are reversible processes: perturbations can be detected, the nature of these causes can be identified, and these causes can be eradicated.	
The reversibility of operations can occur within economically acceptable limits.	
Operations may interact with the actions of the observer.	
To summarize, operations involve	*Projects involve*
Planned actions	Creative actions
Masked actors	Unmasked actors
Process	Praxis
Cooperation	Confrontation
And they are	*And they are*
Rational	Para-rational
Algorithmic	Mosaic
Anhistoric	Historic
Stable and making one feel secure	Rich, ambiguous, unstable

*Entrepreneurial activities are assumed here to be managed using the project "form."

From the table, it is clear that projects as such are learning organizations or learning places. Projects, through the way the project team acts (praxis), are a privileged place for learning: Such project-based learning needs to integrate the two perspectives ("have" and "be" or "operations" and "projects" acting modes), as there is a need for a blend of creative or exploratory learning and application or exploitative learning (Boisot 1998, p. 116). Having in mind the need for efficiency and effectiveness, a project team acts as a temporary structure, generating first information and creating knowledge (adding complexity) with many degrees of freedom, and then applying it (reduction of complexity) in the former stage of a

TABLE 8.3. PUTTING IN PERSPECTIVE PROJECT TEAM, COMMUNITY OF PRACTICE, AND BA.

Community of Practice	Ba	Project Team
Members learn by participating in the community and practicing their jobs.	Members learn by participating in the Ba and practicing their jobs.	Members practice their jobs and learn by participating in the project team.
Place where members learn knowledge that is embedded in the community.	Place where knowledge is created.	Place where knowledge is created, where members learn knowledge that is embedded, and where knowledge is utilized.
Learning occurs in any community of practice.	Need of energy in order to become active.	Need of energy (forming the team) and then learning occurs.
Boundary is firmly set by the task, culture, and history of the community.	Boundary is set by its participants and can be changed easily.	Boundary is set by the task and the project.
Membership rather stable. New members need time to learn and fully participate.	Membership not fixed. Participants come and go.	Membership fixed for the project duration (temporary nature). May vary depending the phases of the project.
Participants belong to the community.	Participants relate to the Ba.	Participants may relate/belong to the project team for the duration of the project but may belong/relate to the operational/functional organization (department, contractors, suppliers, etc.).

project. Of course, the level of knowledge being created will depend of the nature of system project/organization/environment. Some construction projects require a little amount of creativity, while others in a different context will require a lot.

On a larger issue, the notion of knowledge management is so fascinating within projects precisely because all new project teams must solve a unique conundrum: to what degree is the information/knowledge available to complete the project based on past experience, replicable historical processes, and so on, and to what degree must all knowledge and learning be acquired or "emergent" as a result of the unique nature of the project tasks. I am thinking, for example, of the development of the Concorde, when the technical challenges were so new and nonhistorical that new cost estimating methods had to be developed, such as parametric estimation, precisely because all knowledge had to be emergent in regard to these activities.

The consequence at the knowledge management level is twofold. On the one hand, focusing on the "have" side, there is a need of for some form of knowledge—guidance, best

TABLE 8.4. SYNTHESIS OF TWO PERSPECTIVES REGARDING KM, OL, AND LEARNING ORGANIZATIONS.

Epistemology	Positivist—"Have"	Constructivist—"Be"
Main Acting Mode Knowledge management	*Operations* Western approach. Codification. Explicit knowledge. Linear thinking. Knowledge market.	*Projects* "Japanese" approach (and actually French one). Personalization. Tacit knowledge. Dialectical thinking: "synthesizing dialectical thinking," aiming at identifying contradiction and resolving it by means of synthesis or integration, from "compromising dialectical thinking," focusing on tolerating contradiction.
Organizational learning	Single-loop learning. Information theory (knowledge as formal and systematic hard data, codified procedures, universal principles).	Double-loop learning. Information theory (Nonaka 1991, Boisot 1998). System dynamics theory (Senge 1990a, Kim 1993).
Learning organization	Neoclassical learning (N-Learning), knowledge is considered cumulative (Boisot, 1998).	SECI cycle, *Ba*, knowledge assets, needs for a supportive organization (Nonaka, 1991). Schumpeterian learning (S-Learning), change is the natural order of things (Boisot, 1998).

practice, standards, and so on—at the individual, team, and organizational level. The development of professional certification programs, as well as maturity models, are important in this. It's important to recognize that such standards have to be seen as largely social constructs, developed to facilitate communication and trust among those who are adopting them, but their evolution in line with the experiences gained by the users or because of new developments or practices is a vital to avoid any fossilization (Bredillet, 2002). On the other hand, on the "be" side, the need of more creative competence (for example, some professional certifications are incorporating personal characteristics), flexible frameworks (for example, use of meta-rules), and organizational structure to enable the sharing of experience is fundamental.

Consider now the organization of learning and the necessary supporting structures. Each organization running projects has its own characteristics. Each has to build its own learning organization system. Buying some off-the-shelf software or training methods is unlikely to be sufficient, even though they might form the backbone of a learning organization architecture. Being conscious of the specificity of projects, and being clear on the underlying assumptions of the concepts, methods, tools, and techniques available, should, however, certainly help in the design of an appropriate system. In this regard, promising research results have been presented by Morris (2002).

Summary

Projects being the way the organizations implement their strategy and knowledge being the ultimate source of competitive advantage, it seems logical to try to understand how knowledge and learning interact within a project organization. The knowledge management field and the organizational learning field, although relatively young, are nevertheless already well developed in term of concepts, methods, and tools. Thus, they can provide inspiration to organizations aiming at organizational performance and at creating a sustainable competitive position.

To design an effective project learning organization, you must understand the underlying theories and assumptions on which each discipline is grounded. The first choice is thus the choice of a conscious perspective regarding the organizational culture and strategy. Starting from this, it is possible to build a coherent, relevant structure, adapting the methods and tools as appropriate. This construction has to be contextualized, integrating the relevant available explicit knowledge, and providing the necessary and sufficient support for creating, combining, and applying explicit and tacit knowledge.

Project organizations are a privileged place for learning, as there is a need for combining creative and exploitative learning to manage projects efficiently and effectively. There is unlikely to be any panacea or new "one best way" to develop knowledge and learning within any kind of organization, however.

But over all, I strongly believe that at the beginning of any creation of a conscious learning organization, an act of faith is needed: the belief in knowledge as a vector of value for people, organization, and society.

Ordo ab chaos.

References

Ackoff, R. L. 1974. *Redesigning the future: A systems approach to societal problems.* New Jersey: Wiley.

Ansoff H. I. 1975. Managing strategic surprise by response to weak signals. *California Management Review* 18(2):21–33.

Ansoff, H. I., R. Declerck, and R. Hayes, eds. 1976. *From strategic planning to strategic management.* New Jersey: Wiley.

Arcade, J. 1998. Articuler prospective et Stratégie: Parcours du stratège dans la complexité. *Travaux et Recherches de Prospective* 8 (Mai): 1–88.

Argyris, C. 1982. *Reasoning, learning and action: Individual and organizational.* San Francisco: Jossey-Bass.

———. 1990. *Overcoming organizational defenses: Facilitating organizational learning.* Englewood Cliffs, NJ:: Prentice Hall.

———. 1991. Teaching smart people how to learn. *Harvard Business Review* 69 (May–June): 99–109.

———. 1992. *On organizational learning.* Oxford, UK: Blackwell Science.

———. 1999. *On organizational learning.* 2nd ed. Oxford, UK: Blackwell Science.

Argyris, C., and D. Schön. 1978. *Organizational learning: A theory of action perspective*. Reading, MA: Addison-Wesley.

Augier, M., and T. Morten. 1999. Networks, cognition, and management of tacit knowledge. *Journal of Knowledge Management* 3:252–261.

Belout, A. 1998. Effects of human resource management on project effectiveness and success: Toward a new conceptual framework. *International Journal of Project Management* 16(1):21–26.

Boisot, M. H. 1998. *Knowledge Assets: securing competitive advantage in the information economy*. New York: Oxford University Press.

Bomers, G. B. J. 1989. *De lerende organisatie*. Breukelen, Netherlands: University of Nijenrode

Bredillet, C. 1999. Essai de définition du champ disciplinaire du management de projet et de sa dynamique d'évolution. *Revue Internationale en Gestion et Management de Projets* 4(2):6–29.

———. 2002. Genesis and role of standards: Theoretical foundations and socio-economical model for the construction and use of standards. *Proceedings of IRNOP V*. Renesse, Netherlands, May 28–31.

Brooking, A. 1999. *Corporate memory: Strategies for knowledge management*. New York: International Thomson.

Brown, J., and P. Duguid. 1991. Organisational Learning and Communities of Practice. *Organisation Science* (March): 40–57.

Cangelosi, V.E., and W. R. Dill. 1965. Organizational learning: Observation toward a theory. *Administrative Science Quarterly* 10:175–203.

Cleland, D. I. 1994. *Project management: Strategic design and implementation*. New York: McGraw-Hill.

Crawford, L. 1998. Project Management Competence for Strategy Realisation. *Proceedings of the 14th World Congress on Project Management*. Vol. 1. pp. 12–14. Ljubljana, Slovenia, June 10–13.

Cunningham, l. 1994. *The Wisdom of Strategic Learning*. London: McGraw-Hill.

Cyert, R. M., and J. C. March. 1963. *A behavioral theory of the firm*. Englewood Cliffs, NJ: Prentice Hall.

Daft, R. L., and K. E. Weick. 1984. Toward a model of organizations as interpretation systems. *Academy of Management Review* 9:284–295.

Davenport, T. H., 1999. Knowledge management and the broader firm: strategy, advantage, and performance. In *Knowledge management handbook*, ed. J. Liebowitz. Boca Raton, FL: CRC Press.

Davenport, T. H., and L. Prusak. 1997. *Working knowledge: How organizations manage what they know*. Boston: Harvard Business School Press.

De Geus, A. P. 1988. Planning as learning. *Harvard Business Review* 66 (March–April): 70–75.

———. 1997. *The living company: Habits for survival in a turbulent business environment*. Boston: Harvard Business School Press.

Declerck, R. P., J. P. Debourse, and C. Navarre. 1983. *La méthode de direction générale: le management stratégique*. Paris: Hommes et Techniques.

Declerck, R. P., J. P. Debourse, and J. C. Declerck. 1997. *Le management stratégique: Contrôle de l'irréversibilité*. Lille, France: Les éditions ESC Lille.

Deming, W. E. 1993. *The new economics*. Cambridge, MA: Center for Advanced Engineering Technology.

Dierkes, M., A. Berthoin Antal, J. Child, and I Nonaka. 2001. *Handbook of organizational learning and Knowledge*. New York: Oxford University Press.

Dinsmore, P. C. 1999. *Winning in business with enterprise project management*. New York: AMACOM.

Duncan, R. D., and A. Weiss. 1979. Organizational learning: Implications for organizational design. In *Research in Organizational Behaviour*, ed. B. Shaw et al. 75–123. Greenwich, CT: JAI Press.

Drucker, R. E. 1993 *The post capitalist society*. Oxford, UK: Butterworth-Heinemann.

Earl, M. 2001. Knowledge management strategies: Toward a taxonomy. *Journal of Management Information Systems* 18(1):215–233.

Easterby-Smith, M. 1997. Disciplines of organizational learning: Contributions and critiques. *Human Relations* 50(9):1085–1113.

Fincher, A., and G. Levin. 1997. Project management maturity model. *Project Management Institute 28th Annual Seminars/Symposium.* Chicago, September 29–October 1.

Fiol, C. M., and M. A. Lyles. 1985. Organizational learning. *Academy of Management Review* 10: 803–813.

Frame, J. D. 1994. *The new project management: tools for an age of rapid change, corporate reengineering, and other business realities.* San Francisco: Jossey-Bass.

Gareis, R. 1990. Management by projects: The management strategy of the "new project-oriented company. In *Handbook of Management by Projects*, ed. R. Gareis. Vienna: MANZ.

Garvin, D. A. 2000. *Learning in action: A guide to putting the learning organization to work.* Boston: Harvard Business School Press.

———. 1993. Building a learning organization. *Harvard Business Review* 71:78–91.

Grundy, T. 1998. Strategy implementation and project management. *International Journal of Project Management* 16(1):43–50.

Guénon, R. 1986. *Initiation et réalisation spirituelle.* Paris: Editions Traditionnelles.

Hansen, M., N. Nohria, and T. Tierney. 1999. What's your strategy for managing knowledge? *Harvard Business Review* 77(2):106–116.

Hauc, A. 1998. Projects and strategies as management tools for increased competitiveness. *Proceedings of the 14th World Congress on Project Management.* pp. 1–4. Ljubljana, Slovenia, June 10–13..

Hayes, R. H., S. Wheelwright, and K. B. Clark. 1988. *Dynamic manufacturing: Creating the learning organization.* London: MacMillan.

Hayes-Roth, F., D. A. Waterman, and D. B. Lenat. 1983. An overview of expert systems. In *Building Expert Systems*, ed. F. Hayes-Roth, D. A. Waterman, and D. B. Lenat, 3–29. Reading, MA: Addison-Wesley,

Hedberg, B. L. T. 1981. How organizations learn and unlearn. In *Handbook of Organizational Design*, ed. P. C. Nyström and W. H. Starbuck, 3–27. Oxford, UK: Oxford University Press,.

Huang, K. T., Y. W. Lee, and R. Y. Wang, 1999. *Quality information and knowledge.* Upper Saddle River, NJ: Prentice Hall.

Huber, G. P. 1991. Organizational learning: the contributing processes and literatures. *Organization Science* 3:88–115.

Hutchins, E. 1991. Organizing work by adaptation. *Organization Science* 2:14–39.

Kim, D. H. 1993. The link between individual and organizational learning. *Sloan Management Review* (Fall): 37–50.

Kofman, F., P. Senge, R. M. Kanter, and C. Handy. 1995. *Learning organizations: Developing cultures for tomorrow's workplace.* Portland, OR: Productivity Press.

Kolb, D. A. 1984. *Experiential learning.* Englewood Cliffs, NJ: Prentice Hall.

Krogh, G. V., K. Ichijo, and I. Nonaka. 2000. *Enabling knowledge creation: How to unlock the mystery of tacit knowledge and release the power of innovation.* New York: Oxford University Press.

Kuhn, T. 1970. *The structure of scientific revolutions,* Chicago: University of Chicago Press.

Legay, J. M. 1996. *L'expérience et le modèle: Un discours sur la méthode.* Paris: INRA Editions.

Lemoigne, J. L. 1995. *Les épistémologies constructivistes.* Paris: PUF.

Levinthal, D. A., and J. G. March. 1993. The myopia of learning. *Strategic Management Journal* 14:95–112.

Lorino, P., and J. C. Tarondeau. 1998. De la stratégie aux processus stratégiques. *Revue Française de Gestion* 117 (Janvier–Février): 5–17.

Lundin R. A., and C. Midler. 1998. *Projects as arenas for renewal and learning processes.* Boston: Kluwer Academic Publishers.

March, J. G., and J. P. Olsen 1975. The uncertainty of the past: organizational learning under ambiguity. *European Journal of Political Research* 3:147–171.

McElroy, M. W. 2002. Corporate epistemology and the new knowledge management. *IV Conference, Institute for the Study of Coherence and Emergence*. Fort Meyers, FL, December 8.

Meyer, A. 1982. Adapting to environmental jolts. *Administrative Science Quarterly* 27:515–537.

Morgan, G. 1986. *Images of organization*. Beverly Hills: Sage Publications.

Morris, P. W. G. 1997. *The management of projects*. London: Thomas Telford.

———. 2002. Managing project management knowledge for organizational effectiveness. pp. *77–87. Proceedings of PMI Research Conference*. Seattle, July 14–17.

Nonaka, I. 1991. The knowledge-creating company. *Harvard Business Review* 69 (November–December): 96–104.

Nonaka, I. and H. Takeuchi 1995. *The knowledge creating company: How Japanese companies create the dynamics of innovation*. New York: Oxford University Press.

O'Dell, C., and C. J. Grayson, Jr. 1998. *The transfer of internal knowledge and best practice: If only we knew what we know*. New York: Free Press.

Pedler, M., J. Burgoyne, and T. Boydell. 1991. *The learning company*. London: McGraw-Hill.

Peng, K., and S. Akutsu. 2001. A mentality theory knowledge creation and transfer: Why some smart people resist new ideas and some don't. In *Managing industrial knowledge: Creation, transfer and utilization*, ed. I. Nonaka and D. J. Teece, 105–123. London: Sage Publications.

Pinto, J. K., and Prescott, J. 1998. Variations in success factors over the stages in the project life cycle. *Journal of Management* 14(1):5–18.

PMI Standards Committee. 2002. *Project manager competency development (PMCD) framework*. Newtown Square, PA: Project Management Institute.

Polanyi, M. 1958. *Personal knowledge*. Chicago: University of Chicago Press

———. 1966. *The tacit dimension*. London: Routledge and Kegan Paul.

Reich, R. B. 1991. *The work of nations*. London: Simon & Schuster.

Remy, R. 1997. Adding focus to improvement efforts with PM3. *PM Network* (July): 43–78.

Romme, G., and R. Dillen, R. 1997. Mapping the landscape of organizational learning. *European Management Journal*. 15(1):68–78.

Rostogi, P. 2000. Knowledge management and intellectual capital: The new virtuous reality of competitiveness. *Human Systems Management* 19:39–49.

Saures, I. 1998. *A real world look at achieving project management maturity*. Project Management Institute 29th Annual Seminars/Symposium, Long Beach, California, October 9–15.

Schön, D. A. 1971. *Beyond the stable state*. New York: Norton.

———. 1987. *Educating the reflective practitioner*. London: Jossey-Bass.

Senge, P. M. 1990a. *The fifth discipline: The art and practice of the learning organization*. New York: Doubleday Currency.

———. 1990b. The leader's new work: Building learning organizations. *Sloan Management Review* (Fall): 7–23.

Senge, P., A. Kleiner, C. Roberts, R. Ross, G. Roth, and B. Smith. 1999. *The dance of change: The challenges to sustaining momentum in learning organizations*. New York: Doubleday/Currency

Silberston, A. 1967. The patent system. *Lloyds Bank Review* 84 (April): 32–44.

Stata, R. 1989. Organizational learning: The key to management innovation. *Sloan Management Review* 30(3):63–74

Sterman, J. D. 1989. Modeling managerial behavior: Misperceptions of feedback in a dynamic decision-making experiment. *Management Science* 35:321–339.

Sveiby, K. E. 1997. *The new organizational wealth: Managing and measuring knowledge based assets*. San Francisco: Berrett-Koehler.

———. 1999. *The invisible balance sheet: Key indicators for accounting, control and valuation of know-how companies*. Stockholm, Sweden: Konrad Group

————. 1998. Measuring intangibles and intellectual capital: An emerging first standard. Internet version. Updated Aug 5, 1998. www.sveiby.com/articles/EmergingStandard.html.

————. 2001. What is knowledge management? Updated May 17, 2003. www.sveiby.com/articles/KnowledgeManagement.html.

Swieringa, J. and A. F. M. Wierdsma. 1990. *Op Weg Naar Een Lerende Organisatie*. Croningen, Netherlands: Wolters Noordhoff.

Toffler, A. 1980. *The third wave*. London: Collins.

————. 1990. *Power shift*. London: Bantam Press.

Turner, J. R. 1993. *The handbook of project-based management*. London: McGraw-Hill.

————. 1998. Projects for shareholder value: The influence of project managers. 283–291. *Proceedings of IRNOP III, "The nature and role of projects in the next 20 years: Research issues and problems*. Calgary, Alberta, July 6–8.

Ulrich, D., M. A. Von Glinow, and T. Jick. 1993. High-impact learning: Building and diffusing learning capability. *Organizational Dynamic* 22 (Autumn): 52–66.

Voropajev, V. 1998. Change management: A key integrative function of PM in transition economies. *International Journal of Project Management* 16(1):15–19.

Walsh, J. P., and G. Ungson. 1991. Organizational memory. *Academy of Management Review* 16: 57–91.

Weick, K. E. 1979. *The social psychology of organizing*. New York: Random House.

————. 1991. The nontraditional quality of organizational learning. *Organization Science* 2:116–123.

Wenger, E. 1998. *Communities of practice: Learning, meaning, and identity*. New York: Cambridge University Press.

Wenger, E., R. McDermott, and W. M. Snyder. 2002. *Cultivating communities of practice: A guide to managing knowledge*. Boston: Harvard Business School Press.

Wenger E. C., and W. M. Snyder. 2002. Communities of practice: The organizational frontier. *Harvard Business Review* (January–February): 139–145

Wheatley, M. 2001. The real work of knowledge management. *Human Resources Information Management Journal*, 5(2):30–37.

Wiig, K. M. 1999. Introducing knowledge management into the enterprise. In *Knowledge management handbook*, ed. J. Liebowitz. Boca Raton, FL: CRC Press.

Zack, M. 1999. Managing codified knowledge: A framework for aligning organizational and technical resources and capabilities to leverage explicit knowledge and expertise. *Sloan Management Review* 40: 45–59.

CHAPTER NINE

THE VALIDITY OF KNOWLEDGE IN PROJECT MANAGEMENT AND THE CHALLENGE OF LEARNING AND COMPETENCY DEVELOPMENT

Peter W. G. Morris

I was recently invited to speak at two seminars that sought to challenge the prevailing views on what we know about the management of projects. The first stated that it was time to move project management on from its perhaps rather tired and dated positivist, if not to say quasi-normative, origins stemming from its engineering-based background, to reflect a much more complex reality, as characterized, for example, by organizational change-type projects, where more interpretive views of the subject were appropriate. Fair enough, but the implication seemed to be that some of the knowledge we believe we have so carefully been trying to build up over the last 40 or more years since the discipline's emergence in its contemporary form was somehow no longer relevant. This was a position that seemed worth challenging.

The second was a seminar organized on "Managing Projects in the Pharmaceutical Industry" and a research institute's—Fenix's, in Göthenburg—work on product development contrasted against my own research, not least current work on strategy, concurrent engineering, and project-based learning (Morris and Jamieson, 2003; Miranda Lopez and Morris, 2003; Morris, 2002; Morris, Lampel, Jha, and Loch, 2003). Adler, of Fenix, for example, takes three management frameworks for the product development challenge of dealing with uncertainty (Adler, 1999): new product development,; organization theory, and project management. For Fenix, the "dominant model" of project management is control: planning, risk analysis, and monitoring to reduce uncertainty. But who says this is "the dominant model?" Certainly our own research over the years reflects a model of project management that is far broader than project control—what we in fact refer to as "the management of projects." But then again, who says we are right? What *is* "the model" of project management? How generalizable, as well as valid in the first seminar sense, is our knowledge of project management?

Why should there be so much doubt about how best to manage projects?

This chapter argues that there *are* such things as established good practices for managing projects successfully. That there are models, or rules, of project management (i.e., having a predictive ability) that more or less work. But there are indeed problems: Management *is* contextual and there are quite severe limits to our predictive capacity. And within this context, learning is nevertheless possible, though again there are quite significant challenges and limitations.

What Is Project Management?

Project management now occupies an established place in most business bookshops, with dozens of texts telling the reader how to manage projects. There have been literally thousands of books and papers on project management published over the last 40 years, with several hundred being added to the list each year. Most, it is true, talk about planning and control and organization and managing teams (these latter aspects are not included in Adlers' "standard model," incidentally). Increasingly, management of technical issues is also being included—particularly when catering to the systems market, where issues such as requirements management and configuration management figure strongly (e.g., Lientz and Rea, 1999; Forsberg, Mooz, and Cotterman, 1996; OGC, 2002). And if the literature reflects a construction, new product development or defense/aerospace background, procurement will likely figure strongly as well (e.g., DoD, 1996; Marsh, 2001), as the chapters by Venkataraman, Roulston, Milosevic, Morris, and several others in this book attest. All these "popular" writings are trying in effect to suggest norms or good practices of managing projects. But how valid is the knowledge they are seeking to impart? What indeed should be included in the definition of project management? Should the paradigm be an implementation, execution "on time, in budget, to scope" one, or should we be taking a broader view and be including the setting up of the project and the delivering of it to achieve stakeholder satisfaction?

What, first, should the model of project management be? And within this model, how valid is our knowledge?

There are probably at least something like 140,000 or more people—the membership of the Project Management Institute (PMI)—who might be expected to say that project management is as defined in PMI's *A Guide to the Project Management Body of Knowledge* (PMI, 2000).

Though widely accepted, many practitioners, academics, and others, however, believe this model to have serious shortcomings. It contains nothing detailed on project strategy, nothing on project definition, little on value management, nothing on technology management, and little on the linkage with programs and portfolios. There is nothing on leadership and minimal material on team-based development. One begins to suspect, in fact, that it represents an old-fashioned view of project management as tool-based, ignoring the broader context and treating strategy and technology as a given, with people essentially as an interchangeable commodity. All these shortcomings derive largely from its intellectual perspective of project management as primarily an *execution* discipline: of delivering a project "on time, in budget, to scope."

There are many people for whom this perspective is quite adequate. Many firms still separate the project execution end of projects from the project definition and development, labeling the former as project management and the latter as something else (e.g., project development). For others, however, from both a practical and theoretical view, it is not. It misses the whole area of setting up the project objectives (defining scope, budget, and schedule), as well as the linkages with business performance (through portfolio and program management, project strategy, and value management). For many, this really is *the* important area in the successful accomplishment of projects. And for them, seeing the management of the project as a whole brings added benefits of optimization. Project management has to be about delivering business benefit through projects, and this necessarily involves managing the project definition as well as downstream implementation.

Though this is obviously more an owner-oriented view of managing projects, even many contractors need a model of project management that involves managing the front end (commercial, procurement and logistical strategy, technology and design management, etc.), as, for example, in design-build type contracts.

This debate about the intellectual model of project management raises the first and probably the most important philosophical question about project management: What is its remit?

Addressing this question may be approached by looking at the research on project success and failure, as described, for example, in the chapter by Cooke-Davies in this book. The question of success and failure is important because it defines the goals, and hence the activities, of the discipline.

The work by Baker, Murphy, and Fisher (1974, 1988); Pinto and Slevin (1988); and Lechler (1998) assesses success both in terms of project outcome and project management process—did the project meet its mission, were the parent and client groups, and the project team, satisfied with the outcome? For Morris and Hough (1987), however, the measures of success are more explicitly outcome-oriented: What use is it following all the project management processes if the project is not a success? Success measures include both the traditional ones of "on time, in budget, to scope" delivery, but also those most relevant to the different stakeholders. The stakeholders may be many, but the two principal ones are the sponsor—the person funding the project who is responsible for its business success—and the suppliers. Hence, two of Slevin and Pinto's ten key success factors are concerned with consulting with, and selling to, "the client," who is treated almost as an external party, while Morris and Hough consider the client as integral to the project—perhaps even dominant.

Further, Morris and Hough found that the process of *defining* the project was critical to the chances of subsequently delivering the project successfully, and that often external factors such as business-driven changes, geophysical or socioeconomic factors, or technical or supply chain problems, impacted the likelihood of project success more than the control and organization ones typified by the PMBOK-type model. (Admittedly their study was of "major" projects, but interestingly both Pinto and Slevin and Lechler have the management of technology among the top ten factors, while PMBOK ignores this topic altogether.)

The net effect of the Morris and Hough work, I would like to suggest, is to emphasize more the importance of managing the project for a successful outcome and less on the internal processes of project management for their own sake.

Indeed, it is ultimately perhaps more useful to recognize that what really distinguishes projects from nonprojects is primarily their development cycle, as shown in Figure 9.1. The sequence of going from concept into definition, development/execution and into completion/delivery (sometimes including operation, leading to closeout)—or some variant of these words. All projects go through the same sequence, no matter how trivial or complex (only the wording will change). Nonprojects—steady-state operations—do not follow this development cycle. Really, therefore, what we are talking about is *the management of projects*.

Making this recognition leads to a paradigm shift. For the "dominant model" is now far from just project control. It is beyond even PMBOK, but not the APM BoK (APM, 2000), shown in Figure 9.2, or even the IPMA "Competence Baseline" or Japanese ENAA P2M model (Caupin et al., 1998; Engineering Advancement Association of Japan, 2001). It is the management discipline of how one initiates, develops, and implements projects for stakeholder success and includes portfolio and program management, project strategy, technology management, and commercial management as much as the traditional areas of project control and organization. (Crawford concludes on the same point in her chapter on project management standards.)

How we define the paradigm within which we work is of the utmost importance, as Kuhn so clearly showed (Kuhn, 1970). It sets the terms of enquiry; stimulates the areas of research and of professional discourse; and, in a very practical way, as I shall be describing, outlines the challenge of the way we organize, learn, and improve.

Unfortunately, this broader paradigm involves not just a vastly enlarged intellectual framework but, insofar particularly as it requires knowledge leading to successful outcomes, a more contextual understanding where tacit knowledge and judgment are called upon, as much as observation of quasi-normative, positivist PMBOK-type rules.

Defining Our Knowledge about Managing Projects

If there really is a discipline of managing projects that can be described generally, then there needs to be some knowledge about it that can be articulated with a reasonable degree

FIGURE 9.1. THE GENERIC PROJECT DEVELOPMENT CYCLE.

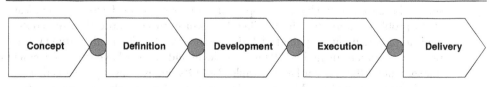

FIGURE 9.2. THE APM PROJECT MANAGEMENT BODY OF KNOWLEDGE (4TH EDITION, 2000).

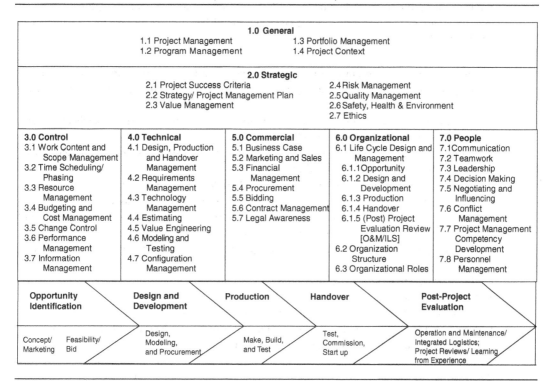

of robustness. One of the challenges implicit in the first seminar was the suggestion that much of the knowledge previously built up in this area is of questionable validity. How valid itself is this suggestion?

All management, as Griseri among many others has argued (Griseri, 2002), is contextual, and hence there will always be limitations to the generalizability of our knowledge of how to manage projects. Yet there have been very serious efforts over the years to document "lessons learned," and indeed I contend that there are indeed some "best practice" guidances that, if applied, should lead to improved chances of successfully managed projects than would be the case were they gone against.

There is nothing wrong in approaches such as ethnography or poststructuralism looking at organizations with fresh eyes, so to speak, to try to elucidate knowledge about how best to manage projects, but there *is* surely knowledge already extant that does have empirical as well as critical validity.

Projects are particularly goal-oriented forms of organization. Indeed, project management is close to an "instrumental rationality" view of the world—"rational action oriented to practical goals" (Weber, 1949). There is every reason why normative rules should apply in some instances; and therefore there is no reason why normative guidance on how to define and deliver projects should not be given (particularly in the more straightforward project management "on schedule, in budget, to scope" model of the discipline).

Thus, for example, I might say that at a general level, good project management will involve the following:

- Aligning the development of the project strategy with the sponsor's (and other stakeholders') business strategy, including reviewing the project strategy at formal review (investment) gates
- Defining the requirements (in a testable manner) so that these lead to specifications and solutions being designed and developed
- Managing design and technology so that innovations are thoroughly examined before proceeding to full project commitment
- Defining and managing the project scope, schedule, resource requirements, and budget (ensuring this represents optimal financing), including limiting changes once the design has been agreed (design freeze); integrating cost, schedule, and scope measurement; and conducting trend analyses on anticipated outturn performance
- Procuring/inducting resources into the project in as positive, cost-effective/value-creating manner as possible
- Building effective project teams
- Exercising leadership
- Ensuring effective decision and efficient communications
- Reviewing lessons-learned after and during the project and feeding other knowledge into the project through formal peer review sessions

The trouble, of course, is determining at what point such knowledge becomes so generalized that it is of limited value, and at what point is it so specific that it is no longer generalizable. Normative rules must inevitably be linked to concrete or generally agreed-to definitions of project management to be useful. Until a definitional basis is established, rules will simply generate more exceptions than use. Further, the more context-specific a discipline, the less normative rules can be generally advanced.

There are at least two problems here. One is the critical realism perspective that reality is stratified (Bhaskar, 1978, 1998; Outhwaite, 1987): There may indeed be causal relationships (laws, event sequences, etc.) discernable at a level of observation, but these are just subsets of what can be observed, and what can be observed is itself a subset of what, at a deeper level of reality, in fact exists. It is inevitable, particularly in the social sciences but in many of the physical ones too, that our knowledge of reality is incomplete.

The other issue is the nature of the knowledge being called upon when talking about how to manage projects. If we accept a framework such as the Morris and Hough/APM Body of Knowledge one, then in order to manage projects successfully, the project management practitioner will need to draw on knowledge, skills, and behaviors in strategic, technical, and commercial management areas, as well as ones in control, organization, and behaviors. The nature of our knowledge in these different areas differs.

Substantively it differs in several ways. The nature of our knowledge about finance, for example, is different from the nature of our knowledge about leadership; similarly production management is different from strategy; and so on. Project management, certainly in the broader paradigm of "the management of projects," covers many branches of management, and to an extent this even holds for the more limited PMBOK model (integration versus time management; quality versus scope).

Even within a topic area, of course, there will be epistemological variations: Our certainty of knowledge about scheduling can be quite robust normatively as regards critical path but more uncertain when it comes to critical chain (insofar as much of the power of critical chain lies in the aggregation of contingencies and in the benefit that management can make of this in motivating teams and juggling grouped contingencies).

Substantively, in fact, as I have already pointed out, it becomes much harder to capture and codify rules relating to the likelihood of certain outcomes emerging. But this is not necessarily the case from a process point of view. Risk management is a good example. The process steps for carrying out a risk analysis can be described easily. But making substantive judgments about which risks are a greater priority, and how best to deal with these, often requires considerable judgment.

In reality, much of our knowledge of how best to manage projects is process-based—how to manage changes, do a risk or value management exercise, prepare a quality plan, form a team, bid a contract, do a design review, prepare a schedule, and so on. This process-based knowledge, however, is then set in a context: how it is done in "our industry," "our company," "our business unit," or even "our project."

It is the process-based knowledge that largely forms the base of project management's generalizable rules. It would be naïve to maintain that these are to a degree also not context-dependent, but they are more generalizable than the substantive knowledge of the context-specific situations that sit above these basic process "good practices." The difference is not quite one of normative, positivist rules versus constructivist insights, but rather one of stratification.

Project Learning

This hypothesis about project management knowledge raises questions about how we can learn the discipline of project management and what relevance this knowledge has in delivering successful projects. Most work on learning in project-based organizations in recent years has, *pace* the preceding, recommended a number of process "good practices"—which themselves begin to resemble valid process rules (laws or event sequences). For example, project performance should be improved through

- the systematic collection of learning on projects (Dixon, 2000);
- periodic project reviews by peer teams (peer reviews and peer assists) (Collison and Parcell, 2001);

- distinguishing between tacit and explicit knowledge (Morris, 2002; Fernie et al., 2003; Nonaka and Takeuchi, 1995);
- identification of key persons as repositories of tacit knowledge and as "owners" of subject matter areas (Ayas and Zeniuk, 2001; Wenger, 1998);
- the use of information management tools to aid in the capture, storing, processing, archiving, looking up, retrieving, and presenting of information (Morris, 2003; Currie, 2003);
- a discipline of accessing knowledge—using checklists or other "look-up" guides by the project teams before beginning a new project task (Brander-Löf, Hilger, and André, 2000);
- having a definition in some way of the knowledge in a particular area: the "Body of Knowledge" (Morris, 2001; Wenger, 1998);
- having a knowledge management program in place (Turner, Keegan, and Crawford, 2000)
- distinguishing between individual, team, and organizational learning (Popper and Lipshitz, 2001);
- a program or programs for using the knowledge/learnings identified (Cross and Baird, 2000; Schindler and Eppler, 2003);
- applying metrics for the usage made of the knowledge and learning;
- implementing a competency development program for updating the knowledge.

But how far does this process view of project management help us in learning how to manage projects better? It is clearly helpful for the more process-oriented type of management challenge—as I have characterized the PMBOK-type model to be—but less so where people are looking for a combination of substantive and process knowledge, where context and judgment is important, and experience and tacit knowledge particularly valuable (Bresner et al., 2003; Fernie et al., 2003). In fact, even at the process level, context is unavoidable.

Many researchers have identified the need for contextual knowledge and have emphasized the importance of tacit knowledge (Dierkes, Antal, Child, and Nonaka, 2001; LeRoy, 2002; Scarbrough et al. 2002; etc.). Loch (2000) and Ferlie and Loch (2001) contend that the process of learning is mediated by both the "content" and the "context" of the process. Boisot (1998) has elaborated a similar idea in terms of his I-Space (information space), where learning takes the form of a spiral (scanning, problem solving, abstraction, diffusion, absorption, impacting) and where the learning process is affected by the degree to which the knowledge is

- codified or uncodified—the business context may be relatively codified (KPIs and strategy) or vague ("our preferred way of working");
- abstract or concrete—for example, a risk management support guide is generally less abstract than an overall project management methodology;
- diffuse or undiffused—the knowledge may be "generally available" or have high company specificity (IPR).

In short, different learning approaches will be required for different types of knowledge. Further, not only are there different learning requirements for different types of (project management) topics, it is generally recognized that the more strategic and cognitive type of learning (more prevalent in the broader model) is more difficult than the more routine and process-oriented learning (more prevalent in the traditional project management model). (This relates to the incrementalist, single-loop theory of learning (e.g., Levinthal and March, 1993; Miner, 1990) compared with the more strategic, cognitive—double-loop—perspective (Argyris and Schön, 1978; Fiol and Lyles, 1985).)

These findings are now being reinforced in research on the rollout of project management best-practice guidance, which suggests that different enabling mechanisms may be appropriate for different project management practices. The thrust of this work, which combines both the model of the spiral of learning of Nonaka et al. (Nonaka and Takeuchi, 1995) and the work on enabling mechanisms of Von Krogh (Von Krogh, Ichijo, and Nonaka, 2000), is that different enabling mechanisms are more relevant in the different modes of knowledge conversion (Socialization-Externalisation-Combination-Internalisation, per Figure 9.3) posited by Nonaka. Further, that within the modes of knowledge conversion

FIGURE 9.3. LEARNING AND SUPPORT MECHANISMS APPROPRIATE FOR DIFFERING STAGES OF THE KNOWLEDGE CREATION SPIRAL.

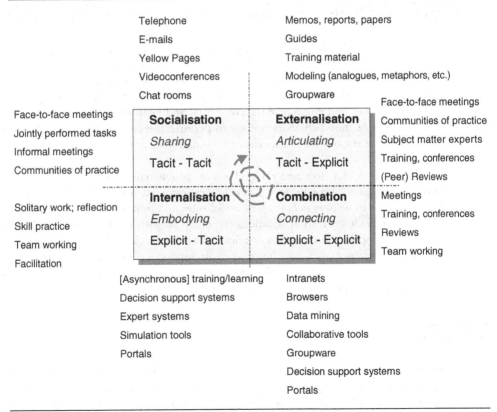

Source: After Nonaka, von Krogh, et al.

there could be a skew in the four quadrants. It may be skewed at different times in the organization, and at different stages of the project (Morris, Loch, Lampel, and Jha, 2003; LeRoy, 2000). For example, the socialization mode appears more dominant where distillation (stories/narratives/word-of-mouth) is the dominant strategic mode (Lampel, Morris, Loch, and Jha, 2003).

And though von Krogh suggests learning/knowledge creation happens best when all quadrants of the spiral are balanced, I have observed that, in a project context at least, this situation may not occur. In fact, all the organizations studied in this research experienced significant learning constraints, even though they might know the theory quite well, due for example to

- strategic intent (perhaps a conscious intent to focus on people rather than tools for knowledge sharing, or from being so preoccupied with creating customer solutions and doing the business that trans-project learning suffers);
- organizational constraints (lack of project orientation, the silo effect);
- virtual working (thus, a focus on explicit ways of working) or the attention it receives or does not receive (wrong priorities and incentives; lack of personal interest).

Implications for Project Management Competence Development

So what does this imply for the development of project management competence? Is there, for example, a place for a generic discipline-based certification program?

Competence is generally taken to be role-specific and to comprise the knowledge, skills, and behaviors needed to perform the role (Boyzatis, 1982). The role specificity of the concept allows us to focus and emphasize the contextual and tacit dimensions of the learning of the knowledge, skills, and behaviors needed to perform the role properly (which, of course, is also one of the reasons why "experience" requirements are often tagged onto definitions of competences.)

The outlook at a more general level is more problematic, however. If the noncontextual, generalized type of knowledge is more process-oriented (more PMBOK-like), then while being of value in helping people learn useful rules and routines, it is less valuable in helping deal with difficult contextual situations requiring judgment. More contextual learning will be more appropriate (such as case studies and team exercises). Professional accreditation or certification is hence at best an inadequate guarantee of professional performance (nice but not necessarily very useful)—which, I would suggest, just about accords with what most practicing project professionals believe to be the case.

Summary

It would be a mistake to ignore the "good practice" guidance and rules that have been built up and formalized as quasi-normative knowledge on project management over the last 40 or more years. Much of this knowledge relates to the processes of managing projects effectively. But even such process knowledge soon requires an understanding of the context in which management operates. This is particularly the case where more strategic and judgmental decisions are being made. The larger "management of projects" framework calls for more of this type of knowledge than the more circumscribed PMBOK-type model of project management.

This has implications for the way we learn about project management and for the value of project management accreditation programs. Incremental-type learning, of rules and practices, should be distinguished from the more challenging strategic and cognitive learning.

References

Adler, N. 1999. *Managing complex product development.* Stockholm: Stockholm School of Economics, EFI.

Argyris, C., and D. Schön. 1974. *Theory in practice: Increasing professional effectiveness.* San Francisco: Jossey-Bass.

Association for Project Management. 2000. *Body of knowledge.* 4th ed. *www.apm.org.uk.*

Ayas, K., and N. Zeniuk. 2001. Project-based learning: Building communities of reflective practitioners. *Management Learning* 32(1):61–76.

Baker, B. N., D. C. Murphy, and D. Fisher. 1974. *Determinants of project success.* NGR 22-03-028.

National Aeronautics and Space Administration. 1988. Factors affecting project success. In *Project Management Handbook.* 2nd ed., ed. D. I. Cleland and W. R. King. 902–919. Hoboken, New Jersey: Wiley.

Bhaskar, R. 1978. A realist theory of science. York, UK: Leeds Books

———. 1998. *The possibility of naturalism.* Atlantic Heights, NJ: Humanities Press.

Boisot, Max. 1998. *Knowledge assets.* Oxford, UK: Oxford University Press.

Boyatzis, R E. 1982. *The competent manager: A model for effective performance.* Hoboken, New Jersey: Wiley.

Brander-Löf, I., J-W Hilger, and C. André. 2000. How to learn from projects: The work improvement review. *IPMA World Congress 2000.* International Project Management Association, Zurich.

Bresner M., L. Edelman, S. Newell, H. Scarbrough,and J. Swan. 2003. Socialpractices and the management of knowledge in project environments. *International Journal of Project Management* 21(3):157–166.

Caupin, G., H. Knöpfel, P. W. G. Morris, E. Motzel, and O. Pannenbäcker. 1998. ICB IPMA competence baseline. International Project Management Association, Zurich. www.ipma.ch.

Collison, C., and G. Parcell. 2001. *Learning to fly.* Oxford, UK: Capstone Publishing Ltd.

Cooke-Davies, T. J. 2000. The "real" success factors on projects. *International Journal of Project Management* 20(3):185–190.

Cross, R., and L. Baird. 2000. Technology is not enough: Improving performance by building organizational memory. *Sloan Management Review* 14 (3, Spring)

Currie, W. L. 2003. A knowledge-based risk assessment framework for evaluating Web-enabled application outsourcing projects. *International Journal of Project Management* 21(3):207–218

Dierkes, M., A. B. Antal, J. Child, and I. Nonaka, eds. 2001. *Handbook of Organisational Learning and Knowledge.* Oxford, UK: Oxford University Press.

Dixon, N. 2000. *The organizational learning cycle: How we can learn collectively.* New York: McGraw-Hill.

DoD. 1996. *Mandatory procedures for major defense acquisition programs and major automated information systems.* Directive 5000.2-R. Washington, D.C.: Department of Defense,

Engineering Advancement Association of Japan. 2001. *P2M: Project and program management for enterprise innovation.* www.enaa.or.op.jp

Ferlie, E., and I. Loch. 2001. Change management and organisational learning in primary care. *Report to NHS Executive R and D (SE Region)*. London: The Management School, Imperial College.

Fernie, S., S. D. Green, S. J. Welleer, and R. Newcomber. 2003. Knowledge sharing: context, confusion and controversy. *International Journal of Project Management* 21(3):177–188.

Forsberg, K., H. Mooz, and H. Cotterman. 1996. *Visualizing project management*. New York: Wiley.

Griseri, P. 2002. *Management knowledge: A critical view*. London: Palgrave.

Kuhn, T. S. 1970. *The structure of scientific revolutions*. Chicago: The University of Chicago Press.

Lampel, J., P. W. G. Morris, P. Jha, and I. Loch. 2003. Projects and the organisation: A strategic learning interface. *Organizational Knowledge and Learning Conference*. Barcelona.

Lechler, T. 1998. When it comes to project management, it's the people that matter: An empirical analysis of project management in Germany. IRNOP III. The Nature and Role of Projects in the Next 20 Years: Research Issues and Problems. Calgary: University of Calgary.

Leintz, B. P. and K. P. Rea. 1999. *Breakthrough technology management*. London: Academic Press.

LeRoy, D. 2002. Knowledge management and projects' capitalization: A systemic approach. *Proceedings of PMI Research Conference*.

Loch, I. 2000. Learning, change, and professional paradigms in primary care. Abstract in *British Academy of Management 2000 Conference Proceedings*. University of Edinburgh.

Lopez Miranda, A., and P. W. G. Morris. Forthcoming. A conceptual framework for maximizing the benefits from implementing concurrent engineering with project management. *International Journal of Agile Manufacturing*.

Marsh, P. 2001. *Contracting for engineering and construction projects*. Aldershot, UK: Gower.

Miner, A. S. (1990) 'Structural Evolution through Idiosyncratic Jobs: the Potential for Unplanned Learning', *Organization Science*. pp. 195-210.

Morris, P. W. G.. 2001. Updating the Project Management Bodies of Knowledge. *Project Management Journal* 32(3):21–30.

———. 2002. Managing project management knowledge for organisational effectiveness. *Proceedings of PMI Research Conference 2002, Seattle*. Newtown Square, PA: Project Management Institute.

———. 1992, 1997. *The Management of Projects*. London: Thomas Telford.

Morris P. W. G., P. M. Deason, T. M. S. Ehal, R. Milburn, and M. B. Patel. 2003. The role of IT in capturing and managing knowledge in designer and contractor briefing. *IT in Architecture, Engineering and Construction* 1(1):1–18.

Morris, P. W. G., and G. H. Hough. 1987. *The anatomy of major projects*. Chichester, UK: Wiley.

Morris P. W. G., J. Lampel, P. Jha, and I. Loch. 2003. Organisational learning and knowledge creation interfaces in project-based organisations. *EURAN 2003*. Milan, European Academy of Management.

Morris, P. W. G. and H. A. Jamieson. 2004. Moving from Corporate Strategy to project strategy. Newtown Square, PA: Project Management Insitute.

Morris, P. W. G., I. Loch, J. Lampel, and P. P. Jha. 2003. A construct for project-based learning: The PROBOL model. *OLK 5 Conference*, Lancaster University Management School, June.

Nonaka, I, and H. Takeuchi. 1995. *The knowledge-creating company*. New York: OUP.

Office of Government Commerce. 2002. *Managing successful projects with PRINCE2*. London: The Stationery Office.

Outhwaite, W. 1987. *New philosophies of social science: realism, hermeneutics, and critical theory*. New York: Macmillan.

Pinto, J. K., and D. P. Slevin. 1988. Critical success factors across the project life cycle. *Project Management Journal* 19(3):67–75.

Popper, M., and R. Lipshitz.. 2001. Organizational learning: Mechanisms, culture and feasibility. *Management Learning* 31(2).

Project Management Institute. 2000. *A guide to the Project Management Body of Knowledge,* Newtown Square, PA: Project Management Institute.

Scarbrough, H., M. Bresnen, L. Edelman, J. Swan, S. Laurent, and S. Newell. 2002. Cross-sector research on knowledge management practices for project-based learning. *European Academy of Management 2nd Annual Conference.* Stockholm.

Schindler, M., and M. J. Eppler., 2003 Harvesting project knowledge: A review of project learning methods and success factors. *International Journal of Project Management* 21(3):219–228.

Shenhar, A. J., O. Levy, and D. Dvir. 1997. Mapping the dimensions of project success. *Project Management Journal* 28(2):5–13.

Turner, J. R., A. Keegan, and L. Crawford. 2000. Learning by experience in the project-based organization. *Proceedings of PMI Research Conference 2000, Paris.* Newtown Square, PA: Project Management Institute.

Von Krogh, G., K. Ichijo, and I. Nonaka. 2000. *Enabling knowledge creation.* New York: Oxford University Press.

Weber, M. 1949. *The methodology of social sciences.* New York: Free Press.

Wenger, E. 1998. *Communities of practice.* Cambridge, UK: Cambridge University Press.

CHAPTER TEN

GLOBAL BODY OF PROJECT MANAGEMENT KNOWLEDGE AND STANDARDS

Lynn Crawford

The definition of a body of knowledge and the development of standards for project management have been significant features of the growing interest in this field of practice as it aspires to recognition as a profession. There has been much debate within the field of project management as to whether it satisfies criteria for status as a profession (Zwerman and Thomas, 2001). One view, however, is that professions can be considered to begin either with the recognition by people that they are regularly doing something that is not covered by other professions or with the formation of professional associations (Abbott, 1988). The impetus behind the formation of professional associations is considered by Eraut (1994, p. 165) to be "derived from the perceived need of a relevant group to occupy and defend for its exclusive use a particular area of competence territory" and that, politically, their interest may lie in claiming that only their members are competent within that territory.

Definition of a distinct body of knowledge and of standards based on that body of knowledge are ways of marking professional territory (Morris et al., 2000; Berry and Oakley, 1994). Assessment and award of qualifications provides a process whereby professionals are recognized as meeting the standards of a profession by demonstrating mastery of the body of knowledge and either minimum or graduated levels of proficiency or competence (Dean, 1997). A body of knowledge, standards, and related assessment and qualification processes can therefore be seen as essential building blocks in the formation and recognition of a profession (see Figure 10.1).

Development and recognition of a distinct profession of project management has certainly been a strong driver in the development of standards. Other related factors that have been significant:

FIGURE 10.1. BUILDING BLOCKS OF A PROFESSION.

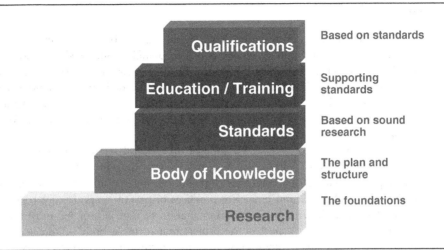

Qualifications	Based on standards
Education / Training	Supporting standards
Standards	Based on sound research
Body of Knowledge	The plan and structure
Research	The foundations

- The need to identify the role and tasks of project managers in an emerging field of practice where neither the project managers, their clients, nor their employers necessarily had a clear understanding of the role
- The need for common terminology
- The need for a common basis for employment and deployment of project personnel, working collaboratively, across functions in multidisciplinary teams; across organizations in strategic alliances and joint ventures; and across continents, in global projects

Project personnel are actively seeking sound guidance for the identification of project management competencies and credentials that will enhance their careers. As project based work takes over from position-based work and careers are defined less by companies and more by professions (Stewart, 1995), project personnel are keen to achieve professional status and independent recognition of their project management competence. If people are to be evaluated, not by rank and status but flexibly according to competence (Stewart, 1995), then evidence of this competence becomes extremely important to individuals as well as to organizations. A distinct body of project management knowledge and standards providing a baseline for project management competence are important building blocks in professional recognition for individuals and for an emerging profession.

This chapter presents an overview of the current principal project management standards and guides for project management knowledge and performance, including a comparison of their content and coverage and an indication of their use in assessment and as a basis for qualifications. Current developments and potential future directions are also reviewed.

Project Management Standards and Guides

Overview

A *standard* is a measure, devised by general consent, as a basis for comparison against which judgments might be made as to levels of acceptability. Standards, to have effect, do not need to be officially endorsed. They can be voluntarily accepted. In the field of project management, there is a very strong link between the definition of a project management body of knowledge and the development of standards, with a number of guides to aspects of the project management body of knowledge being treated as standards for what project management practitioners are expected to know.

It is important to note that although the general territory or coverage of a project management body of knowledge may be defined, the entire body of knowledge, which encompasses tacit and explicit knowledge embodied in published and unpublished material and in the established and emerging practices of practitioners, cannot be captured in any single document. Therefore, any documentation of that knowledge must be considered as a guide to one view or aspect of the project management body of knowledge at a point in time. This distinction is made very clear in the introduction to the Project Management Institute's *A Guide to the Project Management Body of Knowledge* (PMBOK® Guide, Project Management Institute, 2000, p. 3), which is the most widely distributed of a number of body of knowledge guides and is also recognized as a standard by the American National Standards Institute.

Guides and standards have been developed for project management for various purposes, which can generally be classified as the following (Duncan, 1998):

- *Projects.* Knowledge and practices for management of individual projects
- *Organizations.* Enterprise project management knowledge and practices
- *People.* Development, assessment, and registration/certification of people

The most widely known, distributed, and used guides and standards for project management are presented in Figure 10.2, indicating their general focus: projects, organizations, or people. They can be further classified as either focusing on knowledge or on description of practices, the latter being primarily in the form of performance-based competency standards or frameworks intended specifically for assessment and development of project management practice in the workplace. There are a number of standards for aspects of project management that are provided in languages other than English, but these are not included here, since their application tends to be limited to their country of origin. The Japanese P2M is an exception, as an English translation of a significant part of the standard has been widely distributed.

Projects:

Those standards and guides that focus primarily on what project management practitioners need to know (knowledge guides) are also those dealing essentially with management of individual projects. The Japanese P2M stands out as the one exception, specifically extending the focus beyond the management of single projects to management of programs of projects

FIGURE 10.2. SUMMARY OF STANDARDS AVAILABLE THAT FOCUS ON PROJECTS, PEOPLE AND ORGANIZATIONS.

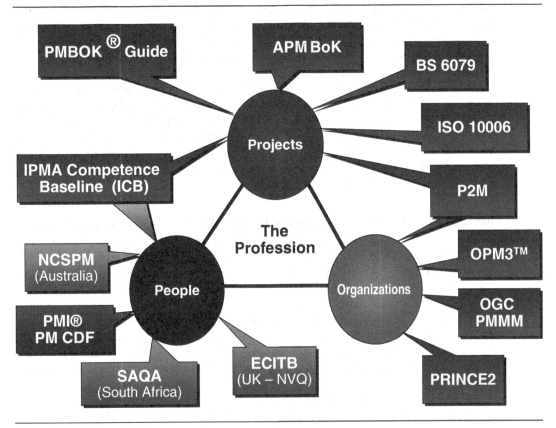

in the context of corporate strategy implementation and enterprise innovation and management. The standards and guides shown in Figure 10.2 as focusing on knowledge and relating primarily to management of individual projects are as follows:

- Project Management Institute's *A Guide to the Project Management Body of Knowledge* (Project Management Institute, 2000)
- *Association of Project Management Body of Knowledge* (APM BoK, UK; Dixon, 2000)
- BS6079 *Guide to Project Management* (British Standards Board, 1996)
- ISO 10006 *Guidelines to Quality in Project Management* (ISO, 1997)
- *ICB: IPMA Competence Baseline* (Caupin et al., 1999)
- *P2M: A Guidebook of Project and Program Management for Enterprise Innovation* (Engineering Advancement Association of Japan Project Management Development Committee; ENAA, 2002)

All of the documents listed are considered to be standards, either formally or informally. All of the documents, except BS6079 and ISO 10006, are used as the knowledge base or standard for professional certification programs. ISO 10006 provides guidelines to quality in project management and is certainly a standard. It is, however, primarily a quality management rather than a project management standard. There has been discussion of the potential for development of an ISO standard for project management, but to date no such standard has been produced.

The P2M was developed by the Project Management Development Committee of the Engineering Advancement Association of Japan with funding from the Japanese government through the Ministry of Economy, Trade, and Industry (METI). The P2M provides the basis for a certification program, and the Project Management Professionals Certification Center (PMCC) was established in May 2002 to manage this certification program and maintain the P2M. All of the other documents have been developed by and are maintained by project management professional associations.

PMBOK Guide. The Project Management Institute (PMI) had been working on defining or mapping what constituted the body of knowledge of project management since the mid-1980s (Wideman, 1986), and its first project management standard was published in 1983 as part of the PMQ Special Report on Ethics, Standards and Accreditation (Project Management Institute, 2002a). The standards portion of the report identified six major project management functions: human resources management, cost management, time management, communications management, scope management, and quality management. The *Project Management Body of Knowledge of the Project Management Institute* was first published in *PM Network* in 1987 and included the six functions identified in the 1983 document as well as two further project management knowledge functions: risk management and procurement management.

The publication, in 1996, of *A Guide to the Project Management Body of Knowledge* marked a major milestone in the development of project management as a field of practice and aspiring profession. It is this document that has been widely accepted throughout the world as a standard for project management knowledge and has been an important factor in the growth of interest in project management, its dissemination benefiting from rapidly increasing use of the World Wide Web. When first published in 1996, the document was made freely available for download from the PMI Web site. It was also made available to all PMI members through PMI chapters, through the PMI online bookstore, and through publication and dissemination by other project management professional associations such as the Australian Institute of Project Management, under generous cooperative agreements. Trainers and consultants were also able to publish and jointly badge the document.

Through this generous promotional program and as a result of the inherent quality of the document, dissemination was rapid, with over 570,000 copies distributed by the end of 2000. There are now over 1 million copies of the PMI Standard in circulation, and all new members receive a copy of the PMBOK Guide on CD-ROM when they join the Institute.

The PMBOK Guide, published in 1996, changed the functions to knowledge areas, adding one more (Integration), and introduced a process focus (see Figure 10.3). The 1996 and 2000 Editions of the PMBOK Guide include nine project management knowledge areas: Integration, Scope, Time, Cost, Quality, Human Resources, Communications, Risk,

FIGURE 10.3. STRUCTURE OF PMBOK GUIDE.

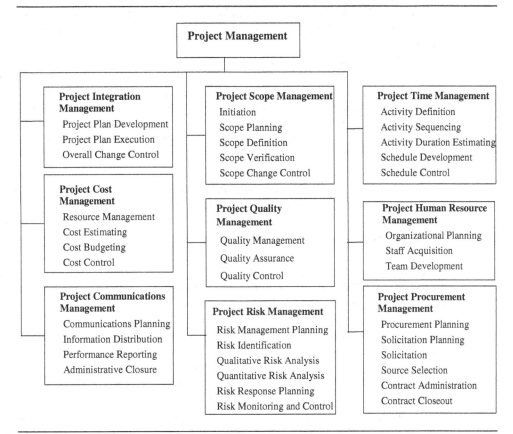

and Procurement, plus a section dealing with project management context and processes, identifying five process groups. A glossary was also added. In the 2000 Edition of the PMBOK Guide, there was "no fundamental revision of the structure or philosophy" (Morris, 2001, p. 23). The major change was a revision and extension of the section relating to risk management. Both editions clearly acknowledge that no single document can embody the whole body of knowledge relevant to project management.

The introduction to the PMBOK® Guide (2000 Edition, p. 3) states that it provides a guide to "that subset of the PMBOK that is generally accepted" in terms of knowledge and practices "applicable to most projects most of the time." The document is not intended to be either comprehensive or all-inclusive. It is used by the PMI to provide a consistent structure for its professional development programs, including but not limited to the following:

- Certification of Project Management Professionals (PMPs®)
- Accreditation of degree-granting education programs in project management

It is intended that *A Guide to the Project Management Body of Knowledge* will be a living document, subject to ongoing review. At the beginning of 2003, the Guide was available in eight languages: Mandarin Chinese, French, German, Italian, Japanese, Korean, Brazilian Portuguese, and Spanish.

In early 1999 PMI was accredited as a Standards Development Organization by the American National Standards Institute (ANSI), and in September 1999, the PMBOK® Guide was approved as an American National Standard (ANSI/PMI 99-001-1999) (Holtzman, 1999). In March 2000, the PMBOK® Guide, 2000 Edition replaced the 1996 Edition as PMI's American National Standard (ANSI/PMI 99-001-2000). The PMBOK® Guide has also been adopted as an IEEE Standard (1490-1998): *IEEE Guide to the Project Management Body of Knowledge* (IEEE, 2000).

Since publishing the PMBOK® Guide, 2000 Edition, PMI has published three other project management standards and guides: *PMI Practice Standard for Work Breakdown Structures*, the *Government Extension to the PMBOK® Guide*, 2000 Edition, and the *Project Manager Competency Development Framework*. PMI has a number of other standards and guides under development (refer to www.pmi.org).

APM BoK. Discussion concerning development of a body of knowledge reference document as a basis for an APM (initially the Association of Project Managers and now the Association for Project Management) certification program began in 1986, led by the Professional Standards Group of the APM (Morris, 2001). The first version of what was referred to as the *APM Body of Knowledge* was published in 1992, updated in 1994 (Second Edition) and 1996 (Third Edition), and significantly revised in 2000 (Fourth Edition). In undertaking what they refer to as a "fundamental revision" (Dixon, 2000, p. 7), the APM sought the assistance of UMIST's Centre for Research into the Management of Projects (CRMP).

The APM Body of Knowledge (Third Edition; APM, 1996) identified 40 key areas that the Association for Project Management considered people involved in project management should have both knowledge of and experience in. The document also listed eight principal personality characteristics that a Certificated Project Manager should display. This edition of the *APM Body of Knowledge* was clearly intended as the reference underlying a certification program, as it included several references to the knowledge, experience, and personality characteristics that would be expected of a Certificated Project Manager. Guidance was given for those seeking certification.

The 40 key areas of knowledge and experience were grouped under four headings:

- *Part 1*. Project Management
- *Part 2*. Organization and People
- *Part 3*. Processes and Procedures
- *Part 4*. General Management

The significantly revised Fourth Edition of the APM BoK has 42 topics listed under 7 headings (see Figure 10.4).

As mentioned earlier, the Fourth Edition of the APM BoK was primarily based on research cosponsored by APM and the industry and conducted at UMIST's Centre for Research into the Management of Projects (CRMP), under the direction of Professor Peter

FIGURE 10.4. APM BOK (FOURTH EDITION, 2000)—SECTIONS AND TOPICS.

1 General
1. Project Management
2. Programme Management
3. Project Context

2 Strategic
4. Project Success Criteria
5. Strategy / Project Management Plan
6. Value Management
7. Risk Management
8. Quality Management
9. Health, Safety and Environment

3 Control
10. Work Content and Scope Management
11. Time Scheduling / Phasing
12. Resource Management
13. Budgeting and Cost Management
14. Change Control
15. Earned Value Management
16. Information Management

4 Technical
17. Design, Implementation and Hand-Over Management
18. Requirements Management
19. Estimating
20. Technology Management
21. Value Engineering
22. Modelling and Testing
23. Configuration Management

5 Commercial
24. Business Case
25. Marketing and Sales
26. Financial Management
27. Procurement
28. Legal Awareness

6 Organisational
29. Life Cycle Design and Management
30. Opportunity
31. Design and Development
32. Implementation
33. Hand-Over
34. (Post) Project Evaluation Review (O&M/ILS)
35. Organisation Structure
36. Organisational Roles

7 People
37. Communication
38. Teamwork
39. Leadership
40. Conflict Management
41. Negotiation
42. People Management

Morris. The aims of the CRMP research (Morris et al., 2000; Morris, 2000) included the following:

- Identifying the topics that project management professionals consider need to be known and understood by anyone claiming to be competent in project management
- Defining what is meant by those topics at a generically useful level.
- Interviewing and collecting data in over 117 companies over a 14-month period

According to Morris (2000, p. 5), the CRMP research "endorsed the breadth of topics in the APM BoK [Third Edition]" and suggested "an even broader scope of topics". Morris (2000, p. 5) points out that the original APM BoK was "strongly influenced by research then being carried out into the issue of what it takes to deliver successful" projects, including work by Baker, Murphy, and Fisher (1988) and undoubtedly by Morris' own work on the subject (Morris and Hough, 1987). Morris et al. (2000) contend that the PMBOK® Guide does not cover all the factors that must be managed to deliver successful projects, claiming that there are other important issues, included in the Fourth Edition of the APM BoK, such as technology, design management, environmental, external, business, and commercial issues. The authors of the PMBOK® Guide would respond by referring to the disclaimer, printed in the Guide, that it is not intended to be either comprehensive or all-inclusive, but to present that subset of the PMBOK that is generally accepted (Project Management Institute, 2000).

The purpose of the Fourth Edition of the APM BoK (2000) has a similar clarity of definition. It claims to be "a practical document, defining the broad range of knowledge that the discipline of project management encompasses" representing those "topics in which practitioners and experts consider professionals in project management should be knowledgeable and competent" (Dixon, 2000, p. 9). It does, however, clearly state that it is not "a set of competencies: (p. 7), although it is suggested that it might be used, in organizations, as "the basis of the project management element of a general competencies framework" (p. 9).

Topics included are those considered to be potentially applicable to all project management situations. They are described at a high level, and readers are referred to texts and other sources for further detail.

A separate *Syllabus for the APMP Examination* (Hougham, 2000) has been published by APM, based on the APM BoK (Fourth Edition). It defines the topics that candidates for the APMP Examination (APM's baseline professional qualification) are expected to know, learning objectives, and a glossary of key terms.

Following the release of the Fourth Edition of the APM BoK, the APM has produced an edited book titled *Project Management Pathways* (Stevens, 2002), which they claim draws upon "current and forward thinking concerning the theory and practice involved in the art and science of project and programme management throughout the world." (APM, 2002, p. 2) and is required reading for the APMP Examination and for all levels of APM professional certification.

ICB (IPMA Competence Baseline). The IPMA (International Project Management Association) has developed the *ICB: IPMA Competence Baseline* (also referred to as the "Sunflower"), which it considers to be a global standard (Pannenbäcker et al., 2002). Work on the ICB was initiated in 1993 and a First Version, in English, French, and German, was presented in June 1998. The primary purpose of the ICB is to provide a basis for certification of project managers. Another important role of the ICB was to bring together or "harmonise" a number of European national project management body of knowledge documents.

The following national project management body of knowledge guides formed the basis of the ICB:

English. Project Management Body of Knowledge, APM, Version 3

German. The Swiss Beurteilungsstruktur, VZPM, Ausgabe 1.0 and The German PM-KANON, PM-Zert, Version 1.0

French. The French Matrice des Projects (AFITEP)

There are 28 core elements and 14 additional elements of project management knowledge and experience (a total of 42 elements) identified from an analysis of the four national documents (see Figure 10.5).

The 28 core elements are presented as a "sunflower" to overcome the difficulties of achieving agreement on a knowledge structure (see Figure 10.6). As Morris et al. (2000) point out, the project management profession has less difficulty agreeing on content of a body of knowledge than they do on a structure for that knowledge.

FIGURE 10.5. ICB: IPMA COMPETENCE BASELINE: CORE ELEMENTS OF PROJECT MANAGEMENT.

Source: Caupin et al., 1999.

FIGURE 10.6. ICB: IPMA COMPETENCE BASELINE: PERSONAL PROFILE.

Attitude	Inventiveness
Common sense	Prudent risk taker
Open-mindedness	Fairness
Adaptability	Commitment

The ICB also includes a section on the expected personality characteristics for a Certificated Project Manager: These are the same as appear in the APM Body of Knowledge (Version 3.0). They were developed in a series of practitioner workshops or meetings conducted by the APM. It is understood that they have no empirical basis.

The *ICB: IPMA Competence Baseline* is intended as the basis of the IPMA Validated Four-Level Certification Program (Pännenbacker, K et al., 1998).The IPMA is a federation of national project management professional associations, and it encourages the national associations to develop their own National Competence Baselines (NCB). The *IPMA Competence Baseline* (ICB) provides the "reference basis" (Pannenbäcker et al., 2002, p. 15) for National Competence Baselines. Each NCB is required, by IPMA, to include all 28 of the ICB core elements and at least 6 additional elements chosen by the nation plus the 8 aspects of personal attitudes and 10 elements of general impression. Up to eight of the additional elements can be eliminated or replaced by new elements, allowing each nation, in developing their NCB, to take into account local requirements or emerging developments in project management. The NCB is used as the basis for national certification programs validated by the IPMA Certification Validation Management Board.

At the end of 2001 there were 21 published NCBs and 5 in preparation. The majority of NCBs have been developed by European countries. An *Egyptian Competence Baseline* was published in 1999 and an Arabic version was in preparation in 2002. The Chinese NCB, also referred to as the C-PMBOK, was published in Standard Modern Chinese in 2001. An Indian NCB was in preparation in 2002.

P2M: A Guidebook for Project and Program Management for Enterprise Innovation. The P2M, *A Guidebook for Project and Program Management for Enterprise Innovation*, was released, in Japanese, in November 2001. An English-language summary version of the P2M (ENAA, 2002) is available covering the total of Parts 1, 2, and 3 and an overview of Part 4, which deals with 11 project segment areas. The Project Management Professionals Certification Center (PMCC) of Japan, a nonprofit organization established in May 2002, has published this summary version. The PMCC is responsible for maintenance of the P2M, for promotion of project management, and for the Certification System for Project Professionals, based on the P2M document.

The Japanese version of the P2M, published in November 2001, is a 420-page document developed following extensive worldwide research, over two and a half years, by the Innovative Project Management Development Committee of the Engineering Advancement Association (ENAA) under the leadership of Professor S. Ohara. The development of the P2M and the associated certification process received support from the Japanese government, primarily through the Ministry of Economics, Trade, and Industry (METI), in recognition that effective project and program management had potential to assist in revival of the Japanese economy.

The focus of the P2M is on "value creation to enterprises, either commercial or public, and a consistent chain from a mission, through strategies to embody the mission, a program(s) to implement strategies, to projects comprising a program" (ENAA, 2002, Preface). Therefore, the document has application to individual projects and also to programs of projects and the wider organizational context. The document is intended to provide a Capability Building Baseline (CBB) for project management and mission-performer professionals, where mission-performer professionals are described as "integration-oriented professionals who perceive complex problems and issues from a high perspective and realize right and optimal solutions" (p. 3). The authors of the P2M align it with PM knowledge guides and standards, claiming that although P2M is "considerably more extensive than existing project management bodies of knowledge or project management competency standards, it does not try to explore every detail of the topics discussed" (p. 2). Rather, it is intended that capability should be developed and expanded through professional experience, as well as related disciplines of science and technology in the context of continuing professional development.

The P2M uses what it calls a Project Management Tower to represent its coverage (see Figure 10.7). Part I: Entry describes how to make a first step as a professional; Part II: Project Management explains the basic definitions and framework of project management; Part III: Program Management introduces management of programs of multiple projects; Part IV: Project Segment Management, considered similar to "Knowledge Areas of Project Management," includes 11 "segments" or knowledge areas of project management that can be used in a "standalone or combined manner for individual tasks and challenges of project management and program management" (ENAA, 2002, p. 17).

People

All of the guides and standards linked in Figure 10.2 with 'People' are primarily concerned with management of individual or stand-alone projects, rather than programs of projects or enterprise project management. With the exception of the ICB (*IPMA Competence Baseline*), which is primarily a knowledge guide, they are generally in the form of performance-based competency standards. Performance-based competency standards are specifically designed for assessment and recognition of current competence, independent of how that competence has been achieved. They also encourage self-assessment, reflection, and personal development in order to provide evidence of competence against the specified performance criteria.

Standards referenced under People in Figure 10.2 that are used as a basis for assessment of competence of project management practitioners and are prepared and/or endorsed by national governments as follows:

FIGURE 10.7. P2M—PROJECT MANAGEMENT TOWER.

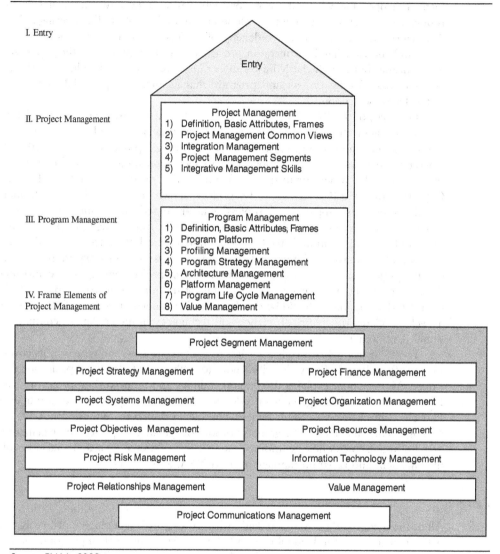

Source: ENAA, 2002.

- National Competency Standards for Project Management (NCSPM) (AIPM [sponsor], 1996) (Australian National Training Authority, or ANTA)
- National Occupational Standards for Project Management (ECITB, 2002) (Qualifications and Curriculum Authority, or QCA)
- National Certificate in Project Management—NQF Level 4 (SAQA) (South African Qualifications Authority, 2001)

All of these standards are formally recognized and provide the basis for award of qualifications within national qualifications frameworks. The National Competency Standards for Project Management also form the basis for award of professional qualifications by the Australian Institute of Project Management.

The Project Management Institute's Project Manager Competency Development Framework (PMI PMCDF; Project Management Institute, 2002c) is intended as a guide to self-assessment and development and is specifically *not* intended as a basis for award of qualifications. The purpose of this document is to describe the "competencies likely to lead to effective project manager performance across contexts," and it is intended for "use in professional development of project managers rather than for use in selection or performance evaluation." Part of the Framework, however, is in a form similar to the performance-based competency standards used by the government-endorsed standards of Australia, the United Kingdom, and South Africa.

The *ICB: IPMA Competence Baseline* (Caupin et al., 1999) is indicated as having application to people, as it is specifically intended to provide a basis for the International Project Management Association's (IPMA) certification program for project personnel. It purports to provide a basis for assessment of knowledge, experience, and personal attitude. The *IPMA Competence Baseline*, as a basis for assessment, includes a guide to what is considered to be the required knowledge and experience for an effective project manager, a taxonomy for identifying the extent of knowledge and experience required at each level of competence and a profile of the personality characteristics and attitudes that are expected in a project manager.

As the ICB has already been discussed under standards relating specifically to single projects, only the following people-focused standards will be discussed in further detail here.

Australian National Competency Standards for Project Management. The first performance-based competency standards for project management were the Australian National Competency Standards for Project Management that were developed through the efforts of the Australian Institute of Project Management and endorsed by the Australian Government on 1st July 1996.

At the commencement of 2003, there were 12 levels in the Australian Qualifications Framework (from school level to Doctorate), and the National Competency Standards for Project Management have been developed at Levels 4 (Certificate IV), 5 (Diploma), and 6 (Advanced Diploma). The standards have been adopted by the Australian Institute of Project Management for their 3 level professional registration process (see Figure 10.8).

The Australian National Competency Standards for Project Management were developed over a three-year period, commencing in 1993 and culminating in the endorsement

FIGURE 10.8. AIPM REGISTERED PROJECT MANAGER PROGRAM LEVELS.

AQF Level	AIPM Recognition Descriptors		Level of Experience Required for Assessment of Competence at This Level
Certificate IV	QPP	Qualified Project Professional	Specialist or Team Member
Diploma	RegPM	Registered Project Manager	Project Manager of well-defined or less complex projects Section Leader of complex projects
Advanced Diploma	MPD	Master Project Director	Project Manager of complex projects Project Manager or Director of multiple projects

of the standards by the Australian Government in 1996. Development was carried out by a consultant working under the guidance of a Steering Committee and Reference Group representing over 50 Australian organizations.

The standards development process is well documented (Gonczi et al., 1990; Heywood et al., 1992) and requires the examination of existing information about the occupation and "analysis of the purpose and functions of the profession and the roles and activities of its members" (Heywood et al., 1992, p. 46) in order to derive the units and elements of competency that provide the structure for the standards. In developing the standards, the team decided to follow the structure of the PMBOK® Guide and, at the same time, recognize the PMBOK® Guide as a contributing knowledge base to the standards.

There are nine units in the standards, described at Levels 4 (Certificate IV/QPP), 5 (Diploma/RegPM), and 6 (Advanced Diploma/MPD), as shown in Figure 10.9.

Level 4 of the standards does not include the unit relating to Integrative Processes. The only specific reference to personal characteristics or attributes in the standards is on page 16 of the Guidelines (AIPM [sponsor], 1996), which, in describing the nature of the project manager, suggests that "technical know-how alone is not sufficient to bring a project to successful completion" and "desirable attributes of a project manager include:

Leadership ability

The ability to anticipate problems

Operational flexibility

Ability to get things done

An ability to negotiate and persuade

An understanding of the environment within which the project is being managed

The ability to review, monitor, and control

The ability to manage within an environment of constant change

By following the structure of the PMBOK® Guide, which is by far the most widely distributed and recognized of the project management body of knowledge guides, the developers

FIGURE 10.9. UNITS IN THE AUSTRALIAN NATIONAL COMPETENCY STANDARDS FOR PROJECT MANAGEMENT.

	LEVEL 4	LEVEL 5	LEVEL 6
UNIT 1	Not applicable at Level 4	Guide Application of Project Integrative Processes	Manage Project Integration
UNIT 2	Apply Skills in Scope Management	Guide Application of Scope Management	Manage Scope
UNIT 3	Apply Skills in Time Management	Guide Application of Time Management	Manage Time
UNIT 4	Apply Skills in Cost Management	Guide Application of Cost Management	Manage Cost
UNIT 5	Apply Skills in Quality Management	Guide Application of Quality Management	Manage Quality
UNIT 6	Apply Skills in Human Resources Management	Guide Application of Human Resources Management	Manage Human Resources
UNIT 7	Apply Skills in Communications Management	Guide Application of Communications Management	Manage Communications
UNIT 8	Apply Skills in Risk Management	Guide Application of Risk Management	Manage Risk
UNIT 9	Apply Skills in Procurement Management	Guide Application of Procurement Management	Manage Procurement

of the Australian National Competency Standards for Project Management ensured that the standards would attract interest worldwide.

The Australian National Competency Standards for Project Management, as endorsed standards within the Australian Qualifications Framework, are the responsibility of Business Services Training Australia. With funding from the Australian National Training Authority (ANTA), Business Services Training Australia commenced a review of the Australian National Competency Standards for Project Management in 2001 and revised standards were finalized in mid-2004. Discussion has been based on the Australian National Competency Standards for Project Management as developed and endorsed in 1996. Information on the current status of these standards is available through the Australian Institute of Project Management Web Site (www.aipm.com.au).

Although the New Zealand Qualifications Authority (NZQA) has a qualification framework, they do not yet have project management standards, and it currently seems unlikely that they will develop their own standards. Mutual recognition arrangements between Australia and New Zealand are more likely to be utilized.

National Occupational Standards for Project Management (UK).

In the United Kingdom, the Occupational Standards Council for Engineering produced standards for Project Controls (OSCEng, 1996), which were endorsed in December, 1996, and for Project Management (OSCEng, 1997), which were endorsed in early 1997. The Construction Industry Standing Conference (CISC), the Management Charter Initiative (MCI), and what was then called the Engineering Services Standing Conference (ESSC), now the Occupational Standards for Engineering (OSCEng), together developed Level 5 NVQ/SVQ competency standards for Construction Project Management. A section of the Management Charter Initiative Management Standards, titled "Manage Projects" (MCI, 1997), provided a further set of competency standards for project management, but in this case, within the general management framework.

As with the Australian National Competency Standards for Project Management, the OSCEng NVQ/SVQs in project management and project controls were developed and tested in conjunction with experienced project management practitioners (OSCEng, 1996, p. 3) from over 50 employer organizations. The standards claimed to be generic and applicable:

- *At Level 4.* To all those who take responsibility for managing projects at the operational level
- *At Level 5.* To those with a strategic role in project management

NVQ/SVQ Levels 4 and 5 do not equate directly to the AQF Levels 4 and 5, as there are 5 levels in the NVQ/SVQ Framework and 12 levels in the Australian Qualifications Framework. For general purposes of comparison, NVQ/SVQ Level 4 equates to AQF Level 5 (Diploma), and NVQ/SVQ Level 5 equates to AQF Level 6 (Advanced Diploma).

Review of the UK OSCEng Standards commenced in mid-2001, and the revised standards were endorsed by the regulatory authorities in August 2002 (ECITB, 2002). These reviewed standards are the responsibility of the ECITB (Engineering Construction Industry

Training Board) and are specifically intended as cross-industry standards. The standards have been written as 51 separate units of competence, each relating to a distinct functional area covering both strategic and operational project management functions. Revision of the standards is claimed to incorporate content of the 4th Edition of the APM Body of Knowledge (Dixon, 2000). Definition of terms within the document is taken from BS6079. Two project management qualifications have been designed based on the 51 units. The Level 4 qualification is essentially for operational project management, and the Level 5 qualification may be considered as strategic project management. Each of these qualifications comprises a total of 20 units that must be completed, of which 11 are mandatory and 9 can be drawn from a selection of options. This allows for variation in the context (range indicators) in which the profession/occupation is performed.

National Certificate in Generic Project Management (Project Administration and Co-ordination) at NQF Level 4 (PMSGB/SAQA). The South African NQF Level 4 standards were developed using a process similar to that followed in both Australia and the United Kingdom. This is the first set of standards and associated qualifications to be produced for project management in South Africa, and work is proceeding on development of standards at other levels. The NQF Level 4 standards are at approximately the same level as the Australian AQF Level 4 (Certificate IV).

The National Certificate in Generic Project Management (Project Administration and Co-ordination) at NQF Level 4 is available with electives offering specialization in one of the following:

* Supervising a project team of a developmental project to deliver project objectives
* Supervising a project team of a technical project to deliver project objectives
* Supervising a project team of a business project to deliver project objectives
* Supporting project environment and activities to deliver project objectives

The qualification is intended to provide recognition of basic project management skills in the execution of small simple projects or providing assistance to a project manager of large projects. The focus is primarily on skills as a project team member but includes working as a leader on a small project/subproject involving few resources and having a limited impact on stakeholders.

There are 15 core titles or units and 4 elective titles that relate to differing contexts or specializations as indicated previously (e.g., developmental or technical projects). The structure and number of units bears more similarity to the revised UK standards (ECITB, 2002) than to the Australian standards.

Organizations

The focus of attention in development of project management standards has been on the management of individual projects. Even those standards that are intended for assessment of the project management competence of individuals are concerned primarily with their ability to manage individual projects and to some extent, at the strategic level, consider the need to manage and report on multiple projects or programs of projects. Management of multiple projects, program management and aspects of enterprise management that foster the effective management of projects have not received the same level of attention as the management of single projects.

P2M: A Guidebook for Project and Program Management for Enterprise Innovation. As mentioned previously, the P2M (ENAA, 2002) can be linked to enterprise or organizational project management through its attention to integration across projects, management of programs of projects, and its declared focus on project and program management for enterprise innovation. The focus, however, is on the development of individuals rather organizational assessment and development.

OPM3. The OPM3, or Organizational Project Management Maturity Model, is a standards development project of the Project Management Institute that has been active since 1998 through a globally representative team of volunteers. The declared purpose of the OPM3™ project is "to develop a global standard for organizational project management," and the vision is "to create a widely and enthusiastically endorsed maturity model that is recognized worldwide as the standard for developing and assessing project management capabilities within any organization" (Project Management Institute, 2002b). At the end of 2003 the Project Management Institute released the OPM3™ as a book and interactive CD-ROM. The OPM3™ has three elements. A Knowledge element presents "Best Practices" and provides guidance on how to use the accompanying material and CD-ROM. An Assessment element is an interactive database on the CD-ROM that enables organizations to evaluate current practices and identify areas for improvement, for which guidance is provided in the third, Improvement, element (Project Management Institute, 2004).

During the long gestation period of the OPM3™ (1998 to 2004), a significant number of proprietary models designed for assessment of organizational project management capability and maturity were developed by commercial organizations, in many cases drawing on either or both of the PMBOK® Guide and the SEI Capability Maturity Model (CMM) (Software Engineering Institute, 1999). In their work on the OPM3™, the team reviewed more than 30 maturity models, including those that could be considered as derived from business excellence and quality models from the SEI's CMM, and others individually or corporately developed (Schlichter, 2002).

OGC PMMM. PMMM is the Project Management Maturity Model owned as a Crown Copyright product by the Office of Government Commerce (UK) (OGC, 2002a). The OGC PMMM was developed in response to requests from both public and private sector organizations for a benchmark or standard against which to assess and demonstrate their corporate project management capability. Its development base is similar to that of the OPM3™, drawing on the approach of the SEI Capability Maturity Model and recognized project management standards and bodies of knowledge as well as OGC experience in providing support for project management improvement. It is intended that organizations can seek assessment against the OGC PMMM by accredited assessors.

PRINCE2. PRINCE, which stands for "PRojects IN Controlled Environments," is a project management methodology or approach to the management of projects that was first developed in 1989 by the Central Computer and Telecommunications Agency (CCTA), now part of the UK Office of Government Commerce (OGC). When first developed, it was intended as a UK government standard for IT project management. PRINCE2 is a development of the original methodology that is intended as a generic approach applicable to management of all types of projects (Office of Government Commerce (OGC, 2002b). There are many project management methodologies commercially available, but PRINCE2 is in the public domain and was developed by a UK government agency, with the specific intention of providing a standard approach to management of projects in organizations. Training in the use of PRINCE2 is available through accredited training organizations, and there is a quality-assured process of assessment and certification in use of the methodology. Training and certification are available in many parts of the world.

PRINCE2 could have been included under "Projects," as it is essentially a methodology for use in management of individual projects. It has been included under the heading "Organizations" because it is a methodology for use within organizations for management of projects and because it relates the management of projects directly back to the business case or organizational justification for each project with a strong concern for project and corporate governance.

Managing Successful Programmes. Program (programme) management, defined by the CCTA as "the co-ordinated management of a portfolio of projects that change organizations to achieve benefits that are of strategic importance" (CCTA, 1999, p. 2) is considered under the heading of "Organizations," as its focus is beyond that of the single project. It is a term widely used in practice and has been the subject of a number of conference papers in recent years. However, program management has so far attracted very little interest in terms of development of guides or standards from professional associations. This may be due to an unstated assumption that project and program management are interchangeable terms or that the practices of project management are equally applicable to the management of programs of projects.

The UK Central Computer and Telecommunications Agency (CCTA), the developers of PRINCE, have provided a considerable amount of material on the management of programs, including "An Introduction to Programme Management" published in 1993 (CCTA, 1993). *Managing Successful Programmes*, now published by the Office of Government Commerce (OGC) (CCTA, 1999), describes the framework and strategies of program management and the delivery of business benefits from a set of related projects. Reference is made in the document to PRINCE2. Although *Managing Successful Programmes* is presented by its authors as a standard, it is not. It does not yet form the basis for an assessment process or qualification (although the OGC has developed a "Successful Delivery Skills Programme"; Office of Government Commerce, 2002b). Therefore, *Managing Successful Programmes* cannot really be considered as a formal standard, although it arguably provides a widely disseminated guideline for the management of programs.

Conclusion

Development of standards and guides for management of single projects and for assessment of the competence of individuals primarily relating to management of single projects has clearly attracted more attention and effort than the development of guides and standards for multiple project, program, organizational, or enterprise project management. It is easy to assume that this is merely a reflection of time and professional maturity. However, it may also be influenced by the nature of boundaries relating to professional territory. Effort to establish a professional territory for project management has understandably begun at the level of the single project, where territorial claims can be won without a great deal of competition, especially in an environment of change.

One way in which new professions are formed is by successfully claiming territory (knowledge and expertise) previously occupied by other professions (Abbott, 1988). Management of single projects by a new profession of project management is effectively an appropriation of territory previously held by a number of professions or disciplines including engineering, architecture, and general management. The cross-disciplinary nature of projects creates confusion between roles of traditional professions, enhancing the opportunities for project management to establish professional claims. As the concerns of project management extend upward in organizations toward programs and enterprise project management, competition is primarily with general management, making it increasingly difficult to define their professional boundaries, especially as general management, itself, is actively fighting for professional recognition.

This section has provided an overview of project management standards and guides relating to projects, people, and organizations. Underlying the development of these standards is agreement on what is rightly included in a project management body of knowledge. The following section focuses on project management knowledge standards, often referred to as project management bodies of knowledge, providing a review of coverage and intent. This is done on the basis that there is general agreement on what constitutes the body of project management knowledge, and that knowledge standards and guides focus on parts of that body of knowledge, each from a slightly different perspective.

Project Management Knowledge Standards

Project management knowledge standards such as the PMBOK Guide, ICB, APM BoK, and P2M identify what is considered as the minimum knowledge coverage required relative to the purpose of each standard. Although they differ in scope of topic coverage, depth of treatment, structure, format, and terminology used, they share a nucleus of core content. This shared coverage, although arranged in different structures in each of these knowledge guides, essentially represents the nine project management knowledge areas outlined in the PMBOK® Guide (Integration, Scope, Time, Cost, Quality, Human Resources, Communications, Risk, Procurement). Further, although the majority of content is common across all guides, there are areas that are not dealt with in some guides or where the degree of emphasis and detail of coverage varies considerably across the guides.

Much of the difference in coverage can be attributed to the difference of purpose of each of the guides, although they do share one purpose: to provide a basis for an assessment and certification program for project management practitioners. It is useful, in considering the difference in content of the guides, to recognize the difference in declared purpose of each of these guides. The purpose to some extent influences the length of the document. Both the APM BoK and the ICB are considerably shorter than the PMBOK Guide and the P2M, providing less detail as they include references to other documents. It is important to note that the content of the P2M, reviewed here, represents only part (Parts 1, 2, and 3, and an overview of Part 4 which deals with 11 project segment areas) of a much larger document. The number of words in each document is therefore included in Figure 10.10 as an indicator of the size and therefore the potential for detailed coverage. Both the APM BoK and the ICB are less than half the size of the PMBOK® Guide and the P2M in terms of length in words. The PMBOK® Guide is the most substantial by some margin, except that only the English Summary Translation of the P2M is included in this analysis. In its complete form (420 pages), the P2M would be significantly more substantial that the PMBOK® Guide.

Structure is another issue that needs to be considered in comparing the content of these four documents. Although the content of documents may be similar, it may appear to be different as a result of the way in which the content is structured. Issues of content are not limited to whether or not a topic is included or excluded but also the degree of emphasis given to particular topics in each document.

Content and coverage of project management knowledge guides/standards and performance-based competency standards is reviewed in the section titled *Content and Coverage of Knowledge Guides and Performance-Based Competency Standards* in this chapter.

Performance-Based Competency Standards for Project Management

Performance-based competency standards describe what people can be expected to do in their working roles, as well as the knowledge and understanding of their occupation that is needed to underpin these roles at a specific level of competence. A valuable aspect of such standards is that they are specifically designed for assessment purposes. At the same time they are developmental in their approach, with assessment being undertaken by registered Workplace Assessors, within a well-defined quality assurance process.

Such standards have been developed within the context of government-endorsed standards and qualifications frameworks in Australia (ANTA),[1] New Zealand (NZQA),[2] South Africa (SAQA),[3] and the United Kingdom (QCA).[4] Although there are some differences in the aims of these national frameworks, common themes are as follows:

[1] ANTA—Australian National Training Authority.
[2] NZQA—New Zealand Qualifications Authority.
[3] SAQA—South African Qualifications Authority.
[4] QCA—Qualifications and Curriculum Authority.

FIGURE 10.10. SUMMARY OF STATED PURPOSES AND WORD LENGTH OF KNOWLEDGE GUIDES REVIEWED.

Guide	Purpose	Approx. no of words
APMBoK	• To provide an overall guide to the topics that professionals in project management consider are essential for a suitable understanding of the discipline. • Scope includes not only specific project management topics, such as planning and control tools and techniques, but also broader topics found to have a significant influence on the success of projects such as social and environmental context, technology, economics and finance, organization, procurement, people, and general management • Topics are described at a high level of generality on the basis that detailed description of the topics can be sourced elsewhere. • Topics included are those that are considered generically applicable.	13,000
ICB	• Guide to knowledge, skills and personal attitudes expected in project managers and project personnel. • Contains basic terms, tasks, practices, skills, functions, management processes, methods, techniques, and tools that are commonly used in project mangement as will as specialist knowledge, innovative, and advanced practices used in more limited situations. • Not intended as a textbook or cookbook. • Based on four existing PM body of knowledge guides in three languages and presented in English, French, and German. • Intended as reference basis for National Competence Baselines of IPMA member countries and eferencing of project mangement competence in theory and practice (Caupin et al., 1999, p. 9).	10,000
PMBOK□ *Guide*	• To identify and describe that subset of the Project Management Body of Knowledge (PMBOK□) that is generally accepted (i.e., knowledge and practices applicable to most projects most of the time.) • To provide a common lexicon within the profession and practice, for talking or writing about project management. • To provide a basic reference that is neither comprehensive nor all-inclusive. Further sources of information are listed and application area extensions are either already available or in production.	56,000
P2M	• Specifically intended to encompass management not only of individual projects but of multiple projects, and programs of projects and the wider organizational context. • Intended as a guide. • Although P2M is *considerably more extensive than existing PMBoKs or PM competency standards, it does not try to explore every detail of the topics discussed* (ENAA, 2002, p. 2).	36,000

- To provide an integrated and consistent system for recognition of learning achievements and qualifications
- To facilitate access to and mobility of progression within education, training, and career paths
- To enhance the quality of education, training and outcomes in the form of employability and productivity
- To contribute to and encourage personal development and career progression through learning

The standards developed within these national frameworks in Australia, New Zealand, South Africa and the United Kingdom can therefore be used for many different purposes. Such purposes include both development and assessment, as well as the provision of a basis for recognition of current competence, regardless of how that competence has been achieved.

Performance-based inference of competence is concerned with demonstration of the ability to do something at a standard considered acceptable in the workplace, with an emphasis on threshold rather than high performance or differentiating competencies. Threshold competencies are units of behavior that are essential to do a job but that are not causally related to superior job performance (Boyatzis, 1982). "Performance based models are concerned with *results* (or "outcomes") in the workplace *rather than potential* competence as indicated by tests of attributes. Even when the underlying competence being tested is not itself readily observable—for example, the ability to solve problems—performance and results in the workplace are still observable and the underlying competence they reflect can be inferred readily. Performance-based models of competence should specify *what* people have to be able to do, the *"level of performance required* and *the circumstances in which that level of performance is to be demonstrated"* (Heywood et al., 1992, p. 23).

The definition of competency, within the context of performance-based or occupational competency standards, is considered as addressing two questions:

- What is usually done in the workplace in this particular occupation/profession/role?
- What standard of performance is normally required?

The answers to these questions are written in a particular format.

Units of Competency

Development of performance-based competency standards begins with an overview of the competency of the overall profession or occupation with an emphasis on the competency levels of particular interest. The overall competency of a profession or occupation is then subdivided into manageable components that are meaningful to practitioners and will be observable in the performance of individuals in the workplace. This first subdivision reflects significant functions of the profession and is generally referred to as a "unit." Each unit of competency describes a broad area of professional or occupational performance.

Elements or Specific Outcomes

As a unit is likely to be too large to be practically demonstrable or assessable for the purposes of recognition of competence of individuals in the workplace, units are usually further subdivided into what are referred to in the Australian system as *elements of competency* and in the South African system as *specific outcomes*. Elements of competency or specific outcomes constitute the building blocks of each *unit of competency*, describing in more detail what is expected to be done in the workplace for each unit.

Performance or Assessment Criteria

While units and elements or specific outcomes describe what is done in the workplace, *performance criteria* (Australia) and *assessment criteria* (South Africa) describe the standard of performance that is required. Performance criteria/assessment criteria specify the type of performance in the workplace that would constitute evidence that the required standard has been achieved. They describe what a competent practitioner would do, expressed in terms of observable results and/or behavior in the workplace. Performance or assessment criteria also specify the evidence, in the form of documentation, from which competent performance in an element of competency or specific outcome would be inferred.

Range Statements

Range Statements (Australia and South Africa) describe the circumstances or context in which competent performance is expected. They add definition to the unit by elaborating critical or significant aspects of the performance requirements of the unit. The Range Statement establishes the range of indicative meanings or applications of these requirements in different operating contexts and conditions. The term "scope" is used in the recently endorsed UK ECITB standards to refer to the same concept.

Underpinning Knowledge and Understanding (UKU)

The Australian standards recognize the need for performance to be underpinned by relevant knowledge and understanding (i.e., underpinning knowledge and understanding, often abbreviated to UKU). This is referred to in the South African standards as "embedded knowledge." The UK ECITB standards refer to both underpinning knowledge and understanding and "specific knowledge" required for each unit.

Key Competencies and Critical Cross-Field Outcomes

The South African standards include *critical cross-field outcomes*. These are outcomes that are useful for, and result from, all teaching and learning.

In the Australian context, standards include reference to *key competencies*, defined as generic skills or competencies considered essential for people to participate effectively in the workforce. Key competencies apply to work generally, rather than being specific to work in a particular occupation or industry. The Finn Report (1991) identified six key areas of competence that were subsequently developed by the Mayer committee (1992) into seven key competencies: collecting, analyzing, and organizing information; communicating ideas and information; planning and organizing activities; working with others and in teams; using mathematical ideas and techniques; solving problems; and using technology (see www.anta. gov.au/gloftol.asp).

Structure and Terminology

A summary of the equivalent structural units of each of the government-endorsed performance-based competency standards for project management introduced in section entitled *People*, earlier in this chapter is presented in Figure 10.11.

Levels

The concept of levels, relating to different roles and levels of responsibility in the workplace, as well as to different levels of academic or workplace achievement, is fundamental to government endorsed performance-based competency standards and qualifications frameworks. Competency Standards are written at a number of different levels corresponding to the demands of occupational roles and/or educational requirements. In National Qualification Frameworks, these levels start at the equivalent of secondary school and move through to postgraduate qualifications, reflecting roles from entry level to chief executive officer. As an example, the Australian National Competency Standards for Project Management have been written at three levels, generally corresponding to the following job roles:

Project Team Member

Project Manager

Project Director/Program Manager

FIGURE 10.11. COMPARATIVE STRUCTURE AND TERMINOLOGY OF GOVERNMENT-ENDORSED PERFORMANCE-BASED COMPETENCY STANDARDS FOR PROJECT MANAGEMENT.

NCSPM (Australia)	ECITB (UK)	PMSGB/SAQA (South Africa)
Unit		Unit
Element	Unit*	Specific outcome
Performance criteria		Assessment criteria
Range statement	Scope	Range statement
Underpinning knowledge and understanding (UKU)	Specific knowledge required for this unit	Embedded knowledge
Key competencies		Critical cross-field outcomes

Note: The level of detail of the units of the ECITB standards is nearer to that of the elements/specific outcomes in the other standards and guides.

Figure 10.12 provides a summary of the approximate equivalency of levels or roles addressed by standards introduced in the *People* section, earlier in this chapter.

Content and Coverage of Knowledge Guides and Performance-Based Competency Standards

Review and comparison of the content and coverage of knowledge guides and performance-based competency standards has been undertaken by the Project Management Research team at the University of Technology, Sydney, to provide the platform for initiatives aimed at development of a framework of Global Performance Based Standards for Project Management Personnel, described in the section of the same name, later in chapter.

In conducting this review and comparison, two main research approaches were taken. First, the content of the documents was mapped at the topic, heading, unit, element, or specific outcome level (high-level structure) to identify those concepts or topics that are covered in all or some of the documents. Second, detailed text analysis was carried out on the full text of the documents under review, using Wordsmith Tools Version 3.0 (Scott, 1999). In this way, it was possible to identify concepts/topics that were not identified at the topic level of knowledge standards or at the unit, element, or specific outcome level of the performance-based standards but were significantly represented throughout the text of the documents. Collocation of words was carefully reviewed to identify the use of words in context. Searches were made on words with similar meaning to counteract the possibility that a concept or topic would be identified as absent from a text because of differences in use of terminology.

FIGURE 10.12. EQUIVALENT LEVELS OF GOVERNMENT-ENDORSED PERFORMANCE-BASED COMPETENCY STANDARDS FOR PROJECT MANAGEMENT.

Standard / Guide	Project Team Member	Project Manager	Project / Program Director	Status
NCSPM (Australia)	AQF Level 4 (Graduate Certificate)	AQF Level 5 (Diploma)	AQF Level 6 (Advanced Diploma)	Govt.-endorsed standard
ECITB (UK)		NVQ Level 4	NVQ Level 5	Govt.-endorsed standard
PMSGB/SAQA (South Africa)	NQF Level 4			Govt.-endorsed standard

It is important to note that in any mapping exercise, it is necessary to select a "spine" to map against. The selection of any one of the existing standards as the "spine" for mapping purposes can be seen as privileging the selected standard. To overcome this difficulty, several mapping exercises were undertaken using the higher-level structure of the different documents as the spine. From this activity, lists of concepts or topics were derived. These lists of concepts or topics were then compared with a list of 44 topics compiled by Themistocleous and Wearne (2000) from the "various systems of Body of Knowledge 'elements' used by the APM, the International Project management Association (IPMA) and the US Project Management Institute (PMI), plus pilot testing on a steering committee" (p. 7). As a result of this comparison, some modifications were made to the concept/topic lists derived from the analysis of the knowledge and performance-based competency standards. These lists, and the Themistocleous and Wearne list, are shown in Figure 10.13. Some differences between the lists have been retained for specific reasons.

First, there were some concepts that were noticeably absent from the Themistocleous and Wearne (2000) list but present in the documents reviewed—or vice versa. Specifically, two additional concepts/topics have been included in the performance-based standards list but do not appear in either the Themistocleous and Wearne list or the knowledge guides. These are *Documentation Management* and *Reporting*. Although these might legitimately be categorized under *Information/Communication Management*, they have a level of significance in their own right in the performance-based standards that warranted their separate listing. For instance, while *Documentation Management* and *Reporting* are well represented in the PMSGB/ SAQA Level 4 standards, other aspects of *Information/Communication Management*, which feature strongly in the other standards, are markedly absent. *Documentation* and *Reporting* are clearly more significant at the level of *practice* than they are as areas of *knowledge*, and it seemed appropriate that this should be highlighted.

Benefits Management was included because, although not represented in the Themistocleous and Wearne list and not well represented either at concept or text level in the government-endorsed performance-based competency standards reviewed, it is a potentially emerging concept and practice. Benefits management was identified as a topic in another research study I did (Crawford, 2002b), which included a wider range of standards and guides, including the OGC *Successful Delivery Skills Framework* (Office of Government Commerce, 2002), where it figures prominently.

Although *Configuration Management* and *Change Control* are included together in the Themistocleous and Wearne list, they are listed separately in the other lists. The characteristics of appearance of *Configuration Management* and *Change Control* are quite different in the performance-based standards. While *Change Control* is strongly represented in the high-level structures and throughout the text of the standards, *Configuration Management* receives very little mention in the government-endorsed performance-based standards. The term *Configuration* appears only three times, and then in only two of the performance-based competency standards reviewed. In the knowledge guides, *Configuration Management* is included at topic level in the APM BoK but is only mentioned once in the ICB and three times in the PMBOK Guide, although *Configurations and Changes* is a topic in the ICB.

Estimating is included as a separate item in the knowledge and performance-based standards lists, as it is quite strongly represented in these documents. Its application is not strictly

FIGURE 10.13. LISTS OF CONCEPTS/TOPICS AS PRESENTED BY THEMISTOCLEOUS AND WEARNE (2000) AND DERIVED FROM MAPPING AND TEXT ANALYSIS OF SELECTED KNOWLEDGE AND PERFORMANCE-BASED STANDARDS AND GUIDES.

	Themistocleous and Wearne (2000)	Knowledge Standards/ Guides	Performance-Based Standards/Guides	
1		Benefits management	Benefits management	1
2	Business need and case	Business case	Business case	2
3	Configuration management and change control	Change control	Change control	3
4		Configuration management	Configuration management	4
5	Conflict management	Conflict management	Conflict management	5
6	Cost management	Cost management	Cost management	6
7	Design management	Design management	Design management	7
8			Document management	8
9		Estimating	Estimating	9
10	Financial management	Financial management	Financial management	10
11	Goals, objectives, and strategies	Goals, objectives, and strategies	Goals, objectives, and strategies	11
12	Industrial relations			
13	Information management	Information/communication management	Information/communication management	12
14	Integrative management	Integration management	Integration management	13
15	Leadership	Leadership	Leadership	14
16	Legal awareness	Legal issues	Legal issues	15
17	Marketing and sales	Marketing	Marketing	16
18		Negotiation	Negotiation	17
19		Organizational learning	Organisational learning (inc. lessons)	18
20	Performance measurement	Performance measurement (inc. EVM)	Performance measurement (inc. EVM)	19
21	Personnel management	Personnel/human resource management	Personnel/human resource management	20
22	(Post-) Project evaluation review	(Post-) Project evaluation review	(Post-) Project evaluation review	21
23		Problem solving	Problem solving	22
24	Procurement	Procurement	Procurement	23
25	• Contract planning and administration			
26	• Purchasing			
27	Program management	Program/programme management	Program/programme management	24
28	Project appraisal	Project appraisal (Options/Modeling Investment/evaluation/analysis)	Project appraisal (Options/Modeling Investment/evaluation/analysis)	25
29	Project closeout	Project closeout/finalization	Project closeout/finalization	26
30	Project context	Project context/environment	Project context/environment	27
31	Project launch	Project initiation/start-up	Project initiation/start-up	28
32	Project life cycles	Project life cycle/project phases	Project life cycle/project phases	29
33	Project management			
34	Project management plan	Project planning	Project planning	30
35	Project monitoring and control	Project monitoring and control	Project monitoring and control	31
36	Project organization	Project organization	Project organization	32
37	Quality management	Quality management	Quality management	33
38		Regulations	Regulations	34

FIGURE 10.13. (*Continued*)

	Themistocleous and Wearne (2000)	Knowledge Standards/ Guides	Performance-Based Standards/Guides	
39			Reporting	35
40	Requirements management	Requirements management	Requirements management	36
41	Resources management	Resource management	Resource management	37
42	Risk management	Risk management	Risk management	38
43	Safety, health, and environment	Safety, health, and environment	Safety, health, and environment	39
44	Schedule management	Time management/scheduling/phasing	Time management/scheduling/phasing	40
45		Stakeholder/relationship management	Stakeholder/relationship management	41
46	Strategic implementation plan	Strategic alignment	Strategic alignment	42
47	Stress management			
48	Success criteria	Success	Success	43
49	Supply chain management			
50	Systems management			
51	Teamwork	Team building/development/teamwork	Team building/development/teamwork	44
52	Testing, commissioning, and handover/acceptance	Testing, commissioning, and handover/acceptance	Testing, commissioning, and handover/acceptance	45
53		Technology management	Technology management	46
54	Value improvement	Value management	Value management	47
55	Work management	Work content and scope management	Work content and scope management	48

limited to any of the other concepts such as Cost, Time, or Resources, and it was therefore considered inappropriate for it to be hidden within other concepts.

From analysis of the content of the documents reviewed, inclusion of separate topics of *Contract Planning and Administration* and *Purchasing*, used in the Themistocleous and Wearne list, did not appear necessary, as they were consistently represented in the context of procurement.

Knowledge Standards/Guides

In Figure 10.13, those concepts/topics that are shaded are those that are represented in the higher level structures of ALL of the standards reviewed. They may be considered CORE concepts or topics. Of these, *Project Environment/Context, Resource Management, Risk Management, Information/Communications Management,* and *Project Life Cycle/Project Phases* are the only topics that are present at this level in all four knowledge guides. *Scope, Time, Cost, Quality,* and *Procurement* are represented at the topic level in all knowledge guides except the P2M, where *Scope, Time, Cost* and *Quality* are included together in *Project Objectives Management* and *Procurement* within *Project Resources Management.*

Overall, topics considered core to the management of individual projects and that are extensively covered in the PMBOK® Guide receive little coverage in the Summary Translation of the P2M. The developers of the P2M gave full recognition to existing knowledge guides, and this may represent an acknowledgment that these topics are well covered in the PMBOK® Guide. They are also more fully covered in Part 4 of the P2M, which has not been translated into English and is not included in this analysis. In any case, the developers of the P2M specifically aimed to provide a knowledge guide that went beyond "delivery-focused traditional project management models and to develop a guide to allow the integration of project business strategy elements and utilization of valuable knowledge created through projects and programs" (ENAA, 2002, p. 1).

Programme/program management is understandably well covered in the P2M, for which the full title is *A Guidebook of Project and Program Management for Enterprise Innovation. Programme/program management* also receives topic-level coverage in the APM BoK but receives only passing mention in the PMBOK® Guide and ICB. Similarly, *Strategy* is covered at topic level in both the APM BoK and P2M, receiving only minor mention at text level in the PMBOK® Guide and ICB. The APM BoK is the only knowledge guide to specifically deal with the *Business Case*. Knowledge relating directly to interpersonal skills of project personnel, including *leadership, problem solving, conflict management,* and *negotiation* are only covered as discrete topics in the APM BoK and ICB.

Contextual issues such as *regulations, environment, health,* and *safety* are specifically covered in the APM BoK and ICB but receive very little mention, even in passing, in the PMBOK® Guide and P2M. Marketing is also covered, at topic level, in the APM BoK and ICB but is barely mentioned in the PMBOK® Guide and P2M. *Legal aspects* are covered at topic level in the APM BoK and ICB, and although not specified at topic level in the PMBOK® Guide and P2M, they are covered in the body of these texts.

The PMBOK® Guide is the only knowledge guide reviewed that mentions *life cycle costing,* while *modeling and testing* are only specifically included in the APM BoK. *Post project evaluation review* is mentioned only in the APM BoK and *reviews* in general receive little attention in any of the knowledge guides.

There are a number of aspects of management that are only covered at topic level in the APM BoK and receive little or no attention in the other knowledge guides. These include *Design, Technology* and *Requirements Management. Requirements Management.* The latter is particularly interesting, as this was one of the handful of topics on which there was less than 50 percent agreement concerning its importance as a knowledge area for competent project managers in the survey undertaken by the CRMP under the leadership of Professor Peter Morris in the review of the APM BoK (Morris et al., 2000). Morris expresses surprise that there is less than 50 percent agreement on the importance of *project success* criteria as a knowledge area for project personnel, yet success criteria are only covered at topic level in the APM BoK and ICB, and are not specifically dealt with in the PMBOK® Guide or P2M.

Value Management is covered only in the APM BoK and the P2M, with a single mention in the ICB and no mention at all even at detailed text level in the PMBOK® Guide, although *Value Engineering* is mentioned in the PMBOK® Guide.

Relationship Management is interesting, as it is a topic area in the P2M but does not appear in any form in the other three knowledge guides. This may be considered as another way of referring to *Stakeholder Management*, but although stakeholders are mentioned at text level in the other guides, and at topic level as Project Stakeholders in the PMBOK® Guide, *Stakeholder Management* is not specifically covered in any of the guides.

In summary, although the APM BoK and the ICB are the shorter of the four documents, they have the widest coverage. The APM BoK has slightly wider coverage than the ICB, but this is understandable, as the APM is a member of IPMA and the APM BoK is effectively the APM's National Competence Baseline (NCB). The PMBOK® Guide is very clearly focused on management of single projects, and the content of the P2M is directed more toward enterprise project management and the role of project management in value creation.

Performance-Based Competency Standards

A number of concepts/topics, not identified as core in Figure 10.13, that do not necessarily appear in the high-level structures and in some cases appear only in Range Statements, Underpinning Knowledge, and Understanding or other parts of the standards, are strongly or clearly represented at a text level in all the documents. These concepts/topics are shown in Figure 10.14.

Competency Models and Personal Competence

Performance-based competency standards represent an approach to competence that is strictly defined within the National Qualification Frameworks of countries such as Australia, New Zealand, South Africa, and the United Kingdom and is essentially concerned with threshold competence or minimum standards of performance required in the workplace. This approach to competence has attracted little interest in the United States, although it is worthwhile noting that Mexico has a National Qualifications Framework based on a performance-based approach to competence that includes a standard of competence for project management. When the term "competence" is used outside the context of National Qualifications Frameworks, it is often used in the sense of the Competency Model or Attribute-Based approach to competence, which is concerned with identification and definition of high-performing or differentiating competencies that contribute to superior performance.

The Competency Model approach is largely derived from the work of McClelland and McBer in the United States, beginning in the 1970s and reported by Boyatzis in the early 1980s (Boyatzis, 1982). Followers of this approach define a competency as an "underlying characteristic of an individual that is causally related to criterion-referenced effective and/or superior performance in a job or situation" (Spencer and Spencer, 1993). Five competency characteristics were defined by Spencer and Spencer (1993). Two of these competency characteristics—knowledge (the information a person has in specific content areas) and skill (the ability to perform a particular physical or mental task)—are considered to be surface competencies and the most readily developed and assessed through training and experience. Three core personality characteristics—motives, traits, and self-concept—are considered difficult to assess and develop.

FIGURE 10.14. CONCEPTS/TOPICS WELL REPRESENTED IN SOME DOCUMENTS ONLY AND/OR BELOW THE ELEMENT/SPECIFIC OUTCOME LEVEL.

Concept / Topic	Comment
Financial management	This concept is distinct from cost management. Cost management focuses on management of costs *within* the project, generally assuming that the necessary funds are available. Financial management is more concerned with interfaces between the project and its context and includes the processes involved in securing funds for the project and ensuring return on investment.
Goals, objectives and strategies	Although the term "goal" is rarely used, "objectives" is one of the most frequently occurring words across all four documents. These words are all used in the sense of goals, objectives, and strategies for the project (internally focused). Relationship to organizational strategy (externally focused) is treated as a separate concept (strategic alignment) and is far less strongly represented.
Legal issues	Legal issues are referred to in all documents but primarily in range statements and underpinning knowledge and understanding requirements.
Organizational learning	This includes the capture and sharing of knowledge between projects.
Performance measurement	Performance measurement appears in a number of ways including reference to 'assessment of performance' and is closely associated with organizational learning. Although earned value management falls within the ambit of performance measurement, it is only mentioned in the range statement at Level 6 of the Australian standards.
Personnel/human resource management	Human resource management is a unit in the Australian standards and is present as personnel management in the UK standards. It is outside the role addressed in the South African Level 4 standards.
Project appraisal	This is a concept applicable in various ways throughout a project and includes options, modeling, evaluation, and analysis, which are not represented in higher-level structures of the standards but appear in association with other concepts.
Requirements management	This is used primarily in the sense of information, procurement, quality, legal, resource, and other project requirements. In some cases (e.g., Australian standards AQF L6) it is used in the sense of quality and end product/stakeholder requirements. It can also be traced through the use of other words such as "needs."

The Competency Model approach, used extensively as the basis for numerous corporate competency development programs worldwide, sees competencies as clusters of knowledge, attitudes, skills, and in some cases personality traits, values and styles that affect an individual's ability to perform. While knowledge standards and performance-based standards as described in this chapter are available for project management, there are no such standards available for the behavioral competencies or personal competencies associated with project management.

A brief listing of attitudes and behaviors that are expected in a project manager are included in both the Association of Project Management Body of Knowledge (APM BoK) (Dixon, 2000) and the International Project Management Association's Competency Baseline (*ICB: IPMA Competence Baseline*) (Caupin et al., 1999) but do not constitute a standard. The Project Management Institute has included Personal Competencies in their Project Manager Competency Development Framework, but this document is very clearly intended as a guide to professional development rather than a standard that may be used as a basis for assessment.

Neither the listing of attitudes and behaviors in the APM BoK and the ICB, nor the Personal Competencies in the Project Management Institute's Project Manager Competency Development Framework, have a strong foundation in research. The Personal Competencies in the Project Manager Competency Development Framework are based on general management competencies presented in the work of Spencer and Spencer (1993). In some cases using methods similar to those presented in the work of Boyatzis (1982) and Spencer and Spencer (1993), a number of organizations have developed corporate competency models for project management, identifying the behaviors that are considered desirable and associated with superior performance within a specific corporate context. The NASA Competency Development Framework, incorporated in the NASA Project Management Development Process (PMDP) (www.nasaappl.com/ilearning/pmdp/pmdp.htm) includes both performance-based and behavioral or personal competencies and was developed by an expert group of NASA personnel.

Gadeken's work (Gadeken and Cullen, 1990; Gadeken, 1991; Gadeken, 1994) remains the most important work on behavioral competencies of project managers, but the results should be addressed with some caution because of the focus on both acquisition and the armed forces. Based on critical incident interviews with 60 US and 15 UK project managers from Army, Navy, and Air Force acquisition commands, the study identifies six behavioral competencies that distinguished outstanding program/project managers from their peers:

- Sense of mission
- Political awareness
- Relationship development
- Strategic influence
- Interpersonal assessment
- Action orientation

A further five behavioral competencies were demonstrated to distinguish outstanding program/project managers at a slightly lower level of significance:

- Assertiveness
- Critical inquiry
- Long-term perspective
- Focus on excellence
- Initiative

Project Management Qualifications

Each of the knowledge standards and guides and the performance-based competency standards reviewed previously are the basis for assessment for award of a project management qualification.

The APM *Body of Knowledge* and the *ICB: IPMA Competence Baseline* and associated National Competency Baselines (NCBs) form the basis for the IPMA's four-level certification program (Pännenbacker, et al., 1998). The entry level to this certification program, Level D, is a knowledge test based on the ICB or National Competency Baseline, which in the case of the United Kingdom is the APM Body of Knowledge.

The PMBOK® Guide is the project management knowledge guide of the Project Management Institute. The PMBOK® Guide is the basis for the Institute's single-level, Project Management Professional (PMP®) certification, which includes a multiple-choice knowledge exam plus project management experience. The PMI Certificate of Added Qualification (CAQ) recognizes knowledge, skills and experience in project management in specific industries. It is intended that Certificates of Added Qualification will be offered in a range of industries, with a CAQ relating to the automotive industry being the first to be made available (Project Management Institute, 2001).

The P2M is the basis for a three-level project management certification program that includes interviews, essay tests and project management experience. Qualifications are conferred by the Project Management Professionals Certification Center (PMCC), which was founded in April 2002.

The qualifications and assessment processes based on the knowledge standards and guides are summarized in Figure 10.15.

The performance-based competency standards discussed in the *Performance-Based Competency Standards for Project Management* section earlier in this chapter have been developed in the context of National Qualifications Frameworks and are therefore specifically designed to provide the basis for assessment and award of qualifications.

Assessment against government-endorsed performance-based competency standards is undertaken by Registered Workplace Assessors. Candidates are required to gather evidence of use of practices in accordance with Performance Criteria specified in the Competency Standards. It is part of the role of a Registered Workplace Assessor to work with candidates, advising and assisting them in achieving recognition of competence. Candidates are assessed either as competent at a particular level against the standards or "not yet competent." If assessed as competent against a set of performance-based competency standards at a particular level, a candidate may be awarded a qualification that is recognized within a government endorsed-qualifications framework.

FIGURE 10.15. EXAMPLES OF KNOWLEDGE-BASED PROJECT MANAGEMENT STANDARDS, ASSESSMENT PROCESSES, AND QUALIFICATIONS.

Standard or Guide	Level	Description	Form(s) of assessment
PMBOK® Guide *(Project Management Institute)*	PMP	Project Management Professional	• Multiple choice exam • Record of experience • Record of education
Industry specific extensions to the *PMBOK° Guide*	CAQ	Certificate of Added Qualification	• Must hold current PMP Certification • Record of industry-specific experience • Examination demonstrating industry-specific knowledge and skills
ICB: IPMA Competence Baseline and National Competence Baselines *(International Project Management Association, and member National Associations, e.g., AFITEP, APM)*	Level A	Programme or Projects Director	• Self-assessment, project proposal • Project report • Interview
	Level B	Project Manager	• Self-assessment, project proposal • Project report • Interview
	Level C	Project Management Professional	• Evidence of experience, self-assessment • Formal examination with direct questions and intellectual tasks • Interview
	Level D	Project Management Practitioner	• Formal examination, direct questions, and open essays
P2M *(Project Management Professionals Certification Center)*	PMA	Program Management Architect	• Interview and essay tests • Experience of at least three projects required
	PMR	Project Manager Registered	• Interview and essay tests • Experience of at least one project required
	PMS	Project Management Specialist	• Written examination

The Australian Institute of Project Management has a professional registration process that is aligned with the Australian National Competency Standards for Project Management and the Australian Qualifications Framework. Requirements for this registration process are available from the web site of the Australian Institute of Project Management (www.aipm.com.au). The equivalent project management role and professional and government recognized qualifications for Australia are shown in Figure 10.16.

In South Africa, the Project Management Standards Generating Body continues to work on development of standards for a range of project management roles. At time of writing, the only standards available for project management were those intended as the basis for award of a National Certificate in Generic Project Management (Project Administration and Co-ordination) at NQF Level 4.

The qualification is intended to provide recognition of basic project management skills in the execution of small, simple projects or providing assistance to a project manager of large projects. The focus is primarily on skills as a project team member but includes working

FIGURE 10.16. EQUIVALENT PROJECT MANAGEMENT ROLE AND PROFESSIONAL AND GOVERNMENT RECOGNIZED QUALIFICATIONS IN AUSTRALIA.

PM Role	Australian National Training Authority Qualification	Australian Institute of Project Management Award Title	Post Nominals
Project Team Member	Certificate IV	Qualified Project Professional	QPP
Project Manager	Diploma	Registered Project Manager	RegPM
Project Director/ Program Manager	Advanced Diploma	Master Project Director	MPD

as a leader on a small project/subproject involving few resources and having a limited impact on stakeholders.

The UK ECITB standards form the basis for project management qualifications recognized in the UK National Vocational Qualifications (NVQ) framework, namely:

- *At NVQ Level 4.* All those who take responsibility for managing projects at the operational level
- *At NVQ Level 5.* Those with a strategic role in project management

Demand for Global Project Management Standards

The possibility of achieving globally recognized project management standards and associated qualifications was a primary topic of interest amongst representatives of 29 countries at a Global Project Management Forum in October 1995 (Pennypacker, 1996). The idea of global standards and certification has continued as a topic for discussion at subsequent Global Forums held in association with major project management conferences, including the biannual World Congress of the International Project Management Association.

The International Project Management Association (IPMA) initiated a series of Global Working Parties at a meeting in East Horsley, England in February 1999. The working parties addressed Standards, Education, Certification, Accreditation/Credentialing, Research, and the continuation of the Global Forum process.

Following the inaugural meeting in February 1999, the Global Working Group: Standards subsequently met independently and in association with Global Forums, with participation by representatives of over 20 countries. The working group identified a number of global standards initiatives. One of these, aimed at development of a global body of project management knowledge, had already been commenced in late 1998, arising from an initiative of the Project Management Institute's Standards Committee. The other project, concerned with development of global performance standards for project management personnel (already mentioned in this chapter) was initiated by the Global Working Group: Standards. A further interest of the Global Working Group: Standards was the possibility of development of an ISO standard for project management, and this has been the subject of a watching brief rather than specific action.

The following are the primary reasons for interest in a global approach to the project management body of knowledge, standards and qualifications:

- Demand by corporations for standards and qualifications that are applicable throughout global operations as a basis for project management methodologies and for selection, development, and deployment of project management personnel
- Demand from practitioners for global recognition and transportability of professional and academic qualifications in project management
- Concern for international competitiveness by nations, corporations, and individuals
- Potential fragmentation of an emergent project management profession because of competition rather than cooperation in development and promotion of project management standards and qualification

Essentially, much energy and investment is wasted by individuals and organizations forced to make choices between competing project management standards and qualifications. A global approach would help to strengthen the image of project management and offer a more effective use of resources.

A brief review of the two primary initiatives relating to global approaches to the project management body of knowledge and standards will be presented here. These two initiatives are as follows:

- Towards a Global Body of Project Management Knowledge (OLCI)
- Global Performance-Based Standards for Project Management Personnel

Towards a Global Body of Project Management Knowledge (OLCI)

Definition of a body of project management knowledge is an essential step in development of a profession, and during the 1990s the key project management associations were all actively engaged in contributing to this development.

In a special edition of the *International Journal of Project Management*, focusing on project management bodies of knowledge, Wirth and Tryloff (1995) claimed that, at that time, at least ten national or international organizations appeared to be writing their own project management body of knowledge documents, while others were waiting to see what happened before attempting to write their own or adopting someone else's.

Although many separate efforts directed toward development of the project management body of knowledge were indicative of a very healthy and growing interest in project management, each organization, having developed its own guide to the body of project management knowledge, tended to become protective of its own content coverage, terminology, structure, and associated qualifications. The result was confusion in the minds of project management practitioners and their employers, particularly those involved in globally distributed operations when faced with decisions concerning which project management "body of knowledge" they should accept as a guide to their practice and which qualifications to support.

In October 1998, a small group of people actively involved in development and review of the PMBOK® Guide, the ICB, and the APM BoK took the initiative of holding an exploratory meeting to discuss ways in which they might work together in the interests of global cooperation. Those present at the meeting, although representing strong vested interests in existing project management body of knowledge guides, agreed to put aside those interests and work together to develop a global framework for project management knowledge. The initiative has come to be known as the OLCI or OLC Initiative, reflecting the initial name of the group that met in Long Beach, California, as the Operational Level Coordination Committee, a subcommittee, at that time, of the Project Management Institute's Standards Committee. The Project Management Institute's Standards Committee was disbanded at the end of 1998 because of a change of governance of the Institute. The Global Body of Project Management Knowledge initiative or OLCI has continued, but it has no formal status and has no affiliation with any project management association or other organization. All material produced by the group is in the public domain.

The work began at the first meeting in Long Beach, California, in October 1998 and was continued in a two-day workshop, hosted by NASA and attended by over 30 globally representative and recognized opinion leaders in Norfolk, Virginia, in June 1999 (Crawford and Pannenbäcker, 1999).

The group realized that they could not start with any of the existing documents but would have to start from a neutral base. Project management texts were nominated by the group in advance of the workshop and the indices of these texts scanned and analyzed. A resulting list of 700 words was reviewed and culled and then increased, at the workshop, to over 1,000 words. The initial list of 700 words and the results of two working parties that made a first attempt at structuring of the words are available in a full report on this workshop at www.aipm.com/OLC.

Prior to the workshop in June 1999, Max Wideman had begun a Glossary of Project Management terms and he has subsequently added the terms agreed at the June 1999 workshop as having a place in the project management body of knowledge. The Glossary continues to be updated and is available on the Web (www.pmforum.org).

Since the first two-day workshop in 1999, further meetings have been held annually, hosted by organizations that recognize the value of a global initiative to advance the development of the body of project management knowledge. The 2000 workshop was held in Norway and hosted by Telenor; the 2001 workshop was held in Lille, France, hosted by ESC Lille; and the 2002 workshop was held in Tokyo, hosted by the Project Management Professionals Certification Center (PMCC) and Japan Project Management Forum. A 2003 workshop is planned for Washington, D.C., hosted by NASA, and in 2004 the Project Management Research Committee has offered to host a workshop in China.

When this initiative was commenced in 1998, it is reasonable to assume that those involved generally envisaged that the outcome would be some form of document that represented a globally agreed body of project management knowledge. At the first two-day workshop in 1999 agreement was reached on the overall range of content of the body of project management knowledge. At this and subsequent workshops, there was realization that

- The existing body of knowledge documents (PMBOK® Guide, APM BoK, ICB, and more recently the P2M) are guides to specific parts of the overall body of project management knowledge.
- If a guide to the "global" body of knowledge were to be produced, it should not be in the form of a document but in a flexible and interactive form that enables stakeholders to view those parts of the project management of knowledge that are relevant to them at a point in time—a Web-based knowledge repository was proposed.
- Although structure is important, many structures are possible and different structures will address different needs and purposes.

The initial purpose of the OLCI, to work toward a global body of project management knowledge, has largely been achieved through shared recognition that the body of project management knowledge exists independently of the various guides representing views of parts of its content. An important role of the work of the OLCI has been to place each of the existing guides or standards in context, not as competing representations of the body of project management knowledge but as legitimately different and enriching interpretations of selected aspects of the same body of knowledge. The OLCI has therefore matured into a high-level and independent global "think-tank" and forum for advancement of the project management body of knowledge.

Global Performance-Based Standards for Project Management Personnel

A project for development of a framework of Global Performance-Based Standards for Project Management Personnel was initiated in mid-2000 at a meeting of representatives of project management and cost engineering professional associations, national standards and qualifications bodies, academic institutions and corporations. A history of this initiative is available at www.globalPMstandards.org.

The initiative responds to a growing recognition that standards for project management need to extend beyond knowledge to the application of that knowledge in the workplace. It provides an opportunity to respond to the needs of industry and project management practitioners interested in transferability of standards and associated qualifications across national boundaries. Governments are concerned with ensuring an internationally competitive workforce and in mutual recognition and transferability of qualifications.

The purpose of this initiative is therefore to develop an agreed framework for Global Performance Based Standards for Project Management Personnel that can be used by organizations, academic institutions, professional associations and government standards and qualifications bodies globally. It is proposed that these standards form a basis for review, development and recognition of local standards that map to a global framework.

In support of this initiative, the Project Management Research Team at the University of Technology, Sydney, conducted a detailed comparative review of the government-endorsed performance-based competency standards for project management of Australia, South Africa, and the United Kingdom, and of the Project Management Institute's Project Manager Competency Development Framework. The results of this research were referred to earlier in this chapter.

This research formed the background material for a working session of this initiative that was held in France in February 2003 and resulted in the development of a first draft of a framework of performance-based standards for the project manager role to which the content of existing standards can be mapped. This first draft was issued for wide stakeholder review in March 2003 and is available from the Global Performance Based Standards for Project Management Personnel Web site (www.globalPMstandards.org) (Global Performance Based Standards for Project Management Personnel). This report also indicates those individuals present and organizations represented at the Working Session.

Based on the 48 concepts identified in Figure 10.13, attendees at the Working Session identified 13 units of activities considered to be applicable to most project managers in most contexts and 4 units of activities considered to be applicable either in a role above that of project manager or only to some project managers in some contexts. The grouping of the 48 concepts into these units is presented in Figure 10.17. Keep in mind that what is presented here is a work in progress and may be expected to change significantly in response to review, stakeholder input, and further work. Further working sessions were held in Sydney, Australia in October, 2003 and in Cape Town, South Africa in May 2004. A further draft of performance-based standards has been developed, extended to cover two levels of project management activity. This and ongoing developments will be made progressively available on the Global Performance Based Standards for Project Management Personnel Web site. In the following Figure 10.17, Units considered applicable only to some project managers in some contexts are shown shaded.

SUMMARY

Standards development in project management over the last decade has made a significant contribution to codification of knowledge and practices as a means of establishing the territory of a project management profession, raising its profile, and making it accessible to a wide constituency. While the most widely distributed standards—the PMBOK® Guide, APM BoK, and ICB—have been instrumental in the growth of interest in project management, they can also be seen as limiting its development and sphere of influence. These standards, and current performance-based competency standards, are essentially reductionist and deterministic in their approach and focus on knowledge and practices applicable to management of single projects. Standards and guides for project management in organizations such as Japan's P2M, the emergent OPM3® and the work of the UK Office of Government Commerce (OGC) such as PRINCE2, *Managing Successful Programmes* and the *Successful Delivery Skills Framework* (OGC, 2002) are beginning to provide some guidance for

FIGURE 10.17. UNITS DEVELOPED FROM 48 CONCEPTS/TOPICS.

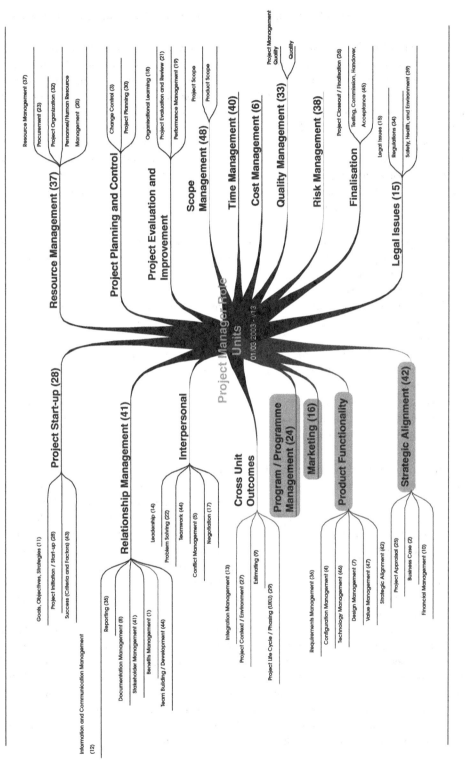

management of multiple projects and programs of projects, the role of the project sponsor, enterprise project management, and provision of corporate support for effective management of projects and programs. These standards and guides, however, enjoy less widespread recognition and appear to have less influence on professional formation than the knowledge and practice standards, largely focused on delivery of stand-alone projects, that form the basis for assessment and award of project management qualifications.

The most recent guide to the project management body of knowledge, Japan's P2M, is the most significant advance toward genuine integration and acceptance of the role of project and program management at the enterprise level. This is the first of the guides, or standards, intended as the basis for assessment and award of project management qualifications, that does the following:

- Develops an approach to enterprise project and program management that starts afresh from the viewpoint of the enterprise rather than drawing on project paradigms developed in the context of large, single, physical projects as the day-to-day business of project-based organizations.
- Directly addresses program management (rather than focusing only on single projects).
- Recognizes and responds to the complexities of fast-moving multistakeholder environments.
- Recognizes and addresses the systemic nature of projects and programs and specifically includes project systems management with reference to both hard and soft systems approaches.
- Addresses integration across programs and portfolios of projects at enterprise level (when other guides and standards mention "integration," in the P2M it is used in reference to integration *within* a single project).

Ensuring that project management develops as a field of practice and aspiring profession that has a strategic influence on the way organizations are managed and results are delivered is hampered by the image of project management that has been defined and presented by the project management knowledge standards and guides developed during the last decade. These standards and guides describe practices and a profession that is firmly lodged at the middle management level of organizations, and research (Crawford, 2002a; Crawford, 2002c) indicates that in the majority of organizations, senior management expects project managers to concentrate on what may be described as first-generation project management, managing efficient delivery of results within established time, cost, and quality restraints (Ohara, 2001, p. 4).

Results of two separate research studies indicate that on one hand less than 40 percent of senior managers surveyed considered strategy, systems, integration, and information management essential knowledge for project managers (Morris et al., 2000), and on the other, increasing levels of use of practices in these areas are associated with decreasing likelihood of being rated by senior management as a top project management performer (Crawford, 2001). This evidence suggests that general managers don't welcome encroachment by "project managers" on what they consider to be their territory. They expect those in project management roles to focus on delivery of their projects to specification, on time and within budget, which is hardly surprising, as this is the image of project management offered by the most widely distributed project management standards.

There is considerable rhetoric, however, promoting the potential contribution of project and program management to corporate strategy implementation (Pellegrinelli and Bowman, 1994), and there is increasing evidence to suggest that the project management approaches enshrined in current standards developed in the context of essentially 'hard' projects in the construction, engineering, defense, and aerospace industries are not uniformly successful when applied to 'soft' projects such as organizational change. Problems in applying hard project management practice to soft projects have prompted rethinking of standards and practices, with a number of writers and researchers turning to systems theory for possible enlightenment (Neal, 1995; Rodrigues and Bowers, 1996; Costello et al., 2002). There is potential for projects in complex, multistakeholder environments to be approached and managed as systemic interventions that engage stakeholders, enable environmental responsiveness, recognize the validity of different viewpoints, and facilitate organizational and individual learning.

Much of this is envisaged in the P2M, but it seems unlikely that it will be reflected in other knowledge and practice guides and standards in the near future. The process for standards development, which is largely a process of making explicit and codifying through consensus the tacit knowledge of experienced practitioners, ensures that standards will remain conservative and will lag behind the cutting edge of both research and practice.

References

Abbott, A. 1988. *The system of professions*. Chicago: The University of Chicago Press.

AIPM (Sponsor). 1996. *National competency standards for project management*. Sydney: Australian Institute of Project Management. www.aipm.com.au/html/ncspm.cfm.

APM. 1996. *Body of Knowledge (Version 3)*. High Wycombe, UK: Association of Project Managers.

APM. 2002. APM's new book. www.apm.org.uk/pub/Pathways%20Flyer%20Final.pdf (accessed February 12, 2003).

Baker, B. N., D. C. Murphy, and D. Fisher. 1988. Factors affecting project success. In *Project management handbook*, 2nd ed., ed. D. J. Cleland, and W. R. King 902–919. New York: Van Nostrand Reinhold.

Berry, A., and K. Oakley, K. 1994. Consultancies: Agents of organizational development: Part II. *Leadership and Organization Development Journal* 15(1):13–21.

Boyatzis, R. E. 1982. *The competent manager: A model for effective performance*. New York: Wiley.

British Standards Board. 1996. *Guide to project management: BS6079: 1996*, London: British Standards Board.

Caupin, G., H. Knopfel, P. Morris, E. Motzel, E. and O. Pännenbacker. 1999. *ICB: IPMA Competence Baseline*, Version 2. Germany: International Project Management Association.

CCTA. 1993. *An introduction to programme management*. London: The Stationery Office.

CCTA. 1999. *Managing successful programmes*. London: The Stationery Office.

Costello, K., L. Crawford, L. Bentley, L., et al. 2002. Connecting soft systems thinking with project management practice: An organizational change case study. In *Systems theory and practice in the knowledge age*, ed. G. Ragsdell. New York: Kluwer Academic/Plenum Publishers.

Crawford, L., and Pännenbacker, O., eds. 1999. Towards a global body of project management knowledge. www.aipm.com/OLC/(accessed March 28, 2000).

Crawford, L. H. 2001. *Project management competence: The value of standards.* DBA thesis. Henley-on-Thames: Henley Management College/Brunel University.

———. 2002a. Senior management perceptions of project management competence. *Proceedings IRNOP V: Zeeland*, ed. J. R. Turner. Rotterdam: Erasmus University.

———. 2002b. *Project management qualification research study.* High Wycombe, UK: APM Group Limited.

———. 2002c. Profiling the competent project manager. In *The frontiers of project management research*, ed. D. P. Slevin, D. I. Cleland, and J. K. Pinto. Newtown Square, PA: Project Management Institute.

Dean, P. J. 1997. Examining the profession and the practice of business ethics. *Journal of Business Ethics* 16(15):1637–1649.

Dixon, M. 2000. *APM Project Management Body of Knowledge*, 4th ed. Peterborough, UK: Association for Project Management.

Duncan, W. R. 1998. Presentation to Council of Chapter Presidents. In *PMI Annual Symposium*. October 10, Long Beach, CA.

ECITB. 2002. *National occupational standards for project management: Pre-launch version September 200.*, Kings Langley: Engineering Construction Industry Training Board.

ENAA. 2002. *P2M: A guidebook of project and program management for enterprise innovation: Summary translation.* Revision 1. Tokyo: Project Management Professionals Certification Center (PMCC).

Eraut, M. 1994. *Developing professional knowledge and competence.* London: The Falmer Press.

Finn, B. 1991. *Young people's participation in post-compulsory education and training.* Canberra, Australia: AGPS.

Gadeken, D. O. C. 1991. Competencies of Project Managers in the NMOD Procurement Executive. Royal Military College of Science.

Gadeken, D. O. C., and B. J. Cullen. 1990. *A competency model of program managers in the DoD acquisition process.* Fort Belvoir, VA: Defense Systems Management College.

Gadeken, D. O. C. 1994. Project managers as leaders: Competencies of top performers. In *12th INTERNET (IPMA) World Congress on Project Management.* pp. 14–25. Oslo, Norway, IPMA.

Global Performance Based Standards for Project Management Personnel. Working Paper No. 1: Report from Working Session 24-26 February, 2003, Lille, France. Sydney: University of Technology, Sydney.

Gonczi, A., P. Hager, and L. Oliver. 1990. *Establishing competency standards in the professions.* Canberra, Australia: Australian Government Publishing Service.

Heywood, L., A. Gonczi, and P. Hager. 1992. *A guide to development of competency standards for professions.* Canberra, Australia: Australian Government Publishing Service.

Holtzman, J. 1999. Getting up to standard. *PM Network* 13(12)(December): 44-46

Hougham, M. 2000. *Syllabus for the APMP Examination.* 2nd ed. High Wycombe, UK: Association of Project Management.

IEEE. 2000. 1490-1998 *IEEE Guide to the Project Management Body of Knowledge.* Adoption of PMI Standard. http://standards.ieee.org/catalog/software2.html#1490-1998 (accessed April 30, 2000).

ISO 1997. *ISO 10006: 1997: Quality management: Guidelines to quality in project management.* Geneva: International Organization for Standardization.

Mayer Committee 1992. *Key competencies: Report of the Committee to advise the Australian Education Council and Ministers of Vocational Education, Employment and Training on employment-related key competencies for postcompulsory education and training.* Melbourne: Australian Education Council and Ministers of Vocational Education, Employment and Training.

MCI 1997. *Manage projects: Management standards—Key Role G.* London: Management Charter Initiative.

Morris, P. W. G. 2000. Benchmarking project management bodies of knowledge. In *IRNOP IV Conference—Paradoxes of Project Collaboration in the Global Economy: Interdependence, Complexity and Ambiguity*, ed. L. Crawford and C. F. Clarke. Sydney, Australia: University of Technology, Sydney.

————. 2001. Updating the project management bodies of knowledge. *Project Management Journal* 32(3): 21–30.

Morris, P. W. G., and G. H. Hough. 1987. *The anatomy of major projects*, Chichester, UK: Wiley.

Morris, P. W. G., M. B. Patel, and S. H. Wearne. 2000. Research into revising the APM project management body of knowledge. *International Journal of Project Management* 18(3):155–164.

Neal, R. A. 1995. Project definition: The soft-systems approach. *International Journal of Project Management* 13(1):5–9.

Office of Government Commerce (OGC). 2002a. Project Management Maturity Model (PMMM): OGC Release Version 5.0, London: The Stationery Office

————. 2002b. *Successful delivery skills framework, Version 1.0.* www.ogc.gov.uk.

Ohara, P.S. 2001. Project management and qualification system in Japan: Expectation for P2M and challenges. *Proceedings of the International Project Management Congress 2001: Project Management Development in the Asia-Pacific Region in the New Century.* November 16–21. Tokyo. Tokyo: Engineering Management Association of Japan (ENAA) and Japan Project Management Forum (JPMF).

OSCEng. 1996. *OSCEng Level 4: NVQ/SVQ in project controls.* London:: Occupational Standards Council for Engineering.

————. 1997. *OSCEng Levels 4 and 5: NVQ/SVQ in (generic) project management.* London: Occupational Standards Council for Engineering.

Pannenbäcker, K., H. Knopfel, and G. Caupin. 1998. *PMA and its validated four-level certification programmes.* Version 1.00. Nijkerk, Netherlands: International Project Management Association.

Pannenbäcker, O., H. Knoepfel, and J. Communier. 2002. *IPMA Certification Yearbook 2001.* Nijkerk, Netherlands: International Project Management Association.

Pellegrinelli, S., and Bowman, C. 1994. Implementing strategy through projects. *Long Range Planning* 27(4):125–132.

Pennypacker, J. S. 1996. *The Global Status of the Project Management Profession.* Newtown Square, PA: Project Management Institute.

Project Management Institute. 2000. *A guide to the Project Management Body of Knowledge..* Newtown Square, PA: Project Management Institute.

————. 2001.Certificate of Added Qualification. www.pmi.org/certification/CAQ/caq.htmwww.pmi.org/certification/CAQ/caq.htm accessed January 17, 2002.

————. 2002a. Introduction to the Project Management Institute (PMI). www.pmi.org/prod/groups/public/documents/info/ap_introoverview.aspgroups/public/documents/info/ap_introoverview.asp (accessed December 29, 2002).

————. 2002b. PMI Standards Open Working Session October 2002. www.pmi.org/info/PP_OWS02.pdf (accessed January 1, 2003).

————. 2002c. *Project Manager Competency Development Framework.* Newtown Square, PA: Project Management Institute.

————. 2004 Organizational Project Management Maturity Model (OPM3) www.pmi.org/info/PP_OPM3ExecGuide.pdf (accessed May 4, 2004].

Rodrigues, A., and J. Bowers. 1996. The role of system dynamics in project management. *International Journal of Project Management* 14(4):213–220.

Schlichter, John. 2002. Organizational Project Management Maturity Model: Emerging Standards. *www.pmi.org* (accessed January 1, 2002).

Scott, M. 1999. *Wordsmith Tools Version 3.* Oxford, UK: Oxford University Press.

Software Engineering Institute. 1999.SW-CMM Capability Maturity Model SM for Software. www.sei.cmu.edu/cmm/cmm.html (accessed February 14, 1999).

South African Qualifications Authority. 2001. General Notice No. 1206 of 2001: Notice of publication of unit standards-based qualifications for public comment: National Certificate in Project Management—NQF Level 4. *Government Gazette* 437 (22846, November 21).

Spencer, L. M. J., and S. M. Spencer. 1993. *Competence at work: Models for superior performance.* New York: Wiley.

Stevens, M. 2002. *Project management pathways.* High Wycombe, UK: Association for Project Management.

Stewart, T. A. 1995. Planning a career in a world without managers. *Fortune* 131 (5, March 20): 72–80.

Themistocleous, G., and S. H. Wearne. 2000. Project management topic coverage in journals. *International Journal of Project Management* 18(1):7–11.

Wideman, R. M. 1986. The PMBOK report. *Project Management Journal* 15 (Special Summer Issue): 102.

Wirth, I., and D. E. Tryloff. 1995a. Preliminary comparison of six efforts to document the project-management body of knowledge. *International Journal of Project Management* 13(2):109–118.

———. 1995b. Preliminary comparison of six efforts to document the project-management body of knowledge. *International Journal of Project Management* 13(2):109–118.

Zwerman, B., and J. Thomas. 2001. Barriers on the road to professionalization. *PM Network* 15 (4, April): 50–62.

CHAPTER ELEVEN

LESSONS LEARNED: PROJECT EVALUATION

J. Davidson Frame

The basic function of project evaluation is to engage in big-picture stock taking, where the most fundamental goals of a project are identified and the extent to which they are being achieved is determined. The implementation of effective evaluations on projects is important for at least three reasons:

- Evaluations force organizations to determine explicitly what it is that they are trying to achieve on their projects. That is, they require managers to identify the core objectives of projects (Locke and Latham, 1990).
- Evaluations supply feedback on project performance, enabling project staff to determine the degree to which the project is on target. This feedback may show that performance is on track (or even "ahead") or that performance objectives are not being attained. Without such information, managers have little or no idea of whether their projects are doing well or are headed toward failure. When evaluations are carried out this way, they are called *summative evaluations*: They summarize actual performance against established performance objectives (Farbey, 1999).
- Evaluations enable organizations to learn what works and what does not. Based on insights gained through evaluations, managers can adjust the organization's processes to improve organizational performance. Thus, evaluations are a core element of *organizational learning* (Symons, 1990). Evaluations that are carried out with a view of providing guidance on future behavior are called *formative evaluations* (Walsham, 1999; Remenyi, 1997).

This chapter focuses on post-project evaluations that are carried out as part of the project closeout process. It examines the function of post-project evaluation, identifies where it fits in the larger closeout effort, and describes how it can be conducted.

The Process of Project Evaluation

Post-project evaluations do not occur in a vacuum. They are in fact the final evaluative action in a process that commences before a project is even selected. There are three distinct types of evaluation that are carried out on projects: pre-project evaluation, mid-project evaluation, and post-project evaluation. (Farbey, et al., 1999). Taken together, they constitute overall project evaluation. The relationships among them are pictured in Figure 11.1.

Pre-project evaluation is also called *project screening, project selection,* and *project appraisal.* With pre-project evaluation, potential candidates for support are examined in respect to project selection criteria (Buss, 1984). Typical criteria include the following:

- *Financial.* What will the project cost? What are the anticipated financial returns?
- *Technical.* What technologies are needed to carry out the project? What technologies will the project create? Do you have the technical capabilities to carry out the project? Will the project enable you to strengthen your technical capabilities?
- *Marketing.* How can you sell your project idea to potential customers? What competitors do you need to contend with? How does the project relate to marketing's Four Ps (i.e., product, price, place, and promotion)?
- *Operational.* What demands will the project place on your operations? Do you have the technical and administrative infrastructure to implement the project smoothly?

FIGURE 11.1. THE PROCESS OF PROJECT EVALUATION.

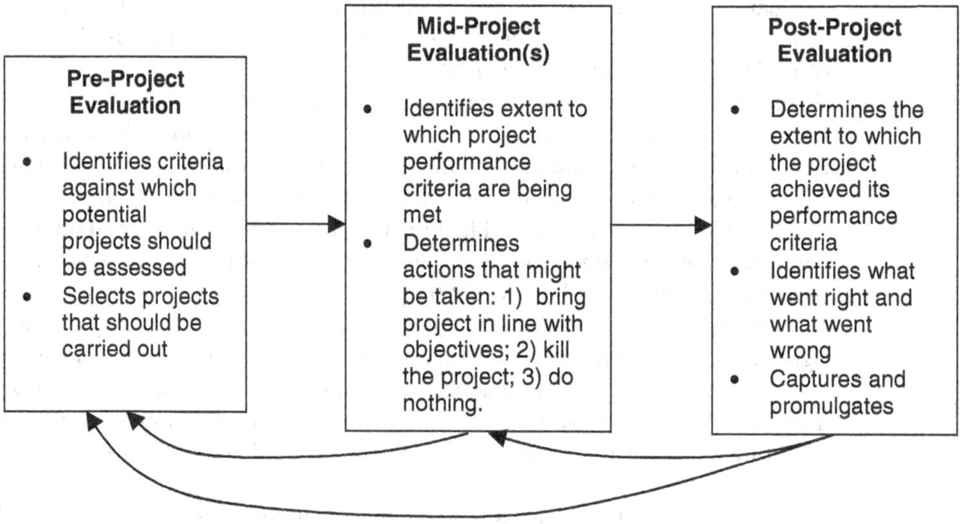

- *Strategic.* To what extent does the project address and support the organization's strategic goals? Does it complement the existing portfolio of projects?
- *Corporate culture.* Does the project align with the organization's overarching culture and goals?

In addressing the questions raised here, performance standards for the project effort emerge. For example, a preliminary budget of $2.3 million may be established for the project. Estimated revenues of $2.6 million may be projected. The need to hire two software designers to serve the project team may be identified. An advertising budget of $55 thousand may be established. Ultimately, these performance standards serve as the basis for determining whether or not the project effort is on target. Note that to answer basic questions raised at the project screening phase, analysts need good forecasting and estimation skills (Frame, 2002). If the projections are off target, then the wrong projects may be chosen.

The importance of effective pre-project evaluation is enormous. When due diligence is not followed in selecting projects, an organization will find that it has committed its resources to pursuing a loser. In this case, project failure has been hardwired into the project before any work has begun.

Mid-project evaluation is carried out periodically during the life of the project. (See also the chapter by Huemann.) Specific evaluative efforts should be established as milestones in the project schedule, requiring the project team to review performance in accordance with an established evaluation schedule. For example, the project plan may suggest that the preliminary design of a database system should be completed by June 12. A formal evaluation of the design effort can be scheduled to be conducted on June 13 and would include a review of the achievement of both technical and business objectives.

Mid-project evaluations address a number of standard questions, including the following:

- Is the project achieving its objectives as planned?
- If not, to what extent is it missing the objectives?
- Are the objectives still worth pursuing, knowing what you know today?
- If the project is not achieving its objectives, should you take corrective action to get it back on track?
- If you decide to bring the project back on track, what actions do you need to take?

A review of these questions shows that mid-project evaluations serve a cybernetic function: They seek *feedback* information for purposes of project *control.*

Mid-project evaluations assume a number of different forms, including the following:

- *Technical evaluations.* These evaluations address the technical performance of the deliverable: Is it meeting the specifications? Technical evaluations usually occur concurrently with system tests in order to assess whether the deliverable is achieving prescribed requirements. On software projects, a popular form of technical evaluation is the structured walk-through. With structured walk-throughs, the project team members step through their work before a panel of outside evaluators whom they have chosen. Because project team members control the whole evaluation process with structured walk-throughs, the level of threat associated with the evaluation is reduced and the team members have little reason to fear surfacing the problems they are encountering. Other commonly encountered technical evaluations include preliminary design reviews and critical design reviews. (See Harpum's chapter on design management.)

- *Performance appraisal reviews.* These evaluations examine the performance of individual employees at predetermined times (e.g., every six months). The principal question they address is this: Are they achieving their performance goals effectively? Individual performance is measured against performance goals established earlier in the project life cycle.
- *Audits.* Mid-project audits are surprise evaluations conducted with little or no advanced warning. Their objective is to see how the project team is functioning at a given moment in time. Through audits, evaluators can see how team members "really" function when they do not expect to be inspected. One reason audits are carried out is to keep project team members on their toes.
- *Managerial reviews.* One of the best-known evaluation methodologies is called *management by objectives* (MBO). Its principal proponent back in the 1950s was the management guru Peter Drucker (Drucker, 1985). With MBO, project teams and their managers negotiate objectives that the team members should achieve at different points in time. Once these points of time are reached, the performance of the team is reviewed to see the extent to which the objectives have been achieved. Unlike audits, MBO reviews eschew surprises— the team members know far in advance what is expected of them during the evaluative review.

Post-project evaluations are conducted after the project effort is completed. The objectives of post-project evaluation are different from those of mid-project evaluation. Clearly, once the project is done, the issue of taking action to get the project back on track is moot. The concerns at this time are as follows:

- At the end of the day, did you do what you said you would do, and did you achieve what you set out to achieve? (Morgan and Tang, 1993).
- What lessons can you learn from the project experience?
- What good practices and results did you encounter that you should attempt to replicate on future projects?
- What troubles did you encounter that you should strive to avoid in the future?

Given the results of the post-project evaluation, managers may determine that organizational processes need to be changed in order to build on strengths and avoid problems in the future. A significant challenge of post-project evaluation is to avoid the "pitfalls of hindsight" (Fox, 1984). That is, for the evaluation to meaningful, it must reconstruct the conditions facing the project as it was implemented in order to avoid critiquing the project on abstract principles that are not linked to reality.

Basic Principles Governing Evaluative Efforts

Effective mid-project and post-project evaluations strive to follow four basic principles: objectivity, internal consistency, replicability, and fairness. Each of these will be discussed briefly.

Objectivity

Evaluations must strive to be as objective as possible. Traditionally, the need for objectivity has been addressed by having evaluators come from outside the group being evaluated. The theory is that by using outsiders, you can avoid conflicts of interest that might arise when you have team members evaluate their own efforts. In other words, you strive to avoid having foxes guard the chicken coop.

While the rationale for objectivity may be solid, experience shows that in practice, relying on outsiders to conduct evaluations can lead to problems. Teams being evaluated under such circumstances often express serious concerns about the process, and their concerns have merit. Commonly encountered problems include the following (Frame, 2002):

- *The outside evaluators are unfamiliar with the circumstances the project team is addressing in its work.* The evaluators may not fully understand where the project team stands technically, what specific instructions it has received from its clients/bosses, how the organization works, what impediments team members have encountered while carrying out their tasks, and so on. Certainly, the evaluators can get up to speed on many of these details, but the effort can be quite disruptive. As the project team is responding to the information-gathering queries of the evaluators, project work may grind to a halt, jeopardizing the team's ability to achieve its goals. This may cause team members to feel hostile toward the evaluative effort.
- *The project team may be suspicious about who selected the evaluators and what instructions the evaluators have received.* Although the point of using outsiders is to maintain objectivity in the evaluation process, it is clear that the process is still susceptible to subjective influences. For example, if the outside evaluators hold a particular ideological perspective, their objectivity is questionable, particularly if their ideology runs counter to the perspective followed by the project at its outset. The assumption of objectivity of the outsiders is also doubtful if their marching orders have been given to them by an executive who is antipathetic to the project's raison d'etre. Project teams understandably want answers to the following questions: Who chose the evaluators? What instructions were they given? What obvious biases—if any—do they hold?
- *The outside evaluators feel compelled to find problems.* An important objective of evaluation is to uncover problems. The earlier problems can be identified, the easier it is to deal with them. If problems are surfaced after they have had time to fester, they can be enormously disruptive. Consequently, you carry out evaluations to discover problems in their infancy, so you can fix them as soon as possible. There is an important point here: Evaluations seek out problems so they can be fixed. Their fundamental rationale should *not* be to identify guilty parties who should be punished for their failings. As soon as evaluations

become associated with punishment, they become exercises in team demoralization and discourage honest reporting of problems. Evaluators should certainly be on the lookout for problems. That is a large part of their job. But they should guard against behaving like traffic police who have a quota of tickets to issue each day.

- *The outside evaluators are not competent.* It occasionally happens that the outside evaluators are not competent to perform their tasks. They may possess skills in the wrong areas or simply do not know what they are doing or what they are talking about. This can be a frightening situation for the project team that is being evaluated, because team members recognize their future lies in the hands of unqualified people.

In recent years, the view that employment of a team of outsiders satisfies the principle of objectivity has become passé. The pitfalls of using "objective" evaluation teams are well recognized, and consequently a number of approaches have arisen to rectify their inadequacies. Two approaches stand out. One is the employment of structured walk-throughs. The other is use of the EISA approach (defined later in the chapter). Key features of these approaches are discussed later.

Internal Consistency

Effective evaluations are conducted in a systematic, logical way. Procedures must be established and followed. Conclusions must map closely to the facts.

Without the employment of internally consistent evaluation procedures, the results of the evaluative effort can be viewed to be arbitrary. This is the predictable result of ad hoc evaluations, where evaluation team members make up the rules as they go along. For example, in tracking budget performance during an evaluation, the evaluation team members may use readily available financial data that, unbeknownst to them, only examines direct costs and leaves out indirect costs. From their review of the incomplete data, they conclude that the project is on target from the perspective of budget performance, when in fact it is experiencing a serious cost overrun. An evaluation team that employs a well-developed evaluation process would not fall into this trap. They would specify what kind of budget data they need to review and would conduct analyses *only* with the proper figures.

Replicability

Ultimately, employment of a consistent, systematic evaluation procedure contributes to the *replicability* of results. A fundamental principle of good science is that results that are achieved through scientific inquiry must be replicable (Garfield, 1987; Merton, 1996). If a scientist makes a seemingly great breakthrough, yet no one can replicate the findings, then the scientific community rejects them (Kuhn, 1996). Results are held to be reliable only after they can be replicated by others.

The same basic principle applies in the arena of evaluation. A properly conducted evaluation is one where the results of the evaluative effort are the same, whether the evaluation is conducted by outside Team A or Team B or Team C. If each evaluation team comes up with dramatically different conclusions, then the evaluative effort is flawed. The people being evaluated justifiably feel that the evaluation results they experience are tied to the luck of the draw—for example, being evaluated by a friendly group rather than an unfriendly one, or by a disciplined group rather than one that follows ad hoc procedures. Project teams that receive poor evaluations owing to the poor conduct of an evaluative effort can become demoralized when they question the fairness of the overall process.

Fairness

In general, healthy people who are in touch with their capabilities have a good sense of when they are doing a good job or bad job. If they are doing a good job, they expect to be praised for their efforts. If they are doing a poor job, they are prepared to experience a measure of criticism. The important thing is for them to believe that the evaluations they experience are conducted fairly. If they feel that they have been unfairly criticized—and if this criticism jeopardizes their job security or bonus status—it can cause them to be very unhappy about their jobs (Adams, 1963; Vroom, 1964).

Being fair can be tricky. For example, if a team is unable to achieve its performance goals on time, within budget, and according to specifications, it appears at first blush that its members deserve a low score on their evaluations. However, it may be that this team inherited a loser project—one that was underresourced and where unrealistic promises were made to the client. In fact, through their heroic efforts, the team may keep cost overruns and schedule slippages modest, as opposed to disastrous. Yet the fact remains that the team has encountered cost overruns and schedule slippages, and there is a danger that they may be punished for this. If they are punished, they may feel justifiable anger at the unfairness of the verdict.

Converting Lessons Learned into Action

It is not enough to go through the motions of conducting a lessons-learned exercise. When lessons are gathered and documented, there is a danger that they will be put on a shelf where they remain unread and gather dust. For the lessons to be truly learned, they must be formulated in such a way that ultimately leads to action. There are a number of approaches people take to promulgate lessons, several of which are described here (Frame, 2002):

- *Share lessons in informal meetings.* In some organizations, project staff meet informally from time to time to exchange project experiences. For example, employees in some companies hold monthly brown-bag lunch meetings where one or two attendees describe recent project experiences. All the people who are present at this meeting are invited to explore what went right and what went wrong on the described projects.
- *Maintain a case study library of project experiences.* Some organizations maintain files describing corporate project management experiences in a case study format. Each case provides a history of a project, from inception to closeout. It focuses attention on special issues and

challenges that arose during the life of the project and the project team's responses to them. It also contains a lessons learned section at the end of the case. A library of such cases can provide employees with valuable insights into the organization's project experiences, enabling them to develop realistic expectations about what they might encounter on future projects and possibly suggesting steps they can take to avoid problems.

- *Change organizational procedures to reflect lessons learned.* The surest way to make certain that the lessons learned are converted into action is to have them incorporated into organizational procedures. For example, a review of problems associated with the execution of a project might lead to the conclusion that a major source of friction in dealing with clients is that project staff take too long to respond to their queries. The conclusion derived from this review might be that project staff must deal with customer queries as quickly as possible. Procedures might be adopted requiring staff to touch base with clients who have questions before the close of the business day in which the query was generated. If this approach is taken, it is important to make sure that the list of procedures is kept lean—each time a new procedure is added, old ones should be examined with a view of throwing out procedures that no longer provide value.

- *Employ captured performance data to establish baseline measures.* A leading cause of struggles on projects is the absence of good estimates that form the basis of project plans. Often, cost, schedule, and other performance estimates tend to be optimistic in order to gain support to move ahead with a project. This is particularly true when trying to win a contract award. Sales staff may promise potential clients that their organizations can deliver incredible deliverables at bargain prices and according to phenomenal schedules. Once a contract is awarded, however, performance often falls far short of the promises, since the promises were based on wishful thinking and not on fact. A factual basis for making estimates can be created if actual performance data is used to develop realistic baseline performance measures. For example, if experience shows that it takes 2.5 days to install a piece of equipment, and that the cost of installation is typically $1,100, then these figures can be employed for schedule and cost-estimating purposes. The trick here is to establish and *enforce* procedures for archiving cost, schedule, resource, and performance experiences.

- *Employ information from the lessons-learned analyses into risk assessments.* In conducting lessons-learned exercises, you will find that most lessons you surface are mundane. For example, you may determine that in order to get reports to clients quickly and reliably, you should ship them using commercial overnight delivery services instead of the national postal service. Or you may find that copies of all correspondence sent to clients should be cc'd to the project manager. Occasionally, you come across a high-impact lesson, and it may be appropriate to embed the lesson into your organization's risk assessment efforts. For example, you may discover that the technical team rarely implements suggestions that arise during the critical design review and that this is a major source of customer unhappiness. Your risk assessment process can be adjusted to audit team performance a week after the critical design review sessions to make sure team members are following up on suggestions.

Conducting Friendly Post-Project Evaluations

Earlier, it was stated that effective evaluations follow four basic principles: objectivity, internal consistency, replicability, and fairness. It was also pointed out that it is often difficult to follow these four principles. For example, in the search for objectivity, you may hire outside evaluators who are ignorant of conditions the project team is facing. Or in a drive to be consistent in your evaluative efforts, you may be unfair in the verdicts you deliver to project teams because you did not take into account the extenuating circumstances facing the team members. To the extent that evaluations are perceived to be unfair and threatening by the people who are being evaluated, they will likely be less than honest in working with the evaluation team. Without honest feedback from the people being evaluated, the whole evaluation effort becomes suspect.

Two "friendly" approaches to evaluation have emerged over the years. What makes them friendly is that the people being evaluated are given a measure of control over the process. Consequently, they develop a sense of trust in the evaluative effort. One approach, developed by IBM in the 1960s, is called the *structured walk-through*. The second is associated with evaluative assessments for such initiatives as ISO 9000 and the Software Engineering Institute's Capability Maturity Model and is called the *EISA approach* (Wilson and Pearson, 1995). Each approach will be discussed in turn.

Structured Walk-Throughs

As mentioned, the structured walk-through approach to evaluation was developed and promoted by IBM in the 1960s (Yourdon, 1988). Managers at IBM recognized that when evaluations are perceived to be unfriendly, it is difficult to gain cooperation from the people being evaluated. So in order to make evaluations friendlier, the people being evaluated should be empowered to run the evaluation effort.

Originally, structured walk-throughs were employed for the purpose of reviewing software code that was being developed. The programming team would walk a panel of evaluators through their work and gain feedback from the evaluators on what they were doing right and what they were doing wrong. Over the years, it grew apparent that the structured walk-through methodology could be expanded to cover a much broader range of evaluations, including design reviews, document reviews, and proposal reviews. This section describes how the structured walk-through can be an important vehicle to conduct post-project evaluation reviews.

As originally conceived, the structured walk-through entails following four rules:

- *Rule 1.* The team being evaluated selects the evaluators.
- *Rule 2.* The team being evaluated sets the evaluation agenda.
- *Rule 3.* The team being evaluated runs the evaluation meeting.
- *Rule 4.* No senior managers are permitted to attend the evaluation session.

Each of these rules will be described briefly.

Rule 1. The Team Being Evaluated Selects the Evaluators. One of the great complaints of people being evaluated is that they are unhappy with the evaluators assigned to judge them. This unhappiness has several roots that were discussed earlier. For example, the team may be concerned about *who* selected the evaluators. Was it someone friendly to the project? Someone unfriendly? Another example: Do the evaluators have any idea of what project team members are doing, or do the team members need to take valuable time getting them up to speed? Still another example: How can project team members deal with evaluators who are fundamentally not qualified to sit in judgment of the project effort?

By selecting their evaluators, the project team members can make sure that they choose people who are sympathetic to their efforts, who are up to speed on what the team is doing, and who are fundamentally competent. An obvious question raised here is this: What's to keep the project team from "rigging" the jury. That is, isn't it likely that they will select friends who will engage in mutual back-scratching? Clearly, this is a possibility. At least two approaches have arisen to deal with it. In one, team members get to choose the evaluators. Then their list of prospects is reviewed by an independent outside panel that will examine the qualifications of the proposed evaluators. In the second approach, the organization maintains a pool of people available to conduct evaluations. The team to be evaluated then selects evaluators from the pool.

Rule 2. The Team Being Evaluated Sets the Evaluation Agenda. With traditional evaluations, the outside evaluators establish the evaluation agenda. They determine what the evaluation will focus on. They define the rules for conducting the evaluation. They even set the time and date for the evaluation review sessions. In this environment, the team finds itself operating according to the vagaries of the evaluators, with minimum input into the process. One commonly heard complaint is that because the evaluators do not fully understand the environment in which the project is being executed, they often focus on the wrong issues and do not ask the right questions. Another concern is that during the evaluation effort, the outside evaluators may try to trick the project team into revealing problems and catching them with a smoking gun. Project team members understandably are concerned about dealing with surprises dealt them by the outside evaluators. A final, more mundane complaint is that activities associated with the evaluation are invariably scheduled at the convenience of the evaluators and not the team members. In fact, it may happen that poorly scheduled evaluative activities actually cause schedule delays on projects.

The structured walk-through empowers project team members to establish the evaluation agenda. They select the topics to be covered. They determine the order in which the topics are treated and who gets to speak on them. They schedule the evaluation at *their* convenience to minimize disruptions to the project effort. Through a process like this, the project team can develop a sense of control over their fate. Anxieties about the relevance of the evaluation and surprise attacks by evaluators disappear.

As with Rule 1, concerns may be raised that by empowering project team members to conduct the evaluation, they may structure the agenda to avoid dealing with real problems. They may do this consciously and cynically keep known problems off the agenda. Or they may do it unconsciously, because they are too close to the work to identify objectively what the evaluation should address. To deal with the possibility of skewed agendas, it is a good idea to let the project team members establish it to the best of their abilities, and then to have it reviewed by an objective outside panel. By following such a process, the team feels empowered while a measure of objectivity can be maintained in establishing an evaluation agenda.

Rule 3. The Team Being Evaluated Runs the Evaluation Meeting. In the spirit of providing project team members with a feeling that they are empowered to run their projects, the structured walk-through has them running the evaluation meetings. If a meeting drifts from the agenda and addresses irrelevant topics, they have the power to bring it back to the agenda. If people speak out of turn, they have the power to insist that speakers stick to the proper protocol.

The principal problem associated with implementing Rule 3 is that project team members seldom have good meeting facilitation skills. In practice, they do not know how to keep the meeting focused on the agenda or how to make sure that speakers stick to their allotted time allocations. If the meeting becomes disordered, the evaluation effort can lose much of its value.

The problem of poor meeting facilitation can be handled by selecting a professional facilitator to run the evaluation session. Professional facilitators have the needed skills to keep the meeting moving forward and focused on the important issues. Because they are objective outsiders, they will not be not be cowed by some of the political dynamics that may arise during the evaluation. It is important when hiring facilitators to be clear that their job is to serve the project team, *not* to serve the organization in some abstract sense. They are like defense attorneys whose responsibility is to defend their clients to the extent possible, not to see that justice is achieved in the abstract. Facilitators *must* view project team members as their clients, in order to maintain the team members' trust in the evaluation process. If the facilitator is presented as a fair arbiter whose task is to serve the organization, it is likely that the project team members will not be completely forthcoming in their participation in the evaluation.

Rule 4. No Senior Managers Are Permitted to Attend the Evaluation Session. Rule 4 is the best-known rule of structured walk-throughs. Its rationale is obvious. How honest will project team members be in describing the problems they are encountering if the people who determine their salaries and career development are sitting in the room? Interestingly, many organizations have extended Rule 4 to cover customers as well. That is, they stipulate that customers should not attend structured walk-through sessions. How honest will project team members be in describing problems if customers are sitting in the room?

Ultimately, senior managers and customers need to be brought into the loop if they are going to have the information they need to function properly. In dealing with updating senior management on project issues, some organizations carry out two walk-throughs. The first—conducted with no senior managers present—entails a tough, honest review of problems. At the end of the session, the project team spends time determining how to deal with problems. Once they have developed solutions to the surfaced problems, a second walk-through can be conducted. Senior managers attend this session. Problems can be discussed frankly. What is good is that in this second walk-through, the team can also present solutions to the problems.

Customers can be brought into the loop in many ways—for example, through progress reports, customer walk-throughs, and customer partnering arrangements. It is important that they not be kept in the dark. However, customers and senior managers should recognize that the driving rationale of structured walk-throughs is to make sure that project team members feel no constraints to being honest.

Employing Structured Walk-Throughs in Post-Project Evaluations

Structured walk-throughs can be employed effectively in post-project evaluations. There is a lot to be gained by empowering project team members to conduct this final project review. Because they had daily exposure to the project, they know better than anyone what worked and what did not. With the structured walk-through, their personal insights can be tempered by the expert views of outsiders. If project team members are asked to select the outside reviewers (subject, of course, to approval by senior managers), you have some assurance that the evaluation team will be composed of qualified people. Furthermore, if the project team establishes the evaluation agenda (subject, again, to approval by senior managers), you have some assurance that the right topics will be addressed.

EISA Approach

EISA is an acronym for External, Internal, Self-Assessment (Wilson and Pearson, 1995). The EISA approach to evaluation has become standard when undertaking major assessments, such as those conducted under the auspices of ISO 9000 reviews, Capability Maturity Model reviews, and Baldridge Award reviews (Software Engineering Institute, 1993a and 1993b). As with structured walk-throughs, the EISA approach combines objective external assessments with major inputs from the group being evaluated.

With the EISA approach, the first round of review is conducted in the form of a self-assessment by the group being evaluated. The group identifies the performance goals it should be addressing in its work, then determines the extent to which these goals are being achieved. Once the self-assessment is complete, independent evaluators from within the organization conduct an internal evaluation. They examine the self-assessment, review the group's performance independently, then make a final judgment on the group's effort. Based on this assessment, deficiencies in the group's performance may be remedied. Finally, outside evaluators are hired to engage in a completely independent assessment of the group's performance.

Employing the EISA Approach in Post-Project Evaluations

As with structured walk-throughs, the EISA approach can be employed effectively in post-project evaluations. Its strength lies in the fact that it relies on input both from the people performing the work that is being evaluated and from independent assessors. Consequently, it likely will yield more meaningful evaluations than those carried out by one or the other party alone.

Customer Acceptance Tests: Built-in Post-Project Evaluation

A standard practice implemented on contracted projects is to have customers conduct a final inspection of the deliverable before accepting it and making final payments. The process of undertaking a final review of the deliverable is called a *customer acceptance test* (CAT). (In the information technology community, the term user acceptance test (UAT) is frequently used.)

The CAT is a form of post-project evaluation. Clearly, it is an important evaluation because it is being undertaken by the players for whom the deliverable is being developed. If they are unhappy with it, then the project has not met the important objective of customer satisfaction, which is central to all projects these days. They may not accept what has been produced and may insist that modifications be made to it.

CATs focus on the functional and technical performance of the deliverable. Is it achieving the functional requirements satisfactorily? For example, will the new data entry forms cut data entry errors in half, as required? Is the deliverable meeting technical requirements? Does the circuit board layout on actual circuit boards correspond 100 percent with layouts drawn on paper?

If a deliverable passes muster with a CAT review, then customers sign a statement that they are satisfied that the deliverable has met project requirements. If the CAT fails to satisfy the customers, things can get dicey. When there is an obvious deficiency in the deliverable, the contractor has a responsibility to fix it. However, the source of customer unhappiness might be tied to an interpretation of requirements that is different from the contractor's. In this case, the contractor and customers need to work out a solution that is satisfactory to both parties.

To close out the contract, a responsible member from the customer organization must attest that the contractor has met its obligations satisfactorily. The customer contract manager will also conduct a final review of the deliverable and payments to date to make sure that the organization has received what it has paid for. When the contract officer signs an acceptance document, final payments are made and the project is officially closed out.

Increasingly, well-managed organizations are adopting the CAT approach on noncontracted, internal projects—for instance, a project carried out by the IT department to update the organization's Web servers. To do this, they need to identify internal "customers" whose needs and wants should be satisfied through the project. This is not always easy to do, since projects always have multiple customers and the customers usually have contending needs and wants. With these internal projects, conducting a CAT is usually the last step taken before the project is closed out.

Post-Project Evaluation and the Learning Organization Perspective

Peter Senge's best-selling management book titled *The Fifth Discipline* stimulated substantial discussion on one of the topics covered in the book: the nurturing of a learning organization

(Senge, 1990). (See the chapters by Bredillet and by Morris.) The learning organization perspective is an extension of a basic cybernetic principle, which holds that for systems to survive they must continually use feedback data to help them adjust to changing conditions in their environments. Thus, learning organizations engage in a process of constantly learning lessons from their environmental experiences and adjusting their behaviors accordingly. A little reflection shows that this is precisely the central concern of post-project evaluation.

Through post-project evaluations, organizations can determine what works and what does not work when their employees execute projects. To the extent that these evaluations cover a broad range of issues—including technical, financial, operational, organizational, and legal—they provide insights that will enable organizations to survive and even thrive in highly competitive business environments. Through a learning processes rooted in the conduct of post-project evaluations, they function as learning organizations.

Organizations that continue to conduct business in the same old way—that live by the motto "If it ain't broke, don't fix it"—are likely to find themselves unable to deal with the surprises thrown at them from out of their ever-changing environment. They are not learning from their experiences, and ultimately this will translate into weak business performance, or worse.

Post-Project Evaluation and the Human Resource Management Perspective

Post-project evaluations can have significant human resource management implications. (See the chapter by Huemann, Turner, and Keegan.) Two are examined here: Post-project evaluations can be tied to an organization's reward system, and they can indicate the adequacy of the organization's staffing efforts. Each will be discussed briefly.

Post-Project Evaluations and Reward Systems

An important element of good management is *accountability* (Frame, 1999). People must be held accountable for their actions. High performers should be recognized for their achievements, while low performers should be alerted to their poor performance and provided guidance on how to improve it. Post-project evaluations can play a significant role in an organization's reward system. For example, they can identify highly successful projects and explain the reasons for success. Individuals who contributed to the success can be recognized. Letters of commendation can be written on behalf of these people and included in the performance appraisal review process. By the same token, they can identify troubled projects and the causes of problems. If the problems are closely tied to the behavior of specific individuals, this information can be incorporated into their performance appraisal reviews.

While rewards are often used to motivate project workers, their shadow side should be recognized. Even as they may stimulate employees to do their best, they may also lead to unhappiness among those who are not recognized (Kohn, 1993).

Post-Project Evaluations and Guiding the Organization's Staffing Efforts

In today's fast-paced world, staffing requirements are undergoing continual change. In developing our information systems, yesterday we needed Cobol programmers, today Oracle database experts, and who knows who we will need tomorrow? A major challenge facing human resource management specialists today is determining whether current staffing arrangements work and predicting future staffing requirements. Information on the adequacy of staffing can be gleaned from post-project evaluations. For example, the evaluation might suggest that a project's performance was hampered by the lack of software testing personnel. If this is deemed an important problem, then the personnel department can work to remedy it by hiring new resources or training existing personnel. To gain the maximum benefit from post-project evaluations, organizations should make sure that human resource management personnel are aware of evaluation findings that have staffing implications.

The Bottom Line: Dealing with the Realities of Post-Project Evaluation

Most organizations do not conduct effective post-project evaluations (Kumar, 1990). A problem they encounter is that these evaluations require the commitment of time, money, and expertise that many organizations are not willing to provide for what they perceive to be an overhead activity. The fact that post-project evaluations are carried out at the end of the project life cycle adds to the problem. By the time most projects end, project funds have already been expended. In fact, managers running projects that face cost overruns are not likely to serve as rigorous advocates for additional funding for post-project evaluations. In their desire to save money, they eschew "nonessential" expenditures, such as expenditures on post-project evaluations.

Beyond the matter of funding, at project's end, its momentum is gone and enthusiasm for project work is often low. The principal item of concern of team members at this time is future job assignments. Higher-level managers usually want to put the finished project behind them and focus attention on new revenue-generating prospects. No one is arguing that effective post-project evaluations must be carried out for the long-term good of the organization.

Because the realities of project termination work against the conduct of effective project closeout, proponents of evaluation need to be willing to take a graded approach. They must recognize that they can't have it all. The graded approach identifies which evaluation activities *must* be conducted and which are nonessential, although nice to have. The following is a list of some *must haves* and *might haves*.

- Processes *must* be in place to capture information on project activities as they are being carried out. Included here is data on actual cost of work effort, actual task durations, actual amount of labor employed, and actual milestones achieved. By archiving cost, schedule, resource, and performance data, evaluators are able to examine project performance on these items retrospectively. The retrospective review enables them to identify exactly when cost overruns and schedule slippages occurred. Or when milestones were achieved early and at what cost. Without archived data, the post-project evaluation is based on anecdote and conjecture and its reliability and validity are suspect.

- Checklists of standard project closeout items *must* be developed. They should address questions such as: Have you protected your deliverables, documents, and processes from an intellectual property perspective (e.g., patents, trademarks, copyrights, trade secrets)? Has equipment been properly reassigned for use on other projects? Have people been reassigned to other work efforts? Have all the items in your statement of work been achieved? Have contractors submitted pertinent deliverables, documentation, equipment, and materials as required by the contract (Frame, 2003)?

- A bare-bones, post-project analysis *must* be carried out and its findings written up in a lessons-learned document. At a minimum, the analysis should describe the project experience in narrative format. Quantitative data on cost, schedule, and resource performance should be included as well. At the end of the document, a statement of key lessons learned should be offered and recommendations made for future projects.

- Procedures for capturing and implementing lessons learned *must* be established. For example, procedures might be developed where formal reviews are conducted biannually on lessons learned from recent project experiences. Then the key lessons might be incorporated into the organization's project and business processes. Another example: The organization might create a library of lessons-learned documents that makes real project experiences accessible to all project staff in the organization.

- Highly structured evaluations employing structured walk-through or EISA processes *might* be adopted. These structured evaluations are important for large, complex, high-impact projects. The hassles associated with conducting them do not usually make them worthwhile for smaller projects.

- Detailed and thorough root cause analyses of sources of project problems *might* be conducted. The goal of these analyses is to understand the underlying causes of problems on a given project. The information gleaned can then be used to improve performance on future projects. These analyses will likely be expensive: Senior managers and key project players need to be interviewed; data needs to be carefully analyzed; a detailed study with well-conceived recommendations must be written up and distributed.

- Steps *might* be taken toward creating a learning organization environment, where lessons learned are continually being incorporated into the organization's business processes and obsolete items are being jettisoned.

In the final analysis, effective post-project evaluation requires sensitivity to organizational realities. However, while it is argued here that evaluators must be willing to live with compromises in the evaluation effort, I am not suggesting that extreme shortcuts be taken or that organizations can afford to abandon post-project evaluations entirely. As noted at the outset of this chapter, post-project evaluations are important for a range of reasons and must be implemented so that organizations can develop an understanding of how their projects are doing, can gather information to enable them to see what works and what does not, and can employ this information to adjust their business processes to ensure better performance.

References

Adams, J. S. 1963. Toward an understanding of inequity. *Journal of Abnormal and Social Psychology* 67: 422–436.

Buss, M. D. J. 1984. How to rank computer projects. *Harvard Business Review* 61:118–125.

Drucker, P. F. 1985. *Management: Tasks, responsibilities, practices.* New York: Harper Business.

Ezingeard, J.-N., Z. Irani, and P. Race. 1999. Assessing the value and cost implications of manufacturing information and data systems: An empirical study. *European Journal of Information Science* 7(4): 252–260.

Farbey, B., F. F. Land, and D. Targett. 1993. *How to assess your IT investment: A study of methods and practice.* Oxford, UK: Butterworth-Heinemann.

———. 1999. Moving IS evaluation forward: Learning themes and research issues. *Journal of Information Technology* 8:189–207.

Fox, J. R. 1984. Evaluating management of large complex projects. *Technology in Society* 16(6):129–139.

Frame, J. D. 1999. *Project management competence.* San Francisco: Jossey-Bass.

———. 2002. *The new project management.* San Francisco: Jossey-Bass.

———. 2003. *Managing risk in organizations.* San Francisco: Jossey-Bass.

Garfield, E. December 1987. Is there room in science for self-promotion? *The Scientist* 1(27):9, 14, 187–188.

Kohn, A. 1993. Why incentive plans cannot work. *Harvard Business Review* 74(5):54–61.

Kuhn, T. 1996. *The structure of scientific revolutions.* 3rd ed. Chicago: University of Chicago Press.

Kumar, K. 1990. Post-implementation evaluation of computer-based IS: Current practice. *Communication of the ACM* 33(2):203–212.

Locke, E. A., and G. P. Latham. 1990. *A theory of goal setting and task performance.* Englewood Cliffs, NJ: Prentice Hall.

Merton, R. K. 1996. *On social structure and science.* Chicago: University of Chicago Press.

Morgan, E. J., and Y. L. Tang. 1993. Post-implementation reviews of investment: Evidence from a two-stage study. *Journal of Production Economics* 30(3):477–488.

Remenyi, D., M. Sherwood-Smith, and T. White. 1997. *Achieving maximum value from information systems: A process approach.* Chichester, UK: Wiley.

Senge, P. M. 1990. *The fifth discipline: The art and practice of the learning organization.* New York: Doubleday & Co.

Software Engineering Institute. 1993a. *Capability maturity model for software, Version 1.1, CMU/SEI-93-TR-24.* Pittsburgh: Carnegie Mellon University.

———. 1993b. *Key practices for the Capability Maturity Model, Version 1.1, CMU/SEI-93-TR-25.* Pittsburgh: Carnegie Mellon University.

Symons, V. 1990. Evaluation of information systems: IS development in the processing company. *Journal of information Technology* 5:194–204.

Vroom, V. H. 1964. *Work and motivation.* New York: Wiley.

Walsham, G., 1990. Interpretive evaluation design for information systems. Chapter 12. *In Beyond the IT Productivity Paradox,* ed. L. P. Wilcox and S. Lester. Chichester, UK: Wiley.

Wilson, P. F., and R. D. Pearson. 1995. *Performance-based assessments: External, international, and self-assessment tools for Total Quality Management.* Milwaukee: American Society for Quality Control Press.

Yourdon, E. 1988. *Structured Walkthroughs,* Englewood Cliffs, NJ: Prentice Hall Professional Technical Reference.

CHAPTER TWELVE

DEVELOPING PROJECT MANAGEMENT CAPABILITY: BENCHMARKING, MATURITY, MODELING, GAP ANALYSES, AND ROI STUDIES

C. William Ibbs, Justin M. Reginato, Young Hoon Kwak

How good are your organization's project management practices? How well do your practices compare with those of your peers in the business world? Are you making the appropriate investments in new project management systems, processes, and practices? These are the questions that few firms can answer directly and accurately. Yet their answers can unlock the gate to superior business performance.

The first step in understanding an organization's project management effectiveness is to determine its Project Management Maturity (PMM). By having a grasp of where a company lies on the PMM spectrum, management can determine its project management strengths and weaknesses, which is enormous value in today's highly competitive, project-oriented marketplace.

Stated simply, a company's PMM is a measure of its current project management sophistication and capability. Knowledge about the most sophisticated project management tools does not necessarily mean that those complicated tools will be used on every project. Rather, appropriate knowledge means that the firm and its managers understand which tool is appropriate for the demands of the project. PMM helps gauge such management wisdom.

Once the PMM is known, it can be used to both understand the company's current standing and to develop a roadmap for future improvements in project management processes and practices. Once on the path to such enlightenment, companies can craft their capabilities and strategy to enhance competitive advantage and wealth creation.

The purpose of this chapter is to describe how PMM benchmarking can help organizations develop that roadmap. We do this by first highlighting the importance of PMM in today's competitive marketplace. Second, we exhibit techniques for determining current levels of PMM and defining a course for PMM improvement. Last, we demonstrate methods to enumerate the value of project management improvement to ensure that investments in project management are reaping the desired returns.

This chapter presents a number of quantitative findings, some of which are cast in a statistical manner. Keep in mind, though, that the subject of this chapter is classical statistical methodologies for "real-world management"—that is, t-tests, levels of significance, and other sophisticated statistical concepts—are not directly pertinent to, or reasonable for, such applications because the real world of management is very complex. These statistical relationships should therefore be seen as general tendencies and not treated as precise correlations and cause-effect relationships.

The Importance Of PMM in Today's Marketplace

Businesses are becoming increasingly projectized. Examples of current successful businesses that are organized around projects include Microsoft (operating and networking environments), Boeing (large commercial and defense aerospace ventures), and Amgen (biotechnology R&D). They are successful because they continually grow revenues and profits and have achieved remarkable stock market capitalization. The vast majority of their revenue and profit sources are from projects.

These firms share common characteristics such as devolved power, strong emphasis on intellectual property, powerful brand identification, as well as a premium on project-driven services and products. They are also very much bottom-line focused, managing themselves in a manner that creates increased shareholder value. These companies also share another trait: As they evolve organizationally, their abilities to deliver projects that advance their corporate strategies also evolve.

As projects become the currency for improved business performance, making project management a core capability of successful organizations in turn becomes paramount. No longer are being on time and on budget the only benefits or goals of strong project management. Additional core benefits of a project-centric focus are sophisticated project management tools that improve organizational effectiveness, meeting quality standards, and fulfilling customer satisfaction (Al-Sedairy, 1994; Boznak, 1988; Bu-Bushait, 1989; CII, 1990; Deutsch, 1991; Gross and Price, 1990; Ziomek and Meneghin, 1984).

But to demonstrate a true competence, project management success cannot be an occasional event. Performance that is good, on average, is not sufficient. Repeatability and relentless improvement must be the standard. Project performance that is, on average, good but erratic is not sufficient because one wayward project ripples through and affects the company's portfolio of projects. When management attention is diverted to the errant project, the loss of attention on all other projects can ripple through to other concurrent or subsequent projects, hurting them.

The first step in determining PMM and appropriate directions for future improvements to project management is to evaluate the benefits of project management in quantified terms. Project management cost, schedule and cost performance, PMM, and the financial return on project management investments (what we refer to as PM/ROI[SM]) are all important quantifiable metrics in evaluating the benefits of project management. Research shows that companies with high levels of PMM can leverage a range of quantified benefits (Ibbs and Reginato, 2002):

- *Companies with more mature project management practices have better project performance.* For example, companies with more mature practices deliver projects on time and on budget, whereas less mature companies may miss their schedule targets by 40 percent and their cost targets by 20 percent.
- *PMM is strongly correlated with more predictable project management schedule and cost performance.* The project portfolios of more mature companies, for instance, have lower standard deviations for schedule performance (0.08) and cost performance (0.11) than companies with lower PMM scores (corresponding values of 0.16 for both schedule and cost).
- *Good PM companies have lower direct project costs than poor project management companies.* Highly mature companies have project management costs in the 6 to 7 percent range, while their counterparts average 11 percent (and in some cases reach 20 percent). Note this is just the direct cost spent on project management. Organizations with low PMM risk other undesirable events such as increased indirect costs, late project deliveries, missed market opportunities, and dissatisfied customers.

Each of these points will be discussed in greater detail later in this chapter. The key point being made is that increased levels of PMM correspond to better project performance and, in turn, can be a key driver for corporate success.

PMM: Concepts and Quantification

So how does an organization measure PMM? The following section demonstrates techniques for determining current levels of PMM, as well as measuring it numerically so that it can be quantifiably valued.

Definition

PMM is the sophistication level of an organization's current project management practices and process (Kwak and Ibbs, 2002). Project management techniques serve businesses in planning, controlling, and integrating time and resource-intensive endeavors. Until recently, there was little quantifiable data, suitable methodologies, or well-defined processes that impartially measure PM practices. PMM models were developed to fill that void.

In addition to measuring the internal level of PMM, corporate executives needed a method to validate the investments they were making in project management. PMM can be used in conjunction with valuation techniques such as financial return on investment because it is a quantifiable value. That is, what profit return will a company realize for every $1 it invests in some new project management system, process, or procedure? Also, as many PMM models become standardized (which is a goal of the Project Management Institute's OPM3 initiative), PMM becomes a yardstick from which companies can compare their internal measures of PMM externally with other peer organizations.

Models

As PMM has been thrust to the forefront of building project management competence, several models for measuring maturity have evolved. One such model is the Berkeley Project Management Process Maturity Model. This model has been developed over a seven-year research period and has been successfully implemented in several industries, nonprofits, and government agencies (Ibbs and Reginato, 2001; Kwak and Ibbs, 2002).

The five-step Berkeley Project Management Process Maturity Model is used to establish an organization's current PMM level. This model demonstrates sequential steps that map an organization's incremental improvement of its project management processes. It is schematically illustrated in Figure 12.1.

The model progresses from functionally driven organizational practices to project-driven organizations that incorporate continuous project management learning. An organization's position within the model can be used to determine its position relative to the other companies in the same industry class or otherwise that have been assessed.

Level 1: Ad Hoc. At the Ad Hoc stage, there are no formal corporate procedures or plans to execute the project. The project activities are poorly defined, and cost estimates are inferior. Project management-related data collection and analyses are not conducted in a systematic manner. Processes are unpredictable and poorly controlled. There are no formal

FIGURE 12.1. BERKELEY PROJECT MANAGEMENT PROCESS MATURITY MODEL.

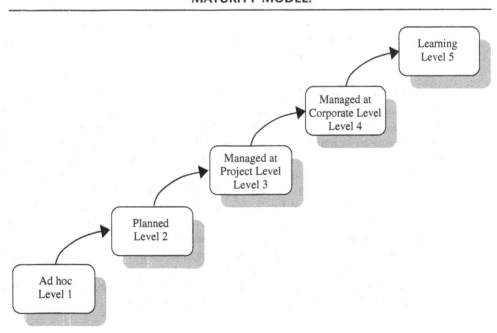

steps or guidelines to ensure continuity of project management processes and practices. As a result, utilization of project management tools and techniques is inconsistent and applied irregularly, if at all, even though the individual project manager may be very competent (Ibbs and Kwak, 2000).

Level 2: Planned. At the Planned level, informal and incomplete processes are used to plan, but not control, a project. Some project management problems are identified, but they are generally not documented or corrected in a systematic manner. Project management-related data collection and analysis are informally conducted but not documented. Project management processes are partially recognized and used by project managers. Nevertheless, planning and management of projects depends largely on individuals.

An organization at Level 2 is more team-oriented than at Level 1. The project team understands the project's basic commitments. This organization possesses strength in doing similar and repeatable work. However, when the organization is presented with new or unfamiliar projects, it likely experiences chaos in managing and controlling the project. Level 2 project management processes are efficient for individual project planning, but not for controlling the project, let alone any portfolio of projects (Ibbs and Kwak, 1997).

Level 3: Managed at the Project Level. At Level 3, PM exhibits systematic planning and control systems that are implemented for individual projects. Project management processes become more robust and demonstrate both systematic planning and control characteristics. The project management team typically works together in an informal setting. For the purposes of project control, most of the challenges regarding project management are identified and informally documented for each project. Various types of analyzed trend data are shared by the project team to help it work together as an integrated unit throughout the duration of the project. This type of organization works hard to integrate cross-functional teams to form a project team.

Level 4: Managed at the Corporate Level. For projects managed at Level 4, management processes are formal, while information and processes are documented informally. The Level 4 organization is fully integrated: It can plan, manage, and control multiple projects efficiently across an organization's project portfolio. A project management process model is probably well defined, with project requirement systems that are in place but not necessarily regularly used. Project-related data and records are formally and systematically collected, reviewed, and distributed to the appropriate parties but are not formally organized. Also, data is collected and analyzed to anticipate and prevent adverse productivity and quality impacts or other trends detrimental to project success. This allows an organization to establish a foundation for fact-based decision making.

In addition to effectively conducting project planning and control for multiple projects, the organization exhibits a strong sense of teamwork within each project and across projects. Project management training is available when needed and is provided to the entire organization, according to the respective role of project team members.

Level 5: Learning. The key characteristic of companies that operate at the Learning stage is that they continuously improve their project management processes and practices. Each project team member spends considerable effort to maintain and sustain the project-driven environment. Training is formally available when needed, presenting lessons learned and other techniques to improve project management on an ongoing basis. Project team members are typically together throughout the entire project duration, and their individual roles are defined based by their strengths and experience. Problems associated with applying project management are fully understood and addressed on an ongoing basis to ensure project success. Project management data are collected automatically to benchmark project management strengths and identify the weakest process elements. These data are then rigorously analyzed and evaluated to select and improve the management processes. Innovative ideas are also vigorously pursued, tested, and organized to improve processes.

Formal comprehensive requirement systems exist and are used regularly. A project management process model is formally defined, distributed, and discussed by all of the team members, using previous project experience as a guideline. Additionally, a project management consulting group is probably created and chartered, and its existence is communicated throughout the organization.

Each level within the Berkeley Project Management Process Maturity Model includes an assessment of PM processes and practices based upon six PM processes and nine knowledge areas, as shown in Figure 12.2. When each organization is assessed along these boundaries, PM strengths and weaknesses are determined. This assessment allows companies to prudently invest in areas to improve upon their weaknesses.

FIGURE 12.2. BENCHMARKED PROJECT MANAGEMENT PHASES AND KNOWLEDGE AREAS.

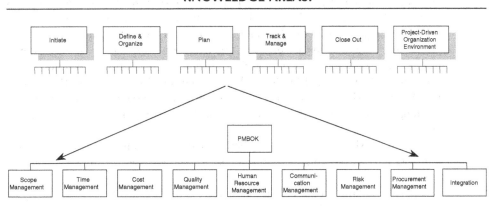

Ideally, an organization evolves smoothly and thoroughly from a less project management-sophisticated organization to a "learning" project-centered organization. However, in practice it is rare that when a company moves to maturity Level N+1, it has implemented *all* the characteristics of Level N. Rather, at Level N+1, an organization has the capability to choose the proper and relevant project management practices or processes that are suitable for a given project.

To illustrate, suppose that the scheduling techniques available to a company range from simple bar charts at the low end to complex simulation for resource optimization at the high end. An organization that has a high level of PMM does not always have to employ the most sophisticated techniques available to them, which in this case would be the simulation. Rather, they enjoy the ability to apply the most appropriate techniques based upon the complexity of the project. This allows for the construction of a broad-ranging PM toolkit as maturity increases.

In addition to the Berkeley Model are other maturity models developed by consultants and practitioners. Among them is the Center for Business Practices Model, Kerzner's Project Management Maturity Model, ESI International's Project Framework, and SEI's Capability Maturity Model Integration. PMI is striving to develop some commonality and consistency among these models through its OPM3 endeavors. For a good summary of maturity models, see Foti's (2002) article.

Measuring PMM

Measuring PMM involves quantifying the internal level of maturity and then comparing it externally to peer organizations. Two interconnected methods—benchmarking and gap analysis—are discussed in the following sections.

Benchmarking. Benchmarking is a process that allows organizations to compare different aspects of current practices against best practices. The basic premise is to improve and learn tools and techniques from other organizations. The purpose of benchmarking is to analyze the internal operation, understand the competition and industry leaders, incorporate best practices, and gain a superior foothold in competitive markets (Camp, 1995).

To assess PMM between different organizations or functional groups within an organization, a rigorous and comprehensive benchmarking methodology must be developed. The methodology adopted by the Berkeley Model involves a detailed, three-part questionnaire for data collection. Part I involves collecting general data regarding each organization, including the size of the organization, personnel structures, and how much it spends on PM per year.

Part II consists of 162 multiple-choice questions. Its intent is to measure the maturity of the organization's standard project management processes. Examples of some such questions are displayed in Figure 12.3.

To calculate the overall PMM, the average score for all 162 questions and their standard deviation are computed for each organization. All questions are weighted equally, so un-

FIGURE 12.3. SAMPLE BENCHMARKING QUESTIONS.

65. Critical path identified

No critical path calculation done. Each subproject identifies critical tasks
independently and sets work priorities.. 1

Critical path based on committed milestone dates. No CPM calculation
performed, or CPM used on individual subprojects.. 2

Key critical tasks identified through nonquantifiable means and used to drive the
critical path calculation.. 3

Critical path calculated through integrated schedule, but only key milestone dates
communicated back to subprojects. 4

All critical tasks identified and indicated in each individual subproject schedule.
Critical path determined through integrated schedule. 5

106. Quality management (QA/QC) system is utilized

No quality management system 1

Informal quality management system, not used... 2

Informal quality management system, hardly used.. 3

Formal quality management system, occasionally used.. 4

Formal quality management system, intensely used 5

128. Project deliverables list reviewed and cross-checked against actual deliveries

No project deliverables list available 1

Deliverables list available, but not reviewed 2

Some informal review of original, approved deliverables list 3

Formal review of approved deliverables list, but with only informal comparison to
actual deliverables 4

Formal review of approved deliverables list with point-by-point comparison to
actual deliverables 5

derlying the assessment is the assumption that all questions are equal indicators to an organization's PMM. Because of industrial and organizational competition, situations arise where some questions are more relevant to an organization than others. However, neglecting such factors allows for achievement of nonbiased circumstances to specific variables.

Part III of the assessment tool collects project-specific data, such as cost and schedule performance, as well as metrics regarding scope and quality attainment. The PMM analysis of Part II can be compared with the project performance data collected in Part III and evaluated as to how project performance improves with corresponding improvements in maturity. It is important that efforts in improving PMM result in increased project perform-ance—improving PMM for project management's sake is not likely to benefit the overall organization's market performance.

As a rough guide based on our experience, we suggest that at least five people partake in any assessment process, with 10 to 15 yielding better results. Less than five lends little to statistical evaluation, and more than 15 becomes difficult to manage. The people partici-pating in the assessment process should be project management professionals that represent typical project managers for the organization. Again, it is important that the people partak-ing in the assessment process be *typical* to the organization to ensure that representative data is being collected.

Process efficacy is further ensured if a manager or organizational decision maker, pref-erably from the vice president level or higher, is involved as a champion of the assessment process. A person at this level can expedite the assessment process, foster buy-in, and help to ensure assessment data quality. Conducting post-assessment interviews with assessment participants can further ensure score stability. Such interviews are particularly helpful when assessment scores exhibit seemingly arbitrary results.

Data collection is highly dependent on the type and size of the organization being assessed. The overarching goal is to collect data that is representative of the entire organi-zation or division being assessed. Obviously, if the group being assessed for PMM is mul-tinational and consists of thousands of employees, then many people from multiple geographic locales should partake in the assessment process. Smaller organizations can, generally speaking, accurately assess their organizations with far fewer participants to the assessment process.

Gap Analysis. Working hand-in-hand with benchmarking is gap analysis. Gap analyses are characterized by the comparison of an organization's current state to its desired state. The current state is defined by current practices, and the desired state is represented by industry best practices (Camp, 1995). The gap between current and best practices serves as the basis for preferred improvement.

Industry best practices are determined by comparing the project management processes of multiple peer organizations. Understanding the project management processes of multiple comparative organizations can be achieved several ways. One common method consists of attending discussions and symposia where other companies discuss their practices and proc-esses. Another common way is to partner with or hire an organization that conducts as-sessments for multiple organizations and hence has access to copious amounts of industry-specific data. The Berkeley database, for example, has project management process data from over 60 organizations in five industries. Other organizations have similar industry-specific best-practice data as well.

Once best practices are understood, the organization attempting to improve its PMM should critically examine which best practices it wishes to adopt. It is important to note that organizations undergoing PMM improvement initiatives should initially select a few areas in which to focus improvement. We suggest "picking the low-hanging fruit" first—that is, choosing the easiest practices to improve at the onset. These practices will provide the least expensive and easiest processes to improve and can serve as a springboard to further process improvement.

Also, once best practices have been determined, implementation should include input from customers, suppliers, subcontractors, operators, and so on. These parties will add further enhancement of the gap analysis by providing insights not attainable by benchmarking alone. Benchmarking is an excellent tool for illuminating best practices, but it is the ability to creatively implement (as opposed to copying) best practices that will allow an organization to become the best-of-class within the project management world.

Maturity's Role in Other Fields

For those readers who are in the software industry, much of the preceding may sound familiar. The Software Engineering Institute (SEI) at Carnegie Mellon University has several models that focus on maturity in the software industry. Research conducted by SEI is widely regarded, and many large organizations require that their software vendors meet certain levels of software capability maturity, usually stipulating that vendors provide continuous improvement over the life of the service contract. This continuous improvement stipulation is not unlike the gains that can be made by systematically improving PMM as shown by the Berkeley PM Process Maturity Model.

PMM'S Relationship to Business Results

Determining organizational PMM should not be an exercise in measurement for measurement's sake. Detailed goals should be outlined, with most, if not all, of those goals tied to business objectives. The following is a discussion of how improvements to PMM can improve overall business processes.

Measurable Maturity Benefits

As previously discussed, the assessment process begins by determining the overall maturity for a group of peer organizations. This step is highlighted in Figure 12.4.

The data presented in the figure were generated by research conducted at the University of California at Berkeley. In this figure companies that were involved in engineering and construction projects are labeled "EC," telecommunications and information management & movement companies are labeled "IMM," information service organizations are "IS," and high technology manufacturing companies are "HTM."

Knowing how an organization stacks up against others is important because the managers of that firm can then ascertain a relative projection of how PM is providing a competitive advantage or stands as a competitive barriers. However, to be most helpful the PMM must be assessed on a "subatomic" level. For example, consider company EC11 in Figure 12.4, as represented by the white bar. Its overall PMM is about 3.70, which puts it in the better half of its peers in terms of PMM.

FIGURE 12.4. OVERALL MATURITY FOR 44 ORGANIZATIONS.

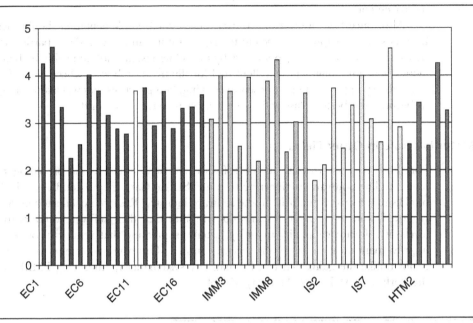

If the managers of EC11 wished to improve the company's PMM, where would they look? A gap analysis would point out specific areas in which EC11 could improve vastly with respect to its peers. For example, EC11 lags well behind its peers in terms of initiating projects, as shown by the white bar in Figure 12.5.

The gap analysis illuminates that EC11, while ahead of most of its peers in terms of overall PMM, can improve substantially by improving the process by which it initiates projects. This analysis allows EC11 to target the most appropriate areas of improvement rather than a hit-or-miss approach that may not improve (or even decrease) overall maturity.

As obvious as this seems, few companies conduct such a diagnostic test on their project management processes and teams before undertaking major project management improvement efforts that they *think* will help their companies. In terms of an analogy, most ailing people would be skeptical of a physician who prescribes a certain treatment before running a full battery of tests. Yet those same rational people routinely spend enormous sums of money, time, and effort on new project management software systems or training without first pinpointing where those improvement efforts would be most helpful.

This methodology can be extended to each of the project management phases and knowledge areas that are represented in Figure 12.2. Table 12.1 displays how a gap analysis can highlight en masse the areas where an organization leads or lags its peers.

FIGURE 12.5. DRILLING DOWN PMM TO THE INITIATING PROJECT'S LEVEL.

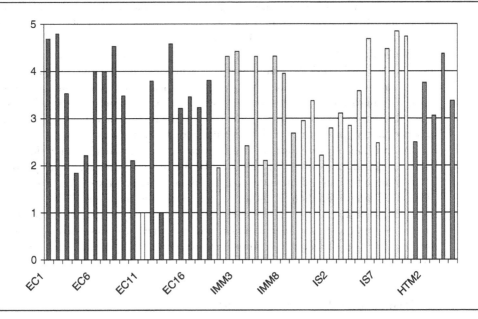

TABLE 12.1. GAP ANALYSIS FOR EC11 AND ITS PEERS.

Process Area	All Companies	Peers	EC11	EC11—Peers
Scope	3.42	4.15	3.84	−0.31
Time	3.37	3.86	3.85	−0.01
Cost	3.47	4.28	4.26	−0.02
Quality	3.00	3.15	3.93	0.78
Risk	2.97	3.35	3.97	0.62
Communication	3.53	3.81	4.10	0.29
Human resources	3.11	3.44	3.88	0.44
Procurement	3.15	4.40	3.97	−0.43
Integration	3.61	3.67	3.86	0.19
Initiate	3.35	4.27	1.00	−3.27
Define/organize	3.65	4.13	3.87	−0.26
Plan	3.21	3.54	3.79	0.25
Track and manage	3.32	3.65	4.40	0.75
Closeout	3.27	3.54	3.33	−0.21
Project-driven organization	2.97	3.63	3.45	−0.18
Overall	3.30	3.73	3.70	−0.03

In the EC11—Peers column, positive numbers represent areas where EC11 surpasses its peers and negative figures show areas of needed improvement. To target areas of improvement, EC11 can simply rank the negative numbers in order of highest to lowest. The largest discrepancies between EC11 and its peers are the areas in which EC11 should focus.

While the table demonstrates that, on average, EC 11 is similar to its peer organizations in terms of *overall PMM*, there are several areas in which it lags behind its peers. The process of benchmarking identifies these areas and allows EC 11 to concentrate and target specific areas for improvement, such as procurement management and project initiation. If EC 11 takes this targeted approach to PMM improvement, then it can more easily adopt best practices and improve its overall PMM above those of its peers.

As mentioned earlier, our initial benchmarking analysis treats each of the 162 questions as being equally important. At any stage of the analysis, though, some questions can be given more importance if the individual company so wishes. Experience shows this must be done with careful forethought, however; otherwise, the managers of the subject company may obtain an inaccurate analysis. For instance, they may think that quality management issues are paramount for their business and ask that such questions be super-weighted, whereas in point of fact their competitors are emphasizing some other aspect of project management.

Project Performance Improvement Benefits

Improving PMM will lead to improvements in project management processes and practices. However, the real goal of project management is not to improve processes and practices per se but to deliver projects more successfully. As discussed in the following section, there is a correlation between improved PMM and improved project performance.

Schedule and Cost Performance. What do companies get for their investments in project management? Our analysis of detailed assessments reveals that companies with higher PMM tend to deliver more of their projects on time and on budget. See Figures 12.6 and 12.7

These figures contrast Schedule Performance Index (SPI) and Cost Performance Index (CPI) against PMM for companies that we have benchmarked over the past six years. SPI and CPI are defined as the ratios of total original authorized duration or budget versus total final project duration or cost, respectfully. That is:

$$\text{Cost Performance Index} = \text{CPI} = \frac{\text{Planned budget}}{\text{Final costs}}$$

$$\text{Schedule Performance Index} = \text{SPI} = \frac{\text{Planned duration}}{\text{Final duration}}$$

It should be noted that our use of the terms CPI and SPI vary from that of common project management vernacular. We use the terms CPI and SPI because it is important to understand that as PMM improves, so do cost and schedule performance. However, the CPI and

FIGURE 12.6. SPI VS. PMM.

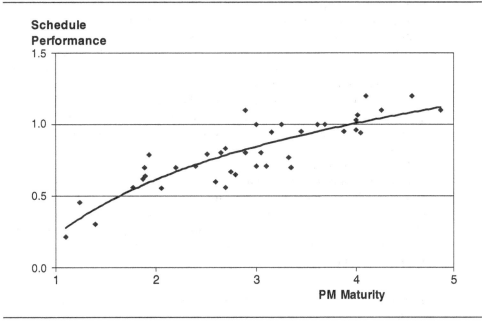

FIGURE 12.7. CPI VS. PMM.

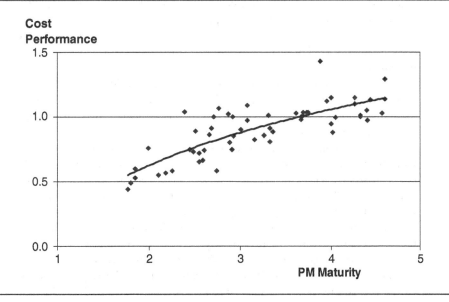

SPI ratios that we have listed are not the same as the ratios by the same name as applied to earned value analysis.

The value to organizations is apparent. In terms of schedule, as PMM increases so does the ability to complete projects on time. An SPI ratio of 1.00 equates to finishing projects in exactly the time that was originally estimated; a number < 1.00 indicates late completion. The ability to accurately forecast the time necessary to complete a project affords senior executives in the firm a powerful tool in meeting time-to-market windows.

Like SPI, CPI increases with higher PMM levels. Also similar to SPI, a CPI value approaching 1.00 signifies accuracy in estimating and delivering projects on budget. Increasing CPI is good because accurate cost forecasts allow companies to confidently and accurately allocate capital.

The R^2 value in these figures is called the *correlation coefficient*. A value $= 1.00$ would mean that the computed regression lines depicted in these figures are correlated perfectly with the actual data. Since the R^2 value for CPI is lower than that for SPI, we can say that cost estimating and control seems to be more erratically performed than schedule planning and control. One possible explanation for this is that companies are more schedule-driven in their projects than cost-driven and therefore are willing in actuality to overspend their projects to meet time commitments.

Schedule and Cost Reliability. At least as important as good SPI and CPI ratios is the reliability of such cost and schedule performance. That is, a PM organization that erratically delivers projects with SPI or CPI $= 1.00$ is not as trustworthy to top management as a team that delivers such reliably.

Our work shows that companies with higher PMM deliver projects with more predictable schedule and cost results. As companies improve their PMM, their individual SPI results tend to deviate less from the overall SPI average. This is seen in Figure 12.8 by examining the standard deviation of SPI over a portfolio of projects.

As also seen with these data, CPI standard deviation decreases as PMM improves. That is, companies with high PMM are less likely to have projects where the budgets escalate out of control. See Figure 12.9.

Budget accuracy is important because it reduces fiduciary risk. For capital-intensive projects, this can lead to a reduction in the cost of capital and large savings for companies that borrow money for project budgets or higher financial ratings for companies that obtain project financing from the capital markets.

A subtle though crucial point that many people overlook is the reliability of SPI and CPI metrics. Many people think that a SPI or CPI that averages more than 1.00 is good, but this is not necessarily the case. It is of little help to a company in estimating project durations if half of its projects have an SPI of 1.25 and the other half 0.75. Such a large variation thwarts effective planning and management of multiple projects.

Similarly, a company that has an average SPI and CPI substantially over 1.00 is being too conservative in its estimates. It may be "leaving money on the table" and not undertaking as many projects as it could with more realistic forecasts.

The data for Figures 12.8 and 12.9 come from a relatively few number of companies, 7 and 7, respectively, but a large number of projects, 46 and 41. In statistical terms this means there is a reliable number of degrees of freedom.

FIGURE 12.8. SPI STANDARD DEVIATION VS. PMM.

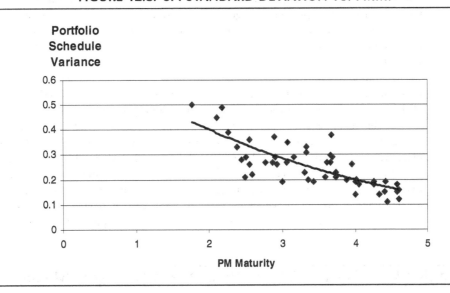

FIGURE 12.9. CPI STANDARD DEVIATION VS. PMM.

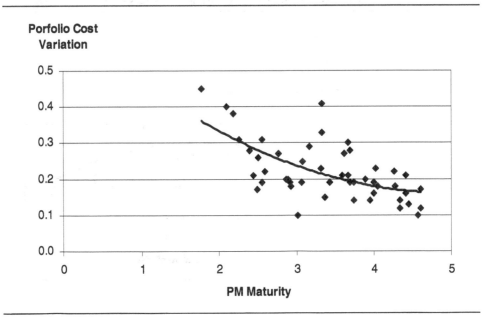

PMM and Project Management Cost Ratio. *Project management cost* entails summing all of the costs incurred by project management to deliver a project. It includes labor and burden costs for direct and indirect project management personnel; hardware, software and communications costs; and training costs; as well as those costs associated with consultants and subcontractors.

In companies that we have studied, this *project management cost ratio* is usually computed by annualizing and dividing all direct project management costs incurred by the total value of the projects executed during that same time frame (see Figure 12.10).

As the regression line in Figure 12.10 displays for N = 32 companies, the project management cost ratio increases until approximately PMM Level 3. From there the project management cost ratio steadily decreases with increasing PMM. This means that companies investing in project management will initially see their investment costs outstrip benefits. Eventually, however, the investments pay off, since mature companies actually pay less, as a percentage of project management costs, to improve their PMM. Economies of scale do hold rewards in project management, just like most other aspects of business.

Bringing It to Closure: The Virtuous Cycle of Project Management

Based on case study interviews and data collected from companies assessed during our research, we have created Figure 12.11 to illustrate what we call the Virtuous Cycle of

FIGURE 12.10. PROJECT MANAGEMENT COST VS. PMM.

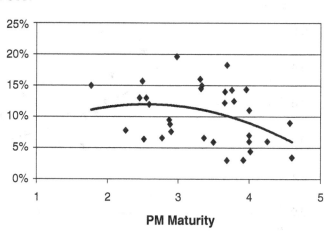

FIGURE 12.11. THE VIRTUOUS CYCLE OF PM.

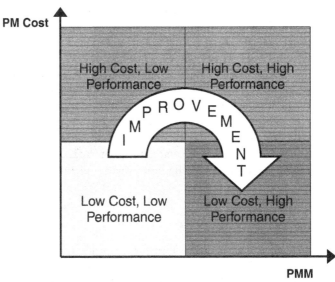

Project Management. This schematic allows organizations to map their PMM and investments in project management to plan and ensure a logical and sustainable progression along the project management cost-effectiveness journey.

The axes of the matrix pit PMM against the project management cost ratio. This diagram was developed by in-depth analysis of project management operations of 24 companies. Quadrants were divided so that each is represented by an equal number of companies to ensure no bias toward any one particular classification. The boundary between "good" and "bad" project management cost ratio % (the vertical axis) is approximately 5 percent to 7 percent; the PMM division is approximately 3.30 to 3.35. Project management growth should, we hope, go through the clockwise cycle depicted on this diagram, though the speed will vary depending on variables such as management commitment, industry circumstances, and the size and dispersion of the company.

Companies in the lower left-hand quadrant are underinvested in project management and are earning low returns, if any. Without adequate investment, both PMM and project performance will continue to lag behind peer organizations.

Continuing to the upper left-hand quadrant is the next step of the progression, and an area where no organization should reside for any prolonged duration. In this quadrant, PM investments have begun to increase, but benefits are not yet being proportionately realized. Organizations whose PM practices exist in this region for extended periods are in danger of paying a steep price for relatively poor PM performance.

In the upper right-hand quadrant, the benefits of improving PM are starting to be realized within the organization, but the cost of those improvements is still relatively steep. While having a high PMM is commendable, the victory is somewhat bittersweet in that PM is still costly for these organizations.

The lower right-hand quadrant is the ideal locale for company-wide PM practices. Companies in this category have best-of-class PMM, low PM cost and very high PM/ROI[SM]. These companies are in "PM nirvana" mainly because they have the highest throughput of projects with respect to their PM investment. For organizations in this arena, PM is a strong organizational competence and even, in some circles, regarded with competitive envy.

Since companies that have high-level PMM are, by definition, companies with high levels of PM learning (see Figure 12.1), they can be self-sustaining and self-improving. This allows them to become pioneering and agile organizations that grow and adapt to changing marketplace challenges, thus offering more value over time.

Summary

Project management can be a key lever for delivering projects. Many companies are interested in, and actively pursuing, initiatives to improve upon their project management processes. Understanding PMM can dramatically aid in the improvement of project management sophistication.

Improving PMM can be efficiently managed with the use of benchmarking and gap analyses. These tools allow for determining an organization's overall level of PM ability, as well as industry best practices. Most important, organizations can utilize benchmarking and gap analyses to make pointed and focused improvements in their PM processes.

Our research has shown that improvements in PMM help deliver three significant benefits:

1. Improvements in cost and schedule performance
2. Improvements in cost and schedule reliability
3. Lower overall project management cost in delivering projects

Large organizations certainly can apply statistical methods to PMM assessment methodologies, but we want to stress that in the dynamic corporate world, such analyses are as much art as science. We have presented methodologies that allow for the combination of statistical methods with experience and judgment to create a methodology that many companies can readily apply within their organizations to improve their PMM.

References

Al-Sedairy, S. T. 1994. Project management practices in public sector construction: Saudi Arabia. *Project Management Journal* (December): 37–44.

Boznak, R. G. 1988. Project management: Today's solution for complex project engineering. *IEEE Proceedings.*

Bu-Bushait, K. A. 1989. The application of project management techniques to construction and R&D projects. *Project Management Journal* (June): 17–22.

CII. 1990. *Assessment of owner project management practices and performance.* Special CII Publication (April).

Camp, R. C. 1995. *Business process benchmarking.* Milwaukee: ASQC Quality Press.

Deutsch, M. S. An exploratory analysis relating the software project management process to project success. *IEEE Transactions on Engineering Management* 38 (4, November).

Foti, R.. 2002. Maturity. *PM Network* 15(9):38–43.

Gross, R. L., and D. Price.1990. Common project management problems and how they can be avoided through the use of self managing teams. *1990 IEEE International Engineering Management Conference.* Santa Clara, CA.

Ibbs, C. W., and Kwak, Y. H. 1997. *The benefits of project management: Financial and organizational rewards to corporations.* Newtown Square, PA: Project Management Institute.

———. 2000. Calculating project management's return on investment. *Project Management Journal* 31 (2, June): 38–47.

Ibbs, C. W., P. W. G. Morris, and J. M. Reginato. 2001. Calculating the value of project management. *Project Management Institute's Annual Seminars & Symposium.* Newtown Square, PA: Project Management Institute.

Ibbs, C. W., and J. M. Reginato. 2002. *Quantifying the value of project management.* Newtown Square, PA: Project Management Institute.

Kwak, Y. H., and C. W. Ibbs. 2002. Project Management Process Maturity Model. *Journal of Management in Engineering* (July): 150–155.

Ziomek, N. L., and G. R. Meneghin. 1984. Training: A key element in implementing project management. *Project Management Journal* (August): 76–83.

CHAPTER THIRTEEN

PROJECT MANAGEMENT MATURITY MODELS

Terry Cooke-Davies

A glance through the contents of this book provides ample evidence that project management is no longer seen as simply being concerned with the skillful and competent management of a single project. There is much more involved in it than that. Organizations undertake many projects, and so require a set of processes and capabilities, of systems and structures, to allow the right projects to be undertaken and supported and to achieve consistent project success. As this recognition has evolved, so has the desire on the part of organizations to assess these systems, structures, processes, and capabilities, and many have turned for help to so-called project management maturity models.

There is no shortage of them; more than 30 were considered as a part of the research leading up to the Project Management Institute's own draft standard OPM3 (Cooke-Davies et al., 2001), and they are supported by claims that an increase in maturity brings organizational benefits (e.g., Kwak and Ibbs, 2000; Pennypacker and Grant, 2003).

The reason for this upsurge in interest is not difficult to understand. As project management has expanded from its origins in the engineering, construction, and defense industries, IS/IT has played an increasingly prominent role in shaping the debate about project management. Well-publicized failures (e.g., Standish Group, 1994) have been accompanied by an increasing focus on developing robust software development and systems engineering processes, as well as improving the management of both software development and business change projects. A significant factor in this development has been the family of Capability Maturity Models (CMM) developed under the leadership of Watts Humphrey by the Software Engineering Institute of Carnegie Mellon University (Paulk et al., 1996).

The principle behind the original CMM is simple: If organizations wish to develop predictability and repeatability in their IS/IT production processes, they need to develop a number of capability areas, each of which consists of families of related processes. In turn, each of these processes needs to develop through a series of stages of maturity from informal at the lower end of the scale to highly routinized and with continuous improvement embedded at the higher end. To prevent the model from becoming excessively complex to understand, the capability areas and process maturity measures are combined into a series of five levels of organizational maturity, into one of which any organization can be categorized.

More and more organizations, in more and more countries, are using the software CMM, and procurers of software are increasingly specifying the level of maturity that must be achieved by would-be suppliers. As a consequence, the general level of maturity of software development organizations has shown significant improvements since the early 1990s (Software Engineering Institute, 2003). The model itself, originally for software development, has since spawned a number of other versions covering such fields as systems engineering, human resources, and, most recently, systems engineering, software development, integrated product and process development, and supplier sourcing in a model known as CMM-I, where the "I" stands for integration.

Since software is developed through projects, it is natural that the concept of organizational maturity would migrate from software development processes to project management, and this has been reflected in an interest in applying the concept of "maturity" to software project management (Morris, 2000; Cooke-Davies et al., 2001). Possibly as a result of this, a number of project management maturity models appeared during the mid-1990s that were more heavily influenced by the thinking of project management consultants and practitioners.

Against this background and recognizing the growing interest in the field, the Project Management Institute in May 1998 initiated a program known as the Organizational Project Management Maturity Model (Schlichter, 2001; Friedrich et al., 2003) to develop a standard for organizational project management processes that would complement the ubiquitous PMBOK Guide, its widely applied standard for the management of individual projects. The first draft of this standard was launched in December 2003.

As often happens in the early days of the development of new concepts, however, the field of maturity models is characterized by a tangle of confused concepts and unclear vocabulary. Several factors contribute to this confusion. In particular:

- There is no universal agreement as to the extent of the practices and processes that are necessary for the successful management of projects.
- Practices and processes are interwoven at many levels simultaneously within the field of "project management," and so it is by no means clear how and to what extent the concepts of "process control," "process maturity," and "capability" can be applied to the whole field.
- "Capability" and "maturity" are words that carry a multiplicity of meanings, some of them technically precise and others more broadly based in common usage. There is no general agreement on how such words, and the concepts that they signify, apply to the general field of project management.

But does this outpouring of creative activity add value to the field of project management, or simply confuse it by offering yet more silver bullets that will ultimately turn out to be illusory? That is the question this chapter addresses.

Before hazarding an answer to this question, however, we must shed light on confused concepts and clarify ambiguous vocabulary. But first of all it is appropriate to review briefly the literature on project management maturity (such as it is) and to describe in a little more detail the most recently developed maturity models that seek to offer themselves as serious candidates to become widely accepted standards.

A Brief Survey of the Literature on Project Management Maturity

This section will inevitably live up to its billing as brief, since there has been comparatively little written about such a new topic. The roots of the concept of process maturity seem to lie deep within the Quality movement and can be clearly traced in the copious writings of its gurus such as Walter Shewhart and W. Edwards Deming. The principle is simple: "a stable process . . . is said to be *in statistical control*. . . . A system that is in statistical control has a definable identity and a definable capability" (Deming, 1986, p. 321, italics in original text). Thus, the efforts of quality improvement are first of all to make the process stable and thus bring it under statistical control, and then to work on improving the capability of the process. A process can thus be said to mature as it passes through the stages from unstable, to stable, and then to enjoying improved capability. The effects can be clearly seen in charts such as Figure 13.1.

FIGURE 13.1. HOW THE PROCESS OF HITTING A GOLF BALL IMPROVES WITH LESSONS.

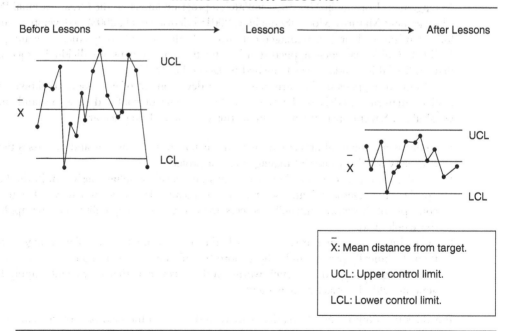

Source: J. Edwards Deming (1982). *Out of the Crisis*. Cambridge: Cambridge University Press, p. 252. Reprinted with the permission of Cambridge University Press.

These principles have been clearly embodied in the Capability Maturity Model for software that was developed by the Software Engineering Institute of Carnegie Mellon University between 1986 and 1993. Integral to the model is the concept that organizations advance through a series of five stages to maturity: initial level, repeatable level, defined level, managed level, and optimizing level. "These five maturity levels define an ordinal scale for measuring the maturity of an organization's software process and for evaluating its software process capability. The levels also help an organization prioritize its improvement efforts." (Paulk et al., 1996, p. 7). The prize for advancing through these stages is an increasing "software process capability," which results in improved software productivity.

As might be expected in a field that is so heavily dominated by practitioners, the literature on project management maturity models is concerned primarily with their practical application rather than with an exploration of the theoretical validity of the concept, or with empirical research to demonstrate their value.

Of the maturity models that have been described in the project management literature, a significant number (e.g., Couture and Russett, 1998; Ibbs and Kwak, 1997; Pennypacker, 2002) show their dependence on marrying two concepts together: "project management" as described in the PMBOK® Guide and "maturity level" as described in CMM (see Figure 13.2). Others, on the other hand, show signs of rethinking the concept of how an organization matures in its ability to manage projects (e.g., Gareis, 2001; Hillson, 2001; Kerzner, 2001). Not surprisingly, each of these develops its own unique description of the path to maturity and its own scope of the practices and processes that are to be assessed.

FIGURE 13.2. FIVE "STAGES" OF MATURITY OF THE SOFTWARE DEVELOPMENT PROCESS.

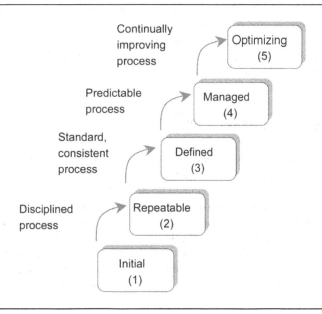

The Project Management Institute's forthcoming OPM3 (Fahrenkrog et al., 2003) has involved at one time or another more than 800 volunteers (Friedrich et al., 2003) drawn from the world's community of project managers, so it is perhaps not surprising that it has begun to influence the development of other custom in-house maturity models, such as that developed for BNY Clearing Services Inc. (Rosenstock et al., 2000).

In addition to descriptions of maturity models, conferences have heard tales of the improvements that can be obtained by individual corporations through their application (e.g., Suares, 1998; Peterson, 2000; Rosenstock et al., 2000). The models have been used in attempts both to assess the state of the art in project management (e.g., Pennypacker and Grant, 2003; Mullaly, 1998) and also (less successfully) to demonstrate the organizational benefits of project management (Ibbs and Kwak, 1997; Ibbs and Reginato, 2002).

Other models, however, are being used within organizations to assess project management maturity as a part of an overall assessment of the quality of business practices (Rosenstock et al., 2000; Cooke-Davies et al., 2001), using models such as the Baldridge National Quality Award (www.quality.nist.gov) or the European Forum for Quality Management's "Business Excellence" model (www.wfqm.org/imodel/model1.htm).

The award-winning article by Jugdev and Thomas (2002) is a refreshing exception to the complaint that little attention has been paid to questioning the fundamental relevance of maturity models to the total scope of managing projects in organizations. The paper examines maturity models (MMs, in the language of the article) from the viewpoint of four different resource-based models in order to assess whether or not the possession of a higher maturity level in project management confers competitive advantage on an organization. The article concludes that MMs possess some but not all of the characteristics of a strategic asset and thus cannot in and of themselves confer competitive advantage. It also asserts that although "MMs are a component of project management[; they are] not a holistic representation of the discipline" (p. 11).

This assertion implies an answer to this chapter's own question about the value of maturity models, about which more will be said later.

Before that, however, it is appropriate to examine two of the most recent additions to the field, each of which could conceivably become, for different reasons, a broadly used model and a widely accepted standard.

OGC's PMMM and PMI's OPM3

Each of these two models has some attributes in common with other models mentioned earlier in this chapter, but it is not the purpose of this chapter to conduct a detailed comparison between individual maturity models. The two models have been selected for special mention for two reasons: First, they are entering the field after sufficient time has elapsed to allow experiences with other models to have been taken into account, and second, each of them is backed by an organization that has demonstrated its ability to establish widely adopted standards—the Office of Government Commerce in the United Kingdom, which has produced both PRINCE2 and *Managing Successful Programmes*, and the Project Management Institute in the United States, publisher of the PMBOK® Guide.

At this writing, neither has yet been formally issued, so the possibility exists of significant change, but the general lines of development of each of them is sufficiently far advanced that their potential contribution to the field can be assessed. The information on PMMM has been based on Version 5.0 of the model, which is available on the OGC's Web site. The information on OPM3 has been taken from papers presented at the PMI Global Assembly 2003—Europe (Fahrenkrog et al., 2003; Friedrich et al., 2003).

PMMM

In the introduction to the Project Management Maturity Model (PMMM), the OGC observes that it has been developed because SEI's experience in the arena of software development between 1986 and 1991 indicated that maturity questionnaires provide a simple tool for identifying areas where an organizations' processes may need improvement. The model is descriptive, with the express intention of providing organizations with guidance to support their process improvement initiatives, and the document describing the model is at pains to point out that the model itself is not to be confused with any questionnaire that may be used to establish an organization's current maturity level.

Each stage is characterized by a discrete set of processes that are definitive of the stage of maturity (see Figure 13.3).

The description of each process includes the process goals and functional achievement, the approach laid down, the deployment that is to be expected, the method of review that

FIGURE 13.3. PROCESSES INCLUDED IN EACH STAGE OF MATURITY—PMMM.

Level 1: Initial Process	Level 2: Repeatable Process	Level 3: Defined Process	Level 4: Managed Process	Level 5: Optimized Process
1.1 Project Definition	2.1 Project Establishment 2.2 Requirements Management 2.3 Risk Management 2.4 Project Planning 2.5 Project Monitoring and Control 2.6 Management of Suppliers and External Parties 2.7 Project Quality Control 2.8 Configuration Definition and Control	3.1 Organizational Focus 3.2 Project Management Success 3.3 Project Training 3.4 Integrated Management 3.5 Lifecycle control 3.6 Interteam Coordination 3.7 Quality Assurance	4.1 Project Metrics 4.2 Organizational Quality Management	5.1 Proactive Problem Management 5.2 Technology Management 5.3 Continuous Process Improvement

is recommended, the way the organization should perceive the process, and the performance measures that should be used. Inherent in the idea of a mature organization is the existence of an organization-wide capability to manage projects based on a set of clearly defined common processes that can be tailored to meet the needs of individual projects. The introduction to the model includes a description of the two extreme states of maturity: immature and mature.

The model can be used for either or both of two purposes:

- To understand the key practices that are part of an effective organizational process to manage projects
- To understand the key practices that need to be embedded within the organization to achieve the next level of maturity

It could be used by any organization wanting to improve its capability to manage projects effectively, by governance bodies and consultancies for the purpose of developing maturity questionnaires, or by accredited service providers in assisting teams to perform project management process assessments or capability evaluations.

OPM3

The Project Management Institute has announced its firm intention to launch OPM3 as a draft standard before the end of 2003. It differs from many of the other models mentioned in this chapter in that it introduces a structure that owes little to the structure of CMM, but rather one that relates explicitly to the PMBOK® Guide (although it is dramatically different in scope); that covers the three domains of portfolio management, program management, and project management; and that explicitly relates the management of projects to organizational strategy.

The basic "building blocks" at the heart of OPM3 are five different kinds of entity:

1. "Best practices" that are associated with organizational project management
2. "Capabilities" that are prerequisite or that aggregate to each "best practice"
3. The observable "outcomes" that attest to the existence of a given "capability" in the organization
4. Key performance indicators (KPIs) and metrics that provide the means of measuring the "outcome"
5. Pathways that identify the capabilities aggregating to the "best practices" being reviewed

The relationships among these are shown in a simplified manner in Figure 13.4.

An example that has been given of a best practice is "Use Teamwork—Cross-functional teams carry out the organization's activities." The four associated capabilities leading up to this are "develop cross-functional training opportunities," "organize project work by functional area," "develop cross-functional teams," and "develop integrated program and project teams." Additional dependencies have been identified between specific "best practices" when

FIGURE 13.4. THE RELATIONSHIP BETWEEN FUNDAMENTAL ELEMENTS IN OPM3.

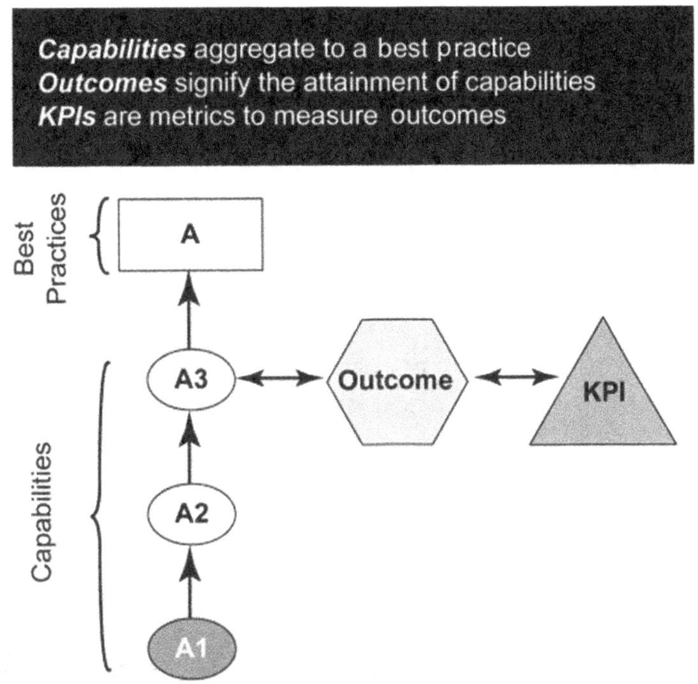

Capabilities aggregate to a best practice
Outcomes signify the attainment of capabilities
KPIs are metrics to measure outcomes

Best Practices { A

Capabilities { A3 ← → Outcome ← → KPI

A2

A1

Source: © Project Management Institute: reproduced by permission.

dependencies exist between one or more capabilities that aggregate to different "best practices."

These basic building blocks are used within a framework of organizational project management processes that identifies five "process groups" (initiating, planning, executing, controlling, and closing processes) in each of three "domains" (portfolio management, program management, and project management).

Each process group in each domain is seen to progress through four stages of "process improvement" (standardized, measured, controlled, and continuously improved) to give an overall framework for the model. (See Figure 13.5.) Each of the "best practices" is mapped onto at least one location within this three-dimensional model, so that OPM3 will tell the user where a "best practice" falls within a "process group," "domain," and stage of "process improvement."

The scope of the model is vast: There are more than 600 "best practices," more than 3,000 "capabilities," and more than 4,000 relationships between capabilities. The finished

FIGURE 13.5. THE PROCESS CONSTRUCT OF OPM3.

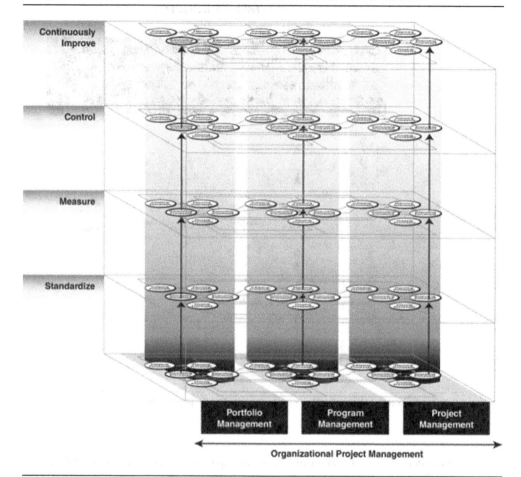

product is likely to take the form of both a book and a CD-ROM, and to contain both means of self-assessment and assessment by external consultants.

The model is designed to be used by an organization for any or all of four purposes:

• To understand what practices and processes have been consistently found to be useful by organizations seeking to undertake "organizational project management," which is defined as "the application of knowledge, skills, tools, and techniques to organisational and project activities to achieve the aims of an organisation through projects" (Fahrenkrog et al., 2003, p. 2).

- To assess its ability to implement its high-level strategic planning at the tactical level of managing individual projects and groups of projects.
- To drive business improvement.
- To integrate organizational practices and processes in the domains of portfolio management, program management, and project management.

Untangling the Vocabulary and Distinguishing Relevant Concepts

Earlier in this chapter, three areas of confusion and ambiguity were identified, each of which adds somewhat to the entangled nature of conversations about project management maturity models. This section attempts to reach toward answers to three fundamental questions: "What is the extent of practices and processes that are necessary to the effective and efficient management of projects?" "What is meant by the words 'practice' and 'process' as they apply to the management of projects?" and "What is meant by the words 'maturity' and 'capability' as they apply to the management of projects?"

What Topics Are Covered by the "Management of Projects"? Unless the scope of the topic can be agreed, it is unlikely that it will be possible to agree on what the "management of projects" might look like in its "perfected end-state," so this first conversation is fundamental to the topic of organizational project management maturity. (Similar discussions occur and points are made in the chapters by Crawford on project management standards and by Morris on the validity of project management knowledge.)

Perhaps the place to start in considering this question is with a review of the "bodies of knowledge" that are produced by several of the world's project management professional associations. Both the longest-established and the most widely distributed is undoubtedly the PMBOK® Guide produced by the Project Management Institute. First produced in 1976, and most recently updated in 2000, this document, which had over 270,000 copies in circulation in September 2001 (Crawford, 2002), seeks "to identify and describe the knowledge and practices that are applicable to most projects most of the time" (Duncan, 1996, p. 3). It recognizes that the management of projects also requires general management and specific application area (i.e., industry, market, or technology) knowledge and practice, but restricts itself to the knowledge and practices that are generally applicable to the management of individual projects.

The Association for Project Management in 1986 developed the framework for what was to become the *APM Body of Knowledge*, which is now in its fourth edition, having been updated in 2000 (Dixon, 2000) on the basis of research carried out by the Centre for Research in the Management of Projects (Morris, 2001). A much broader range of topics is covered by the *APM Body of Knowledge*, in line with the findings of research into project success, which suggests that a much broader range of factors is critical to project success than the knowledge and practices contained within PMBOK® Guide. (Baker, Murphy, and Fisher, 1974; Morris and Hough, 1987; Lechler, 1998; Pinto and Slevin, 1998; Crawford, 2000; Crawford, 2001; Cooke-Davies, 2001; Cooke-Davies, 2002).

Following the development of "bodies of knowledge" by various European professional associations, the International Project Management Association in 1998 published in French, German, and English the "International Competency Baseline" (Caupin et al., 1999), offering a coordinated set of definitions to the terms used in the Swiss, German, French, and UK documents. Other professional associations (e.g., AIPM, PMISA) have their own "bodies of knowledge" and/or competency standards, usually resembling some combination of those that have already been discussed.

Most recently, a three-year joint academic/government/industry study in Japan has resulted in the production of an innovative standard for project management known as P2M (Project Management Professionals Certification Center, or PMCC, 2002). This is remarkable for the thoroughness with which it reexamines and redefines the practices, processes, and competencies that are necessary to deliver innovation through strategies, programs, and projects.

It has been argued forcefully and cogently (Morris, 2001; Morris, 2003; Crawford, 1998; Crawford, 2001; Crawford, 2002) that the absence of global standards works to the detriment of the practice of managing projects in multinational or global organizations. Precisely the same argument can be used with regard to maturity models—to the extent that enterprises are seeking to assess their organizational capability for managing projects, the absence of a generally accepted definition of what is involved inevitably inhibits the value of any maturity model to the whole of an organization.

What Do "Practices" and "Processes" Mean in Connection with Project Management?

The second fundamental area that causes some confusion is precisely what is meant by the terms "process" and "practice." The 1980s and 1990s saw an emerging fashionable focus on adopting a "process" view of organizations—defining a process as "a specific ordering of work activities across time and place, with a beginning, an end, and clearly identified inputs and outputs: a structure for action" (Davenport, 1993, p. 6). The term is used in much the same way in the PMBOK® Guide, where the process groups and the "knowledge areas" are defined in terms of individual processes, described in terms of their inputs, outputs, and mechanisms.

The practice of identifying, describing, and then improving business processes lies at the heart of the quality movement, as has been seen earlier in this chapter. It accounts for the emphasis given to processes in models such as the EFQM "Business Excellence" model and the Baldridge award, to which reference has already been made. In parallel with the quality movement, the fashion for viewing organizations as collections of interlocking processes gave rise to disciplines such as business process reengineering (e.g., Hammer and Champy, 1993) and benchmarking (e.g., Camp, 1989). These movements didn't simply focus on making processes repeatable and predictable, but rather they sought to identify the specific work flows and working practices that lead to improved process performance. Camp's working definition of benchmarking as "the search for industry best practices that lead to superior performance" (1989, p. 12) makes this clear. And process performance is measured by physical characteristics such as throughput time, efficiency measured by unit of output produced for unit of input, or cost per unit of throughput.

Thus, in terms of a "process view" of work, a *process* describes how the organization's inputs are converted into outputs, and *practices* describe how the processes are carried out.

However, the process view is not the only way of describing work. It could be argued that projects, for example, are a specialized subset of processes—those that are carried out only once so as to produce a unique product, service, or beneficial change. Most of the definitions of a project make reference to the unique nature of each of them, whereas the essence of the process view is the repetitive nature of the work carried out.

Clearly the management of projects involves a large number of processes, as described in each of the bodies of knowledge, but equally clearly, different organizations employ their own practices to undertake these processes. And as research shows, different organizations in different industries recognize very different areas of practice as being appropriate to managing projects (Toney and Powers, 1997; Turner and Keegan, 1999; Turner and Keegan, 2000; Cooke-Davies and Arzymanow, 2002; Morris, 2003).

Outside of the process view of an organization, considerable fluidity characterizes the way literature on the management of projects uses the words "process," "practice," and even "discipline." For example, in a discussion of the "topic" of project management, the *APM Body of Knowledge* states that "Project Management is *the discipline* of managing projects successfully," and then on the following page includes a diagram with the caption "this diagram illustrates *the project management process*" (2000, pp. 14, 15; italics mine).

In contrast to the use of the term "practice" by proponents of benchmarking and other approaches to process management to denote how "processes" are carried out, PMBOK® Guide states that "Part II, the Project Management Knowledge Areas, describes project management *and practice* in terms of its component processes" (2000b, p. 7 "italics mine"). Thus, in some usage by project management practitioners, practice is a characteristic of processes, and elsewhere processes are components of practice.

Once more, as in the prior discussion about what is involved in the management of projects, the absence of generally accepted definitions for these two key terms creates confusion for the application of maturity models.

What Do "Maturity" and "Capability" Mean in Connection with Maturity Models? In Collins Dictionary, the adjective "mature," from which the noun "maturity" is derived, has a number of different meanings in common usage. It can, for example, mean "(1) fully-developed or grown up; (2) of plans or theories it can mean that they are fully considered, perfected; (3) of insurance policies or bills it can mean due or payable; and (4) of fruit, wine or cheese it can mean ripe or fully aged." The last two of these do not offer any obvious link to the world of project management (unless through projects that are described as "pear-shaped" when they are in serious trouble), so it is in either meaning (1) or (2) that the word is used in the term "maturity models."

According to SEI's CMM "a software process can be defined as a set of activities, methods, practices, and transformations that people use to develop and maintain software and the associated products (e.g., project plans, design documents, code, test cases, and user manuals). As an organization matures, the software process becomes better defined and more consistently implemented throughout the organization." (Paulk et al., 1996, p. 3) In other words, maturity is used in meaning (2) as a technical description of the state of definition and consistency of implementation of an end-to-end process.

There is a clear implication that this can be accomplished only through the willful application of process improvement effort over time, and this would seem to link the use of the term to its more general meaning: (1). However, a working definition of a mature organization or a mature process would seem to be one that has reached what is, to all intent and purposes, a perfected state—one that is capable of delivering the requisite outcomes consistently, efficiently, and effectively.

This working definition also builds a link to the second word being considered here: "capability." Once more, there are two rather different meanings in the dictionary: "(1) the quality of being capable, ability; and (2) potential aptitude".

Once more CMM states that "software process capability describes the range of expected results that can be achieved by following a software process. The software process capability of an organization provides one means of predicting the most likely outcomes to be expected from the next software project the organization undertakes" (Paulk et al., 1996, p. 3). This sounds more like meaning (2)—dealing with a potential rather than an actual quality possessed. The correlation is made more explicit when the explanation of important concepts goes on to define "software process performance" as being the actual results accomplished by following a software process.

All this looks logical and clear, but all is not as it seems. Having spoken about what happens to "the" software process as an organization matures, the CMM introduces as a fundamental concept "software process maturity"—the development through five stages from definition, through management, measurement, and control to effectiveness. The term "maturity" and the five stages of development are thus applied *both* to the software process and to the organization that is undertaking it.

Within CMM, where each term is tightly defined, this does not appear to cause difficulties. Indeed, the descriptions in the model and the assessment methods are sufficiently tight, and little room is left for ambiguity.

As the terms and concepts are translated into the world of project management, with all of its inherent uncertainties, ambiguities, and disagreements, the distinction between the maturity of a process and the "maturing" of an organization becomes more problematical. It seems reasonably intuitive and logical to describe the "maturity," the "capability," and the "performance" of a single process within the field of project management, such as "activity duration estimating," using precisely the same definitions as are used in CMM. It becomes less so when considering the "maturity," the "capability," or the "performance" of an organization that undertakes many projects and programs of different categories (such as engineering, marketing, or business development) in many different business units, for many different purposes, using many different criteria for success. And it is this consideration that gives rise to the fundamental tacit assumption behind project management maturity models: that there is an underlying development path or trajectory that must be followed by organizations as they seek to improve their ability to manage projects successfully.

But is there?

This question will be examined in more detail a little later in the chapter, when the potential contribution of maturity models is considered in relation to the field of project management. Before that, though, it is appropriate to review the "state of play" on the field of maturity models in the light of these three fundamental questions.

Maturity Models: The "State of Play" Reviewed

Starting with the most recent two models first, it would appear that both PMM and OPM3 avoid some of the inherent difficulties that have been sketched previously, but not all of them. Of course, by the time they are launched as potential standards, they will certainly undergo changes, which may alter the conclusions about them.

OPM3 is by far the largest and most complex of the project management maturity models, and it might well turn out to be the most comprehensive. It recognizes the heritage of maturity models in the quality movement and acknowledges that "practices" are components of processes or process groups. It is also clear in its recognition that a process can be described as "mature" when it has achieved its "perfected end-state," has a measured and defined "capability," and is subject to process improvement initiatives. The identification of both outcomes and KPIs for each "capability" represents a substantial achievement.

In itself, however, OPM3 is far from "mature" in its own terms. The choice of the terms "best practice" and "capability" for two of the five elemental components of the model is not the happiest, in view of the semantic confusion that already exists. In spite of extensive market research (mainly conducted within the community of practitioners that are familiar with PMBOK® Guide), neither the 600 or so "best practices" nor the 4,000 or more dependency paths that the model contains can be demonstrated empirically to describe the essential trajectory (or trajectories—the model incorporates great flexibility) to organizational maturity, as measured by the successful implementation of strategy through projects, which is the stated goal of the model.

PMMM is very different in terms of its detail (67 pages compared with what might turn out to be more than 800 for OPM3), its focus (on government departments undertaking projects with a high IT content, or other organizations embracing PRINCE2, rather than any organization undertaking projects), and its derivation directly from CMM. The 21 processes described in the five stages incorporate many of the broader areas of the field of project management that are included in the *APM Body of Knowledge*, and also the principles of quality management. The terminology of "maturity," "capability," and "process" is very close to that contained in CMM.

On the other hand, there are at present some surprising omissions. For example, neither program management nor portfolio management is included in the model. There is also, more than in OPM3, an implicit single "development path" or "trajectory" toward the state of "maturity" to which, by implication, most organizations should aspire.

To turn back to models that were considered in the literature review, it is clear that different categories of model have different strengths and weaknesses that can be reviewed briefly. The earliest maturity models that combine the concept of CMM's five stages of maturity with the PMBOK® Guide's project management processes (e.g., Couture and Russett, 1998; Ibbs and Kwak, 1997; Pennypacker, 2002) fail to distinguish between organizational maturity and process maturity (the organization is mature when every process is mature) and also omit from their consideration the extensive areas of practice that contribute to the successful management of projects that are not covered by PMBOK® Guide.

The CMM family of models itself and its derivatives, of course, are useful in terms of organizations to whom the software process is an important component of what APM's Body of Knowledge refers to as "technology management". Those organizations that are seeking to improve their overall excellence, using the Baldridge or EFQM models, certainly cover the whole field of practices necessary for the management of not only projects, but of everything else as well. They contain neither implicit nor explicit process or capability elements to assist organizations that are seeking explicitly to improve the maturity of their project management.

Each of the remaining models that has been considered contains its own assumptions about the processes that need to be added at each stage of maturity, and thus implies its own hypothesis about the appropriate "development path" that leads to maturity. Until empirical project management research is in a position to demonstrate the validity of one or more of these development paths to project management maturity, or the correlation of project management maturity to consistent project success, then the adoption of a particular model remains largely an act of faith. But is such an act of faith reasonable? Do maturity models, in their present form, add value to the field of project management, or do they simply add to the confusion? That is the question to which we can now hazard an answer.

Maturity Models: Silver Bullets or Unhelpful Distractions?

The purposes for which an organization might seek to use a maturity model have been variously described as the following:

1. To understand what practices and processes have been consistently found to be useful by organizations seeking to undertake organizational project management
2. To drive business improvement, for example, by understanding the key practices that need to be embedded within the organization to achieve the next level of maturity
3. To assess its ability to implement its high-level strategic planning at the tactical level of managing individual projects and groups of projects
4. To integrate organizational practices and processes in the domains of portfolio management, program management, and project management

Several limitations to the various different types of maturity model have already been described, but these do not necessarily prevent the models from providing value, or from helping organizations to accomplish any or all of the objectives in the list.

As demand has grown for the effective management of an increasing number projects, it has been helpful to the advancement of project management to identify those processes that are "applicable to most projects, most of the time" and to help a growing number of practitioners learn what those are. Maturity models, in a sense, seek to do for organizations seeking to implement strategy through projects what "bodies of knowledge" have done for individual practitioners seeking to improve their ability to manage projects.

But three factors make the practice of "organizational project management" considerably more problematical than the management of individual projects. First, it has long been recognized that the related product-oriented processes "are typically defined by the project life cycle, and vary by application area" (Project Management Institute, 2000a, p. 30). For example, the product-oriented processes involved in the development of new pharmaceutical products, with their inherent technical risk and consequent uncertainties of scope, will inevitably differ in many important respects from those involved in the construction of a new building or the development of software. Thus, different organizations will have a bias toward projects with a certain set of characteristics, and the strategies that lead to business success in different "application areas," industry, or markets may differ radically from each other. Morris' review (2003) of the differences between project management as practiced in four different industries (construction, information technology, defense/aerospace, and pharmaceuticals) describes these differences very clearly. These differences may be so great as to raise doubts about the concept of a single "development path" toward a "perfected end-state."

Second, regardless of the type of "application area," organizations themselves operate in many different industry and market environments. Banks, for example, operating in the financial services market, undertake construction projects, information technology projects, business development projects, process reengineering projects, and so on. The environmental differences can be characterized on a grid as shown in Figure 13.6. Project management

FIGURE 13.6. DIFFERENT "ENVIRONMENTS" FOR PROJECTS.

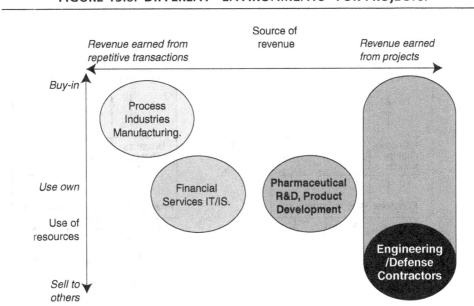

clearly has a different "voice" in the traditional project-based engineering environment, displayed on the right-hand side of the diagram, than it does in those nearer to the left. The vertical axis also plays a part—the more the revenue of an organization depends on the efficient and effective use of its own resources on projects (i.e., the nearer to the bottom of the diagram), the greater the commercial pressures on that organization to develop a world-class corporate capability in project management. This perhaps accounts for the evidence that engineering suppliers to the process industries have the highest project management "maturity" of any industry (Cooke-Davies and Arzymanow, 2002). It also suggests that there may be no common path to "project management maturity" in different industries and for different types of projects.

The third factor that complicates the question of "organizational project management maturity" is that in order to compete strategically in their markets, organizations have to pay attention to two very different aspects of business: "business as usual," in terms of extracting value from their current products and markets, and "business change," in terms of preparing the organization to compete in the future in new markets or with new products. Projects can play their part in either of these aspects, as illustrated in Figure 13.7. For example, one of the gurus of the quality movement, J. M. Juran, has maintained that quality of products or services can be improved only through projects (Anbari, 2003), and it is

FIGURE 13.7. THE RELATIONSHIP BETWEEN "BUSINESS AS USUAL" AND "BUSINESS CHANGE."

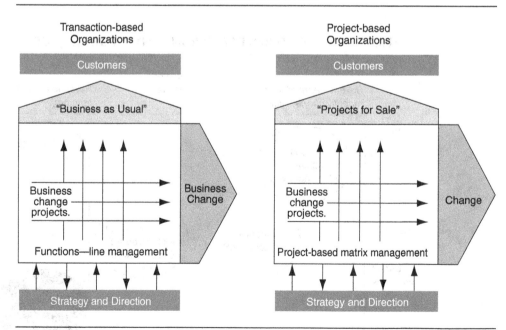

accepted wisdom among project management practitioners that project management is essentially the management of business change.

The reason that this complicates the question of project management maturity is that, as Mintzberg (1989) has argued, business strategy is "crafted" by managers in a way that is analogous to how a potter crafts a pot from the clay with which he or she is working. The strategy is a function of the present situation, with its inherent possibilities and the skills of the organization. But the present situation that organizations find themselves in varies widely depending on the nature of the organization's structure and culture, and the relative pressure felt to emanate from each of the two strategic aspects. It is these three factors that establish the unique environment of each organization as it seeks to accomplish any of the four purposes of a maturity model: to know what "good practice" looks like, to identify sensible improvement initiatives, to implement strategy more effectively, or to integrate the management of all the projects in the organization.

By definition, any model is a simplification of reality. If that were not the case, the model would serve no purpose, since it would be as complex and as opaque as reality itself. Where project management maturity models are concerned, however, there remains a question mark as to whether it is possible to strike a balance between a model that is an essentially accurate representation of the complex reality that covers the management of all projects by all organizations, and one that is simple enough to offer practical help to most organizations, most of the time.

Until that question is answered, each organization that is considering adopting a maturity model should ask itself a number of questions:

- Does the scope of the model cover all those areas of the management of projects that we believe to be strategically important in our current competitive environment?
- Does the definition of maturity—the perfected end-state described by the model—look like a state to which it is strategically important that our organization should aspire?
- Does the cost of assessing our current state of maturity and identifying our desirable path of development toward maturity look as if it is justified in the light of the benefits we could expect?
- Is the use of a maturity model the most cost-effective way that we can assess our current status and identify our future development needs?

The answers are likely to vary considerably from model to model, and from organization to organization.

Summary

There are three general conclusions that can be drawn from this discussion. First, Thomas and Jugdev's assertion (2002) that maturity models are not a holistic view of the field of project management appears to be justified. The field is broader than can be embraced simply within the process view that the models embody, and projects possess distinctive features that distinguish them from processes.

Second, maturity models make a positive contribution to the field because they

- at best broaden the discussion from a narrow definition of project management toward the the management of projects;
- recognise that developing a mature capability to manage projects requires an organizational commitment to the development of incremental capabilities along some kind of a development path, as well as the continual improvement of associated processes;
- bring into the field of managing projects core values that have proven themselves to be valuable in the field of quality management.

Third, maturity models are unlikely to be the silver bullet that some hope for because they

- lack an important precondition in terms of a well-researched and theoretically grounded understanding of just what is involved in the management of projects;
- are built, in many cases, on an unproven assumption that there is an ideal development path toward maturity that most organizations must follow most of the time, regardless of application area, project, and market environment or competitive strategy;
- must steer a perilous path between the "Scylla" of over-simplification and the "Charybdis" of excessive complexity.

Regardless of these conclusions, project management maturity models are a visible feature of the current landscape of project management. Their use as a means of comparing capability across organizations and industries seems likely to ensure their longevity.

The real question that project management practitioners, consultants, and academics should be asking is this: "Will they simply remain an interesting phenomenon of limited relevance and application, or will they provide the means of transforming the success rate of projects for which organizations are searching?" Any answer to this question remains largely a matter of conjecture at this time. In the long run, it will depend on whether or not the considerable investment of effort involved in assessing, adopting, adapting, and implementing a project management maturity model in an organization translates into sufficient value from improved results to justify its continued place in the armory of project management practices. The increase in the adoption of the software CMM suggests that IS/IT organizations are finding the investment worthwhile, but it does not follow that this will also hold true for the much more complex and diffuse world of organizational project management. Watch this space!

References

Anbari, F. T. 2003. Strategic implementation of six sigma and project management. *PMI Global Congress 2003—Europe*. Newtown Square, PA: Project Management Institute.

Baker, B. N., D. C. Murphy, and D. Fisher. 1974. *Determinants of project success*, NGR 22-03-028. National Aeronautics and Space Administration.

Camp, R. C. 1989 *Benchmarking*. Milwaukee: ASQC Quality Press.

Caupin, G., H. Knoepfel, P. W. G. Morris, E. Motzel, and O. Pannenbäcker. O. 1999. *ICB. IPMA Competence Baseline*. Monmouth IPMA.

Cooke-Davies, T. J. 2001. *Towards improved project management practice: Uncovering the evidence for effective practices through empirical research*. www.dissertation.com.

———. 2002. The "real" success factors on projects. *International Journal of Project Management* 20(3): 185–90.

Cooke-Davies, T. J., and A. Arzymanow. 2002. The maturity of project management in different industries. *Proceedings of IRNOP V. Fifth International Conference of the International Network of Organizing by Projects*. Rotterdam: Erasmus University.

Cooke-Davies, T. J., F. J. Schlichter, and C. Bredillet. 2001. Beyond the PMBOK Guide. *Proceedings of the 32nd Annual project Management Institute 2001 Seminars and Symposium*. Newtown Square, PA: Project Management Institute.

Couture, D., and R. Russett. 1998. Assessing project management maturity in a supplier environment. *Proceedings of the 29th Annual project Management Institute 1998 Seminars and Symposium*. Newtown Square, PA: Project Management Institute.

Crawford, L. 1998. Standards for a global profession—project management. *Proceedings of the 29th Annual project Management Institute 1998 Seminars and Symposium*. Sylva, NC: Project Management Institute.

———. 2000. Profiling the Competent Project Manager. *Proceedings of PMI Research Conference*. Newtown Square, PA: Project Management Institute.

———. 2001. Towards global project management standards. *Proceedings of the International Project Management Congress 2001, Project Management Development in the Asia-Pacific Region in the New Century*. Tokyo, November 18–21. Tokyo: ENAA and JPMF.

———. 2002. Developing project management competence for global enterprise. *Proceedings of ProMAC 2002 Conference*. July. Singapore: Nanyung Technical University.

Davenport, T. H. 1993. *Process innovation: Reengineering work through information technology*. Boston: Harvard Business School Press.

Deming, W. E. 1986. *Out of the crisis*. Cambridge, UK: Cambridge University Press.

Dixon, M. 2000. *Project Management Body of Knowledge*. 4th ed. High Wycombe, UK: Association for Project Management.

Duncan, W. R. 1996. *A guide to the Project Management Body of Knowledge*. Newtown Square, PA: Project Management Institute.

Fahrenkrog, S., C. M. Baca, L. M. Kruszewski, and P. R. Wesman. 2003. Project Management Institute's Organizational Project Management Maturity Model (OPM3). *PMI Global Congress 2003—Europe* Newtown Square, PA: Project Management Institute.

Friedrich, R., F. J. Schlichter, and W. Haeck. 2003. The history of OPM3. *PMI Global Congress 2003—Europe*. Newtown Square, PA: Project Management Institute.

Gareis, R. 2001. Assessment of competences of project-oriented companies: application of a process-based maturity model. *Proceedings of the 32nd Annual project Management Institute 2001 Seminars and Symposium*. Newtown Square, PA: Project Management Institute.

Hammer, M., and J. Champy, J. 1993. *Reengineering the corporation: A manifesto for business revolution*. London: Nicholas Brealey.

Hillson, D. 2001. Benchmarking organizational project management capability. *Proceedings of the 32nd Annual project Management Institute 2001 Seminars and Symposium*. Newtown Square, PA: Project Management Institute.

Ibbs, W. C., and Y. H. Kwak. 1997. *The benefits of project management. Financial and organizational rewards to corporations.* Newtown Square, PA: PMI Educational Foundation.

Ibbs, W. C., and J. Reginato. 2002. Can good project management actually cost less? *Proceedings of the 33rd Annual project Management Institute 2002 Seminars and Symposium.* Newtown Square, PA: Project Management Institute.

Kerzner, H. 2001. *Strategic planning for project management using a project management maturity model.* New York: Wiley.

Kwak, Y. H., and W. C. Ibbs. 2000. Calculating project management's return on investment. *Project Management Journal* 31(2):38–47.

Lechler, T. 1998. When it Comes to project management, it's the people that matter: An empirical analysis of project management in Germany. *IRNOP III. The Nature and Role of Projects in the Next 20 Years: Research Issues and Problems.* Calgary: University of Calgary.

Mintzberg, H. 1989. *Mintzberg on management: Inside our strange world of organisations.* New York: Free Press.

Morris, P. W. G. 2000. Researching the unanswered questions of project management. *Proceedings of PMI Research Conference 2000.* Newtown Square, PA: Project Management Institute.

———. 2001. Updating the project management bodies of knowledge. *Project Management Journal* 32(3): 21–30.

———. 2003. The irrelevance of project management as a professional discipline. *17th World Congress on Project Management.* Moscow: IPMA.

Morris, P. W. G., and G. H. Hough. 1987. *The anatomy of major projects: A study of the reality of project management.* London: Wiley.

Mullaly, M. 1998. 1997 Canadian project management baseline survey. *Proceedings of the 29th Annual project Management Institute 1998 Seminars and Symposium.* Newtown Square, PA: Project Management Institute.

Paulk, M. C., W. Curtis, M. Chrissis, and C. V. Weber. 1996. *Capability Maturity Model for Software, Version 1.1.* Technical Report: CMU/SEI-93-TR-024: ESC-TR-93-177: February 1993. Pittsburgh: Software Engineering Institute, Carnegie Mellon University.

Pennypacker, J. S. 2002. Benchmarking project management maturity: Moving to higher levels of performance. *Proceedings of the 33rd Annual Project Management Institute 2002 Seminars and Symposium.* Newtown Square, PA: Project Management Institute.

Pennypacker, J. S., and K. P. Grant. 2003. Project management maturity: An industry benchmark. *Project Management Journal* 34(1):4–11.

Peterson, A. S. 2000. The impact of PM maturity on integrated PM processes. *Proceedings of the 31st Annual project Management Institute 2000 Seminars and Symposium.* Newtown Square, PA: Project Management Institute.

Project Management Institute. 2000. *A guide to the Project Management Body of Knowledge.* 2000 ed. Newtown Square, PA: Project Management Institute.

Project Management Professionals Certification Center (PMCC). 2002. *P2M: A guidebook of project and program management for enterprise innovation. Summary translation.* Revision 1 August 2002 ed. Tokyo: PMCC.

Rosenstock, C., R. S. Johnston, and L. M. Anderson. 2000. Maturity model implementation and use: A case study. *Proceedings of the 31st Annual project Management Institute 2000 Seminars and Symposium.* Newtown Square, PA: Project Management Institute.

Schlichter, F. J. 2001. PMI's organizational project management maturity model: Emerging standards. *Proceedings of the 32nd Annual project Management Institute 2001 Seminars and Symposium.* Newtown Square, PA: Project management Institute.

Software Engineering Institute. 2003. Process Maturity Profile. www.sei.cmu.edu/sema/presentations.html (accessed May 14, 2003).

Standish Group. 1994. Chaos. www.standishgroup.com/sample_research/chaos_1994_1.php (accessed May 16, 2003).

Suares, I. 1998. A real world look at achieving project management maturity. *Proceedings of the 29th Annual project Management Institute 1998 Seminars and Symposium.* Newtown Square, PA: Project Management Institute.

Thomas, J., and K. Jugdev. 2002. Project management maturity models: The silver bullets of competitive advantage? *Project Management Journal* 33(4):4–14.

Toney, F., and R. Powers. 1997. *Best practices of project management groups in large functional organizations.* Newtown Square, PA: Project Management Institute.

Turner, J. R., and A. Keegan. 1999. The versatile project-based organization: Governance and operational control. *European Management Journal* 17(3):296–309.

———. 2000. Processes for operational control in the project-based organization. *Proceedings of PMI Research Conference.* Newtown Square, PA: Project Management Institute.

CHAPTER FOURTEEN

PROFESSIONAL ASSOCIATIONS AND GLOBAL INITIATIVES

Lynn Crawford

Communities of practice (Wenger, 1998) are formed when people doing similar things realize they have shared interests. They recognize that there are opportunities to improve both their practices and their performance by sharing knowledge and experience. The project management professional associations as we know them today began in this way: as informal gatherings and forums for networking and exchanging ideas and information.

INTERNET, now known as the International Project Management Association, IPMA, was initiated in 1965 (IPMA 2003; Stretton, 1994) as a forum for European network planning practitioners to exchange knowledge and experience. The Project Management Institute originated in North America in 1969, as "an opportunity for professionals to meet and exchange ideas, problems and concerns with regard to project management, regardless of the particular area of society in which managers function" (Cook, 1981). The UK national project management association began in 1972 as the Association of Project Managers and was subsequently renamed the Association for Project Management. The Australian Institute of Project Management was initially formed as the Project Managers Forum in 1976. The early focus of these project management professional associations on exchange of knowledge and experience between practitioners clearly illustrates their origins as communities of practice.

Recognition of shared interests results in fairly informal gatherings, often referred to as a forum for meeting, networking, and exchange of ideas. At some point members of this community of practice express a need or desire to define their areas of common interest or practice. They begin to think of themselves as a community, and then sometimes, as in the case of project management, as a profession, and to attempt to define and delineate that profession in order to make it visible and acceptable to those outside the community. This is the point when the community begins to put in place the building blocks of a profession, which include the following (Dean, 1997):

- A store or body of knowledge that is "more than ordinarily complex"
- A theoretical understanding of the basis of the area of practice
- An ability to apply theoretical and complex knowledge to the practice in solving human and social problems
- A desire to add to and improve the stock or body of knowledge (research)
- A formal process for handing on to others the stock or body of knowledge and associated practices (education and training)
- Established criteria for admission, legitimate practice, and proper conduct (standards, certification and codes of ethics/practice)
- An altruistic spirit.

From this list it is easy to identify a number of the issues that have preoccupied the various national project management professional associations as they have emerged from the more informal forums of practitioners with shared interests. Primary preoccupations have been as follows:

- Definition of a distinct body of knowledge
- Development of standards
- Development of certification programs

In focusing on these areas, project management professional associations have tended to develop proprietorial or vested interests in the products they have produced. This has resulted in a proliferation of competing project management standards and certification programs, largely local in their origin, if not in their application. By the second half of the 1990s, members of the project management community began to realize that the nature of their community was changing and that a more unified approach would be needed to promote project management as a practice and profession and to meet the needs of increasing numbers of corporate adopters of project management with increasingly global scope to their operations.

From Local to Global

Modern project management may be considered to have had its genesis in the international arena when, in the 1950s (Stretton, 1994; Morris, 1994), companies such as Bechtel began to use the term "project manager" in their international work, primarily on remote sites. Communities of project management practice, however, with their focus on interactions between people, developed locally, becoming formalized in national project management professional associations. Even international projects were considered as endeavors conducted by national organizations, offshore. During the period in which project management professional associations were emerging, the focus was on techniques for planning, scheduling, and control. Practitioners who were forming these local and essentially national communities of practice were primarily involved in major projects in the engineering, construction, defense, and aerospace industries, and the interests of these practitioners strongly influenced the activities and focus of project management professional associations from their emergence in the 1960s through to the early to mid-1990s.

A change of focus became evident in the early 1990s. For some time the application of project management had been spreading beyond its traditional origins, to a wider range of application areas, particularly information technology. In a rapidly changing and responsive environment, where more and more of the endeavors of organizations are unique and could benefit from being identified and managed as projects, interest in project management grew progressively stronger, extending to project management as an approach to enterprise management (Dinsmore, 1996). However, these "projects" are often internal, without physical end products or clearly identified "clients," although they will often have many interested stakeholders both internal and external to the parent organization. In an era of networking, alliances, and partnerships, there may not even be a single or clearly identified parent organization, and resources are often shared across multiple projects and, in some cases, multiple organizations. Further, the application of project management extended beyond international projects, managed offshore by nationally based companies to use by global corporations through globally distributed operations and projects. Hence, the communities of project management practice, formalized in primarily local or national professional associations, faced, and continue to face, a dual challenge.

One aspect of this challenge is that the expertise that underpins the development and definition of project management practice is founded primarily in the management of clearly recognized and defined stand-alone projects such as those in engineering and construction. These practices were first transferred and minimally transformed in application to information systems and technology projects, but they retain a strong and identifiable legacy from their origins in major engineering projects. They do not recognize or offer a response to the systemic and complex nature of business projects, including implementation of corporate strategy, management of organizations by projects, and enterprise innovation.

The other aspect of the challenge facing project management as a community of practice, with aspirations to being a recognized profession, has been fragmentation through internal competition between the professional associations as the formal manifestations of the community, which have tended to remain locally focused and proprietorial about the knowledge created by their communities. In contrast, aided by the development of information technology; easier, faster, and less expensive travel; and an increasingly global economy, informal communities of project management practice have developed through online communities of practice, attendance at an increasing number of project management conferences and other initiatives for exchange occurring outside the official channels. This development has a distinctly global dimension, encouraged by global corporations and facilitated by global communication technologies.

The movement toward globalization, like modern project management, is considered to have begun in the 1950s, and by 1996, international business had become "the fastest-growing field in the business world, just behind technology" (Lenn, 1997, p. 1), leading to the suggestion that "to be successful in the twenty-first century, global professionals will require an education that guarantees their competence in their individual country and competitiveness in an international marketplace" (Lenn and Campos, 1997, p. 9). Indeed, the North American Free Trade Agreement (NAFTA, 1993) and the World Trade Organization's General Agreement on Trade in Services (GATS, 1994) removed many barriers to professional mobility and required the "development of policies that evaluate professional competence based on fair, objective criteria and transparent (publicly known) procedures" (Lenn, 1997, p. 2). These agreements have put pressure on established professions and their professional associations to consider mutually acceptable standards in cooperation with other countries and to actively plan for reciprocal recognition as a minimum.

By the mid-1990s, these wider external economic and social pressures, as well as internal pressures from the increasing number and range of users of project management, lead to the emergence of a number of initiatives aimed at enhancing the global communication and cooperation between the professional associations as formal manifestations of the project management communities of practice and providing a more globally relevant and rational framework for the definitions of practice and recognition of professional competence through standards and certification.

This chapter briefly identifies and describes key project management professional associations and provides a review of initiatives relating to enhancement of global communications, research, education, standards, and certification in project management.

The Professional Associations

There was relatively little growth in membership and in number of project management professional associations from their emergence in the mid-1960s and 1970s through to the start of 1990s, which heralded a decade of unprecedented growth. The most significant membership growth was that of the Project Management Institute, which experienced a growth from 8,500 members in 1990, located primarily in the United States, to nearly 110,000 members in 2003, of which 69 percent were located in the United States, 11 percent in Canada, and 20 percent in other parts of the world (Project Management Institute, 2003a).

At the start of the twenty-first century, the majority of people who want to participate in project management professional associations can do so through national associations, many of which are members of the International Project Management Association, which describes itself as an "international network of national project management societies" (International Project Management Association, 2003); through one of the 207 Chartered and 52 Potential PMI Chapters located in 125 countries (Project Management Institute, 2003a); or through one of a number of online project management communities.

Only a few of the project management professional associations will be briefly described here: the two organizations that purport to be global or international in their reach (PMI and IPMA), a few of the more influential or active of the associations that are neither PMI chapters nor members of IPMA (AIPM, PMSA), and a small number of other national associations and/or PMI chapters that have characteristics of particular interest or have made specific contributions to the promotion and development of project management practice.

International Project Management Association (IPMA)

The International Project Management Association began as a discussion group comprising managers of international projects and has evolved into a network or federation comprising 30 national project management associations representing approximately 20,000 members, primarily in Europe but also in Africa and Asia (International Project Management Association, 2003). The International Project Management Association has developed its own standards and certification program (see my chapter earlier in the book), which maintains a central framework and quality control process but encourages development of conforming national programs by national association members. The International Project Management Association and member national associations promote their standards and certification program in competition with those of others, primarily the Project Management Institute. The IPMA is hampered by its structure as a federation, by vested interests and priorities of its national association membership, and by lack of funds available for international and global development, which is a particular issue regarding the large number of member associations representing transitional economies who require subsidization of their membership and services.

As membership of the IPMA is subject to change, anyone interested in current membership should refer to the IPMA Web Site—*www.ipma.ch.*

Project Management Institute

The Project Management Institute began as the national project management association for the United States. By the late 1990s the Institute realized that with over 15 percent of its members and a number of chapters located worldwide, it was rapidly becoming an international organization. In September 1997 the PMI Board established a Globalization Subcommittee and then a Globalization Project Action Team (PMI Globalization Project Action Team, 1998) to assist the board in establishing its position on globalization. It has subsequently refocused its activities as a global organization rather than a national association with international or offshore chapters. PMI's headquarters continues to be located in the United States (in Philadelphia, PA), and the organization remains subject to the law of that state. However, in May 2003 the Institute held its first Global Congress in Europe (The Hague) and in June 2003 opened a PMI Regional Service Centre for Europe, Middle East, and Africa (EMEA Regional Service Centre) in Brussels, Belgium (Project Management Institute, 2003b).

The PMI approach, positioning itself as a global organization, plus its significant membership—109,117 individual members as of July 2003 (Project Management Institute, 2003a)—suggests that it is the organization that provides the primary representation and voice for the project management community, globally. The Institute itself claims to be "the leading nonprofit professional association in the area of Project Management" (www.pmi.org, June 2003). However, while headquarters and nearly 70 percent of membership remain located in the United States, its products and services, despite recent efforts to increase globally representative involvement are, or are perceived to have been, primarily developed in the United States to suit the needs of that market, and there is considerable reluctance on the part of project management professionals in some countries outside the United States to relinquish their independence and genuinely national representation. A further issue is economic. Practitioners in many countries cannot afford professional membership fees that are acceptable in the United States. In many cases it has been necessary to establish fully national associations in order to meet the needs of local jurisdictions and/or to provide a more affordable alternative. A notable example is South Africa, where project management practitioners were for many years represented by a PMI Chapter. The PMI South Africa Chapter continues to exist, but a separate national association (Project Management South Africa) was established in 1997 to satisfy local economic and regulatory requirements.

American Society for the Advancement of Project Management (asapm)

While project management practitioners in some countries outside the United States prefer to retain local representation through national associations rather than having their professional interests addressed through a local PMI chapter, a number of those in the United States took the view that if the Project Management Institute is a global organization, there is no longer any national association representation in the United States. As a result, the American Society for the Advancement of Project Management (asapm) was formed in July 2001.

PMINZ: PMI New Zealand Chapter

The Project Management Institute, New Zealand (PMINZ) is an example of a PMI chapter that has been accepted, to date unopposed, as the national project management association. It was established in 1994, as a chapter of the Project Management Institute and by 2003 had 850 members (www.pmi.org.nz, June 2003).

Association for Project Management, UK (APM)

The Association for Project Management (UK), although a member of the IPMA, deserves mention in its own right, as it has more members (14,000) than any of the other member organizations of IPMA and has done considerable influential work in definition of the project management body of knowledge and development of certification programs. The *APM Body of Knowledge* was one of the key documents referenced in writing of the *ICB: IPMA Competence Baseline*, as well as in the development of certification programs.

A PMI chapter was established in the United Kingdom in 1995 and reported membership of 2,000 by 2003, claiming that many of their members "are also members of local UK project management associations and groups" (UK PMI Chapter, 2003)

Australian Institute of Project Management (AIPM)

The Australian Institute of Project Management, begun in 1976 as the Project Managers Forum, is the Australian national project management association and by 2003 had 4,000 members distributed over eight state and territory chapters. As AIPM is not a member of the International Project Management Association, it has been well placed to offer an independent voice and in some cases act as an intermediary between the Project Management Institute and the International Project Management Association in the interests of global cooperation. The AIPM is also notable for having secured government support for development of performance-based competency standards for project management, recognized within the Australian Qualifications Framework (see my earlier chapter). Another role of AIPM has been to encourage development of national project management associations in the Asia Pacific Region and cooperation among them. AIPM has cooperative agreements with a number of Asian professional associations and has participated in a fairly loose Asia Pacific Forum for some years. In 2002 the AIPM took a leading role in the formation of the Asia Pacific Federation for Project Management (APFPM), an umbrella organization of national project management institutes.

The Australian Institute of Project Management remained unopposed as the national project management association until 1996, when the first of a number of PMI chapters was chartered in Australia. By 2003 there were PMI chapters established in Sydney, Melbourne, Canberra, Adelaide, Queensland, and Western Australia (Project Management Institute, 2003c), with a total membership of 1,500, with 700 of these being members of the Sydney Chapter (Project Management Institute, 2003d). Relationships between the Australian Institute of Project Management and the Australian PMI Chapters varies from time to time, and from state to state, between friendly cooperation and active competition.

Project Management South Africa (PMSA)

Project Management South Africa (PMSA) has already been mentioned in discussion of the Project Management Institute. It is worthy of separate mention because, like the Australian Institute of Project Management, it is not a member of the IPMA and therefore also has the opportunity to offer an independent voice in the global arena. However, because the PMI South African Chapter was formed first (1982) and was very active for a long time before formation of PMSA and because PMSA was essentially formed by members of the PMI South Africa Chapter (which continues to exist), there is a far closer and more consistently cooperative relationship between the two organizations. Membership of PMSA increased from 400 at formation in 1997 to over 1,200 in 2003.

The drive to create PMSA came from a need for a cross-sector forum for practitioners to meet and work together and for a national body to work with local organizations and the South African government in developing effective project management within South Africa. Another argument in favor of formation of the national association was that PMI membership fees had become prohibitive for many in South Africa with the decline in the value of the South African rand relative to U.S. currency.

PMSA has taken an active role in working with the South African government in development of performance-based competency standards for project management, similar in format to those developed in Australia and recognized within the South African National Qualifications Framework.

Japan Project Management Forum (JPMF)

Japan is another country that embraces both a national association, the Japan Project Management Forum (JPMF) and a PMI chapter (Tokyo), both established in 1998. The JPMF is a division of the Engineering Advancement Association (ENAA), which was founded in 1978 as a nonprofit organization based on corporate rather than individual membership, dedicated capability enhancement, and promotion of the Japanese engineering services industry. ENAA enjoys government support through the Ministry of Economy, Trade and Industry (METI), and membership encompasses 250 engineering and project-based companies. Since inception, ENAA has had a Project Management Committee, and 46 member companies are involved in this committee. While ENAA engages and addresses the needs of industry and corporations, JPMF acts as the professional association for individual practitioners.

To advance the use of project management approaches in Japan, the ENAA was commissioned by the government (METI) over a period of three years from 1999 to develop a "Japanese style project management knowledge system," which has been developed and published as a standard guidebook under the *title P2M: Project and Program Management for Enterprise Innovation* (ENAA, 2002). Subsequently, in April 2002, the Project Management Professionals Certification Center (PMCC) was established, to widely disseminate the P2M in Japan and beyond and to establish a related certification process (see my earlier chapter).

Project Management Research Committee, China (PMRC)

Established in 1991, the PMRC was the first, and claims to be the only national, cross-industry project management professional association in China, supported by over 100 universities and companies and 3,500 active individual members form universities, industries, and government.

In 1994 the PMRC initiated, with support from China Natural Science Fund, the development of a *Chinese Project Management Body of Knowledge* (C-PMBOK), which was published together with the *China-National Competence Baseline* (C-NCB) in May 2001. Over 160,000 copies of this document had been issued by the start of 2003.

PMRC supports both the IPMA, of which it has been a member since 1996, and the Project Management Institute. It helped introduce the PMP certification into China and started an IPMA certification program in July 2001 (Yan, 2003).

In 2003, there were three PMI Potential Chapters in China (Project Management Institute, 2003c). Also active in China is the Project Management Committee (PMC), established in November 2001 as a branch of the China Association of International Engineering Consultants (CAIEC), an organization guided by the Ministry of Foreign Trade and Economic Cooperation, China (MOFTEC). The PMC promotes project management training and certification among the 255 member companies of CAIEC (Yan, 2003).

Society of Project Managers, Singapore

The Society of Project Managers, Singapore is worthy of separate mention because of its active role in the Asian region, strong links with other parts of the region, notably Japan, and as it purports to be "an amalgam of learned society and a professional body." The Society was formed in 1994 by a group of professionals as a vehicle for advancing the development of project management. It has held a number of well-regarded research-based conferences in the region (The Society of Project Managers, 2002).

Global Initiatives

As outlined in the previous section, there is strong evidence of considerable and increasing interest in project management from individuals, corporations, and governments. Responses have originated at the local level, giving rise to many competing membership opportunities, conferences, standards, and certifications, not only between countries but often within a single country. Two key organizations have attempted to achieve a more unified and global approach: the International Project Management Association, as a federation of national associations, and the Project Management Institute in taking an active stance as a global organization. To a large extent this has merely increased the dilemma of choice for individuals and organizations, especially as national governments such as those in Australia, South Africa, the United Kingdom, and Japan have recently begun to take an active role in support and recognition of project management standards and certifications. The International Project Management Association, on behalf of member associations, promotes the *ICB: IPMA Competence Baseline* and aligned National Competence Baselines, along with its 4-Level Certification Program. At the same time, the Project Management Institute promotes its PMBOK Guide, PMP Certification, and a range of other standards and certification products, globally. Individuals and organizations must therefore not only decide between one national and one "global" or international membership and set of standards and association certifications, but between several products available at both the national and global levels.

A frustration for the "users" of the products of professional associations, primarily in the form of standards and certification, has been the apparent unwillingness or inability of the project management associations, despite the signing of "cooperative agreements," to really work together to enhance global cooperation and communication and to bring a sense of global unity to the profession, resolving the dilemmas of decision in terms of investment in project management association membership, standards, and certification. As a result, efforts toward global communication, cooperation, and alignment of standards and certification have primarily been informal and unfunded, occurring outside the project management professional associations. Some of the primary initiatives are reviewed here.

Global Project Management Forum (GPMF)

A response to the independent and largely local development of standards and certification, as well as research and education, first came to a head in 1994. At the PMI Symposium in

Vancouver, Canada, there was a meeting of representatives of PMI, IPMA, APM, and the AIPM, at which "formal cooperation on several global issues, including standards, certification and formation of a global project management organization or 'confederation'" were discussed (Pells, 1996, p. ix). Another, informal meeting, of individuals from about a dozen countries, also gathered, intensely discussing cooperation and communication among project management professionals around the world and formulating a declaration of intent. The first of a series of Global Project Management Forums, held in association with the PMI Symposium in New Orleans in 1995, was the result.

There were nearly 200 attendees, representing over 30 countries, at the first Global Project Management Forum. There were high hopes that the energy and enthusiasm evidenced at this meeting would be "an opportunity for the world's leading project management associations to take another major step towards achieving agreements on international standards, recognition of project management certifications, and development of a global core Project Management Body of Knowledge" (Pells, 1996, p. x). Breakout sessions were held to discuss five topics:

1. International project management standards
2. Globally-recognized project management certification
3. Global communication among project management professional organizations
4. Global cooperation and organization of the project management profession
5. Development of a global core project management body of knowledge

This first meeting of the Global Project Management Forum was a high point for this initiative and will be a lasting and important memory for those privileged to be there. However, by the time the thirteenth Global Project Management Forum was held in Moscow in June 2003, little real progress had been achieved. The GPMF had evolved into an informal association with a slogan "Toward the globalization of project management" and a stated mission "To advance globalization of the project management profession by promoting communications and cooperation between and among project management organizations and professionals around the world" (www.pmforum.org). The forums have been held as a one-day event associated with the annual or biannual conferences of project management professional associations. Until 1999 the key venues were the annual Symposia of the Project Management Institute and the biannual IPMA World Congress, but from 1999 onward, the Project Management Institute declined to provide for the forums in association with their symposia, preferring instead to offer a PMI Global Assembly aimed primarily at representatives of their globally distributed chapters.

The agenda for each forum has become relatively predictable, with presentations and breakout sessions on Research, Standards, Education, and Certification. The initial ideal that the initiative would bring together people from all over the world in an open forum to keep touch with developments in the field of project management has been fulfilled, but it became clear within a couple of years of the first GPMF meeting that meaningful cooperation between the project management professional associations was far from being realized. Informal cooperation and lip service were possible. Formal cooperation and real progress in the interests of a strong and unified project management profession were hampered by political issues and vested proprietorial interests.

Recognizing that real achievements were required to maintain the momentum begun by the GPMF, the IPMA established and convened a series of Global Working Groups that first met in East Horsley, UK, in February 1999 (IPMA, 1999).

Global Working Groups

The Working Groups were established in six areas, namely (IPMA, 1999):

- Standards
- Education
- Certification
- Accreditation/Credentialing
- Research
- The Global Forum

Certification and Accreditation/Credentialing were recognized as related and were merged. The subsequent five Working Groups were tasked with delivering a progress report at the next Global Project Forum, the last held in association with a Project Management Institute Symposium, in Philadelphia in September 1999. These five Working Groups have continued to present progress reports at each of the following Global Forums. The Working Group on the Global Forum confirmed that it should remain informal, independent of established professional associations. Because of their nature, Education and Research have generated considerable interest, and breakout sessions have provided excellent opportunities for sharing ideas. A research project concerning the benchmarking of the degree of project-orientation of societies was originated in 1999 and has been furthered under the auspices of the IPMA through Projekt Management Austria (PMA) (IPMA, 2003). The Certification Working Groups and GPMF breakout session have facilitated ongoing discussion of global issues surrounding certification and credentialing, but little progress is possible without formal involvement of the professional associations that "own" the existing certification products and processes. A global approach to certification also needs, as a prerequisite, a global approach to standards.

Progress arising from the Working Group: Standards is reviewed in detail in my previous chapter. The most active and promising of the initiatives generated by the Global Working Group: Standards is the development of a global framework of performance-based standards for project management personnel (see my earlier chapter and http://www.globalPMstandards.org).

Toward a Global Body of PM Knowledge (OLCI)

This initiative, which began in 1998 by bringing together those working at the time on various representations of the body of project management knowledge (*ICB: IPMA Competence*

Baseline, reviews of the *APM Body of Knowledge* and the PMBOK Guide), has progressed through a series of annual workshops hosted by organizations, including NASA, Telenor, ESC Lille and the Project Management Professionals Certification Center, and JPMF. This initiative is discussed in detail in my earlier chapter.

International Research Network on Organizing by Projects (IRNOP)

IRNOP, the International Research Network for Organizing by Projects, is somewhat different from the initiatives outlined previously, as it is not directly associated with project management professional associations but is an important expression of the informal interactions of the global communities of project management practice and has been an important initiative in achieving global communication and cooperation in project management research. It has no formal organization but comprises a loosely coupled group of researchers. It was initiated in 1993 to support and enhance efforts aiming at the development of a theory on temporary organizations and project management. Its main activity is the facilitation of research conferences, and these have been held in Sweden (1994), Paris (1996), Calgary (1998), Sydney (2000), and the Netherlands (2002). The next conference will be held in Turku, Finland, in 2004.

Prior to staging the first Project Management Research Conference by PMI in Paris in 2000, IRNOP provided the only real impetus and opportunity for students, researchers, and academics in project management to come together to share ideas and information about their research.

Summary

By referring to the formal manifestations of project management communities of practice as "professional associations," we assume that project management is in fact a profession. Although there is a strong sense of aspiration among project management practitioners and their representative associations to professional status, this remains a matter of debate and has been powerfully questioned by Zwerman and Thomas (2001), who have highlighted the "barriers on the road to professionalization." They maintain that although project management has been moving toward satisfying various criteria indicative of professional status, it is still some distance away and achievement will require significant effort on the part of the professional associations and members.

Even recognition of project management as an occupation or field of practice is vulnerable, as many see it merely as an aspect of general management, and there is a growing view that project management should form part of a wide range of skill sets. Much of the knowledge base of project management is shared with or has been annexed from bodies of knowledge of other professions, and Turner (1999) suggests that in order to be a mature profession, project management must develop a sound theoretical basis indicating that considerable further research is required to establish a sound foundation for professional status.

A challenge that professional associations are likely to face is the increasing involvement of government in defining practice standards for project management as evidenced by the project management standards forming part of National Qualifications Frameworks in Australia, South Africa, and the United Kingdom; government support of project management standards development in Japan and China; and active leadership by of the Office of Government Commerce in the United Kingdom in developing standards and certification programs for individuals and organizations involved in projects and programs. Influenced by trade agreements such as the World Trade Organization's General Agreement on Trade in Services (GATS, 1994), governments are motivated to seek mutual recognition in areas of standards, accreditation, and certification. Further, unless project management professional associations take the lead, global corporations will do so in order to satisfy their own needs.

Clearly, there is considerable work to be done to establish project management as a profession, and this suggests the need for project management's formal representation to adopt a globally unified stance rather than pursue the fragmentation and internal competition that has characterized development to date. It is interesting to note that the majority of activity promoting global communication and cooperation, the free sharing and exchange of ideas among members of the global community of project management practice, is occurring outside the formal structures, through online discussion groups and through informally structured initiatives.

References

Cook, D. L. 1981. Certification of project managers: Fantasy or reality? In *A decade of project management: Selected readings from* Project Management Quarterly—*1970 through 1980,* ed. J. R. Adams and N. S. Kirchoff. Newtown Square, PA: Project Management Institute.

Dean, P. J. 1997. Examining the profession and the practice of business ethics. *Journal of business ethics* 16(15):1637–1649.

Dinsmore, P. 1996. On the leading edge of management: Managing organizations by projects. *PM Network* (March): 9–11.

ENAA 2002. *P2M: A guidebook of project and program management for enterprise innovation: Summary translation.* Revision 1. Tokyo: Project Management Professionals Certification Center (PMCC).

International Project Management Association. 2003. About IPMA. Available at www.ipma.ch/ (accessed June 26, 2003).

IPMA 1999. Documentation of Meeting: Global Working Groups. East Horsley, UK, February 27, 1999.

IPMA. 2003.Research and development. www.ipma.ch/ (accessed June 26, 2003).

Lenn, M.P. 1997. Introduction. In *Globalization of the professions and the quality imperative: professional accreditation, certification and licensure,* ed. M. P. Lenn and L. Campos. Madison WI: Magna Publications, Inc.

Lenn, M. P., and L. Campos. 1997. International organizations. In *Globalization of the professions and the quality imperative: professional accreditation, certification and licensure,* ed. M. P. Lenn and L. Campos, 9–10. Madison, WI: Magna Publications, Inc.

Morris, P. W. G. 1994. *The management of projects.* London: Thomas Telford.

Pells, D. L. 1996. Introduction. In *The global status of the project management profession,* ed. J. S. Pennypacker, ix–xii. Sylva, NC: PMI Communications.

PMI Globalization Project Action Team. 1998. Project definition. www.pmforum.org/featindex.htm/ gpatp1.htm (accessed April 13, 2003).

Project Management Institute. 2003a. The PMI Member FACT sheet—July 2003. www.pmi.org/prod/groups/public/documents/info/gmc_memberfactsheet.asp. (accessed August 28, 2003).

Project Management Institute. 2003b. PMI opens regional service centre in Brussels, Belgium. www.pmi.org/prod/groups/public/documents/info/ap_news-emeaopen.asp (accessed June 28, 2003).

Project Management Institute. 2003c. PMI Chapters Outside the United States. www.pmi.org/info/GMC_ChapterListingOutsideUS.asp#P128_1923. (accessed June 26, 2003).

Project Management Institute, Sydney Chapter. 2003d. PMI Sydney Chapter. Available at http://sydney.pmichapters-australia.org.au/ (accessed June 26, 2003d).

Stretton, A. 1994. A short history of project management. Part one: The 1950s and 60s. *The Australian Project Manager* 14(1):36–37.

The Society of Project Managers. 2002. The society. www.sprojm.org.sg/web/socpm/ (accessed June 29, 2003).

Turner, J. R. 1999. Project management: A profession based on knowledge or faith? Editorial, *International Journal of Project Management* 17(6):329–330.

UK PMI Chapter. 2003. Welcome to the UK PMI Web page. www.pmi.org.uk/ (Accessed June 26, 2003).

Wenger, E. 1998. *Communities of practice: Learning, meaning and identity*. Cambridge, UK:Cambridge University Press.

Yan, Xue. 2003. PMRC and other organizations in China. E-mail from Xue Yan (xue_yan@cvicse.com.cn) to Lynn Crawford (Lynn.Crawford@uts.edu.au). January 2, 2003.

Zwerman, B., and J. Thomas. 2001. Barriers on the road to professionalization. *PM Network* 15 (4, April): 50–62.

INDEX